The Analysis of
Drugs in Biological Fluids
Second Edition

Joseph Chamberlain
Former Head of Analytical Department
Hoechst Pharmaceutical Research Laboratories
Milton Keynes, Buckinghamshire, England

CRC Press
Boca Raton   New York   London   Tokyo

Library of Congress Cataloging-in-Publication Data

Chamberlain, Joseph.
  The analysis of drugs in biological fluids / by Joseph
Chamberlain. — 2nd ed.
    p. cm.
  Includes bibliographical references and index.
  ISBN 0-8493-2492-0
  1. Drugs—Analysis.   2. Body fluids—Analysis.   I. Title.
RB56.C48   1995
615'.1901—dc20   95-19022
                                                     CIP

No claim to original U.S. Government works
International Standard Book Number 0-8493-2492-0
Library of Congress Card Number 95-19022
Printed in the United States of America 1 2 3 4 5 6 7 8 9 0
Printed on acid-free paper

PREFACE

When the first edition of this book was completed over a decade ago, the future of analysis of drugs in biological fluids appeared to be reasonably clear cut. Although the analyst was, as ever, being called on to design more sensitive assays, it was envisaged that he would be up to the task using natural developments of a whole armory of established techniques, and perhaps adding new automated methods and computer routines to make his work more productive. In general, the analyst would expect to be measuring new compounds (but no really novel structures), applying the appropriate technique as suggested by his experience. To some extent this is what has happened, so that literature descriptions of analytical method for new drug entities have become so predictable that such papers tend to be of archival interest only, breaking no new ground in analytical technology.

Luckily, however, life is not so boringly predictable. The first edition suggested there might be considerable advances in automation, but this has not happened. Machines are more reliable and materials such as column packings are also more reproducible; these features have improved productivity rather than the takeover of laboratories by robots. In the actual analytical techniques, there has been a considerable revival of older physicochemical methods, such as mass spectrometry and nuclear magnetic spectroscopy, both previously considered too insensitive for bioanalytical work. These techniques have been considerably helped by advances in instrument design, as well as by the use of computers enabling the use of routines that would not be possible otherwise. One of the major features of the use of these physicochemical methods is the ability to skip the separation step, a procedure almost unthinkable a decade ago when chromatography was preeminent. Now that such techniques as tandem mass spectrometry are widely used, at least one author has suggested that these are the methods of choice. The question of sensitivity has been addressed by a swing back toward biological end points, with radioimmunoassay and its derivatives leading the way.

The present edition includes the basic principles of the major techniques and describes the latest advances and examples of their application. The sections on automation, however, have been shortened on the basis that the automated laboratory is not specific to bioanalytical fluids and more specialized works can be consulted. Good Laboratory Practice is now a well-established part of all laboratories involved in bioanalytical work and the detailed review in the first edition has been omitted — the subject no longer being considered of special interest in such laboratories.

Chirality has become an issue in drug analysis over the last few years. The first edition was not alone in similar works of its time in not including reference to this important point, but this edition redresses the balance with descriptions of the methods now available for chiral analysis, as well as discussions on its impact.

Finally, the new edition delves deeply again into the practical aspects of bioanalytical work; despite the considerable reliability of materials and instruments mentioned above, there will always be that extra touch that laboratory analysts will need to bring to their work.

THE AUTHOR

Joseph Chamberlain, Ph.D., F.R.S.C., is currently the Editor of the *Journal of Pharmacy and Pharmacology*, and founding Editor of the rapid publication journal *Pharmaceutical Sciences.*

Dr. Chamberlain graduated in 1960 from Liverpool University with a B.Sc. honours degree in Chemistry and obtained his Ph.D. in 1963 from the University of Birmingham, working on the analytical problems involved in the study of progesterone metabolism. Following post-doctoral work at Harvard (1963–1965), Oxford (1965–1966), and Charing Cross Hospital (1966–1969) continuing his research interests in steroid metabolism, Dr. Chamberlain's subsequent career was mainly in pharmaceutical industrial research with Syntex, Searle, Hoechst and Merck, Sharp, and Dohme.

During his research career Dr. Chamberlain published over 50 research papers, mainly in the field of hormone and drug analyses.

Since 1991, Dr. Chamberlain has been the full-time Editor of the *Journal of Pharmacy and Pharmacology*, and in 1995 was the founding editor of *Pharmaceutical Sciences*, but continues a research interest in the theoretical aspects of drug metabolism and pharmacokinetics.

TABLE OF CONTENTS

1 Why Analyze Drugs in Biological Fluids?

CONTENTS

1.1 INTRODUCTION

Any parent, or teacher of young children, is accustomed to the barrage of questions from enquiring and curious minds: "Who is that?", "What is this?", and "How many of those?" But the most persistent query, and the one most difficult to answer, is the question "Why?" As well as being the hardest to answer, this is most probably the most important of life's many questions. In trying to answer the child's "Why?", we may often be forced into thinking more deeply about the point being discussed. It does not matter, for example, *when* Martin Luther made his break with the Church,* but the reasons for the break are very important to clearly understand that particular turning point in European culture. It does not matter who was the first man

---

*   1518

1

on the moon,* but to establish *why* we thought it necessary to undertake such a project is a fascinating excursion into our minds and into the complexities of human motivation. Often, our immediate answer to a child's "Why?" is the first reasonable reply that comes to mind, but if the child continues persist with one "Why" inevitably leading to another, then our own comprehension and acceptance of the ideas will also change.

This book discusses the analysis of drugs in biological fluids. It is, apparently, an important subject; anyone who takes even a passing interest in current literature on the subject is aware of the torrent of papers that are published each year. A rough estimate in 1983 revealed that about 5,000 papers a year were being published which had as the main topic the analysis of drugs in biological fluids; in 1993, this figure was still being maintained. There are thousands more that depend on an analyst somewhere producing the raw analytical data for clinicians, pharmacokineticists, biochemists, and toxicologists. Many papers and books are devoted to the problems of performing such analyses, and historical reviews will tell you who first applied a particular technique, and when. Some authors will explain at great length how one method is superior to another in terms of sensitivity or specificity, but will omit to explain why such superiority is needed. In this opening chapter, then, the author will describe the various situations in medicine and research where the analysis of drugs in biological fluids is considered desirable. The answer to the question "Why?" will not always be the same for the diverse situations and the realization of this should lead to the acceptance of different criteria for deciding on the analytical method to be used in the different situations.

Figure 1.1 shows the various stages in the research, development, and application of medicinal products. All these stages require some analytical input. The most obvious situations where this need arises are in the use, or more commonly the abuse, of drug substances. As a result, the author hopes you will forgive him for ignoring the chronology set up in Figure 1.1, and discuss these aspects first.

1.2   DRUGS IN USE

1.2.1   FORENSIC TOXICOLOGY

The author begins with this subject because it is to the layman perhaps the most fascinating, and so beloved of detective story writers. If we ask why these analyses are necessary the simple answer is that the investigator needs to know if a crime has been committed by examining relevant samples from the victim, usually post mortem, for evidence of a poison. In this situation, the investigator needs evidence primarily regarding the presence of a foreign compound and, if so, of its identity. The actual concentration of the substance is a question which is a distant third. Thus this type of work is often qualitative rather than quantitative; any quantification is a gross estimate, rather than a precise measurement required in some of the situations described later. In forensic work the analyst is usually presented with an ill-defined sample, often of dubious origin — a general unknown — and he does not know what he is supposed to find. The approach used has to be a series of more or less logical

---

*   Neil Armstrong

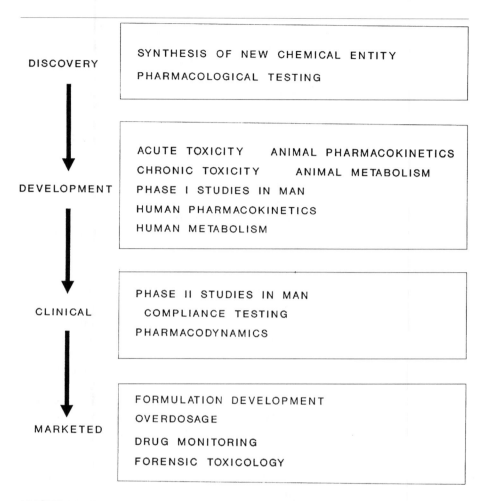

FIGURE 1.1 The stages of research, development, and application of medicinal products requiring analytical input.

tests beginning with a widely cast net. First, the analyst will apply a screening test to establish whether there are any components of the sample not normally present — no easy task if the sample is an old blood or urine specimen. From the results of the first screen he will decide on the next screen or series of screens looking for specific suspected constituents and finally he will apply confirmatory tests aimed at specific drugs. This type of work can cover the whole range of sophistication in analytical tests, from classical spot tests in the primary screens, via simple chromatography for provisional identification, to chromatography coupled to mass spectrometry or infrared spectrometry for unequivocal identification. Even for primary screening the raw specimen may need some prior extraction and the choice of solvent for these procedures is an art requiring considerable experience when one considers the diversity of samples and possible drugs (and other toxins) in such samples. Overall the forensic toxicologist would like to identify the presence of a particular poison in the sample and state whether the amounts are compatible with fatal poisoning or more in accordance with therapeutic doses. The toxicologist is not normally called on to give precise concentrations, nor is the sensitivity a problem because the investigation is presumably only necessary when large doses are suspected. However, this last aspect is becoming more of a problem as forensic toxicologists are being asked to analyze drugs at therapeutic concentrations.[1] Quantification is also important in situations where it may be necessary to extrapolate plasma levels to a previous point in time, for example when investigating alcohol levels in traffic offenses. This has been a matter of some debate and calls for a measure of interpretation — which is not necessarily part of the analytical chemist's expertise.

Not all biological samples presented to the forensic analyst are derived from murky sources. I have already mentioned alcohol in blood and in this case the sample may be obtained in well-controlled conditions to yield a clean serum or plasma sample. The same is true for analysis of drugs in blood of suspects.

1.2.2   OVERDOSAGE

The monitoring of drugs in cases of overdose is one of the oldest areas of drug measurement in clinical medicine.[2] In cases of overdosages with drugs, the prime reason for an analysis is to establish which dose has been overdosed in order to apply the appropriate remedy. The need for an analysis will arise if the patient is comatose and cannot tell the physician, and there is no other evidence (e.g., medicine bottles, known history of drug taking) to suggest the immediate cause. It is important to recognize that in most cases the patient recovers with the help of intensive support therapy although the case team may have no knowledge of the poison ingested or its concentration in biological fluids. The difficulty and dangers of obtaining appropriate samples must be balanced against the immediate clinical problem and the likelihood that the knowledge obtained will be crucial to the course of action. The sample, however, once obtained, will be more amenable to analysis than that usually provided to the forensic toxicologist and in addition the range of possible intoxicants is less. In fact, there is a general trend in the type of compounds that are overdosed and a knowledge of the current patterns as well as local conditions enables one to identity the overdosed compound using a limited number of tests aimed at specific compounds or groups of compounds. Once the intoxicant has been identified, appropriate treatment can be initiated. Although the primary purpose of the analysis is to identify the drug rather than its concentration, a degree of quantification may be desirable. For example, the level of the drug and knowledge of its elimination characteristics in humans may enable the physician to decide whether to remove the drug, or whether to allow the episode to run its course, allowing the body's own mechanisms to complete the elimination process. Alternatively, if treatment is used, then it may be useful to monitor progress by taking further blood samples for reanalysis. In this case, it may be more appropriate to use a different analytical method, one which has been designed for routine, rapid, quantitative analysis as described in later sections of this book. This may be typified by glutethimide, where knowledge of the levels of both glutethimide and its 4-hydroxy metabolite are useful in following the clinical course of respiration and coma.[3] Prescott[2] has considered the drugs for which toxicological monitoring may be required in overdose cases and has concluded that such monitoring is important for the proper management of poisoning with paracetamol, salicylate, iron, phenobarbitone, lithium, paraquat, and methanol and of very severe intoxication with a variety of other drugs and poisons when measures for their active removal are considered.

1.2.3   DRUG ABUSE

Drug testing has become a new feature of the workplace in the last decade, particularly in the United States, where testing laboratories are part of a $600 million business. Drug abusers, the argument goes, are more likely to need time off for illnesses, more likely to require greater healthcare coverage, and more likely to be less efficient, or more dangerous in their jobs, because of their drug habits. Thus, drug testing has routinely become a part of the candidate selection process. Monitoring of existing employees seems to be less prevalent, although presumably the same arguments apply. It is said that marijuana is the drug most tested for in such screens, despite there being no evidence of biological damage among relatively heavy users. This contrasts with tobacco and alcohol, both of which are proven harmful agents, but with less importance attached to the problem of abuse, although the U.S. Department of Transportation does require alcohol and drug tests for anyone in safety-sensitive jobs, such as pilots, drivers, and sea captains.

Because such tests need to be noninvasive, urine is the most common test substance, and for initial screening, radioimmunoassay, or other type of immunoassay, is usual. However, such nonspecific assays must never be accepted as conclusive and any well-managed scheme needs to include a definitive and confirmatory test, such as gas chromatography-mass spectrometry in its program.

Thus, testing for drugs of abuse needs both a method of rapidly screening large numbers of urine samples and a very sophisticated method for definitively identifying the drug claimed to be present. Recently the National Research Council in Washington D.C. has released a report which suggested that the relationship between alcohol and other drug use by employees and their liability to accidents was inconclusive.[4]

1.2.4   DRUGS IN SPORT

The use of drugs in sport is a topic almost as contentious as politics in sport. There are those who maintain that drugs should be banned completely from all competition, in animals as well as in humans. There are two separate standards that seem to apply to the world of animal sport (dog and horse racing) on the one hand and human sport on the other. In the animal sports, the overall intention of the rules is that the animal is drug free at the time of competition, drugs generally being the substances that fall in the categories shown in Table 1.1.[5] The drug list for horses is very comprehensive, with the aim being that the animals are competing totally without the help or hindrance of chemical modifiers introduced by unscrupulous owners or those likely to profit by the horse's performance in racing. Certainly the doping of racehorses goes back a long way with the use of exciting substances being recorded as banned from races in Worksop as early as 1666.

TABLE 1.1
Prohibited Substances Under the
Rules of the Jockey Club

Drugs that act on the autonomic system
Drugs that act on the cardiovascular system
Drugs that act on the central nervous system
Drugs that affect the coagulation of blood
Drugs that affect the gastrointestinal system
Drugs that affect the immune system and its responses
Analgesics, antipyretic, and anti-inflammatory drugs
Antibiotics, synthetic antibacterial, and antiviral drugs
Antihistamines
Antimalarial and other parasitic agents
Cytotoxic substances
Diuretics
Endocrine secretions and their synthetic counterparts
Local anesthetics
Muscle relaxants
Respiratory stimulants
Sex hormones, anabolic steroids, and corticosteroids

By contrast, the original intention of the Medical Commission of the International Olympic Committee was to ban those drugs which were likely to be harmful when misused, the intention being that competitors undergoing a legitimate course of treatment should not be banned from competing. Of course, if the treatment should have effects other than those for which they are prescribed, but are desirable for the athlete's performance, then the emphasis shifts to the question of unfair advantage. An example would be the β-antagonist, used to treat certain heart ailments, but which has the effect of steadying hand tremor, a useful advantage for such sports as rifle shooting. In recent years, the question of unfair advantage, and even the ethics of drug taking in sport, has been more discussed than the potential harm, particularly where the evidence of harm is not always so incontrovertible.

If the drugs are banned in animals and humans, then any such use is obviously covert and the authorities are faced with devising methods to detect their use. The obvious way is to sample biological fluids from the suspected competitors and assay them for drugs. Such methods generally need to be noninvasive and urine is almost always the fluid of first choice, although for some drugs which are not excreted in urine, blood samples may be essential. Because of the nature of the problem (i.e., the identity of the drug is unknown), laboratories engaged in analyzing for drugs in sport will employ screening analyses in the first instance, followed by confirmatory methods such as gas chromatography-mass spectrometry.[6]

1.2.5   THERAPEUTIC DRUG MONITORING

To understand the reasons for performing drug monitoring in patients it is necessary to appreciate the concept of benefits and risks as applied to drugs. One of the goals of drug research is to produce a drug which is efficacious yet completely nontoxic. Some would argue that this is impossible, because by its very nature — the ability to alter biological events — any drug is potentially harmful. Nevertheless, the aim is to separate the beneficial effects from the harmful effects by developing new compounds which are effective at doses considerably lower than the doses at which toxic manifestations occur. For example, the initial success of benzodiazepines lay, in part at least, in their relative nontoxicity compared with barbiturates used for the same indications.[7] Despite intensive research in many therapeutic areas, however, it is still necessary to persevere with drugs where the so-called therapeutic ratio (the ratio of toxic levels to therapeutic levels) is low and dosing must be carried out with extreme care. In the simplest cases, this merely means keeping the dose at the correct recommended level. However, it is not the dose itself which is important, but the concentration of the drug at the pharmacological sites, and for individuals this is not always easily predictable from the dose. Koch-Weser[8] has reported that of 200 ambulatory patients prescribed the usual therapeutic dose of phenytoin, only 28.5% achieved plasma levels in the range considered appropriate for effective anticonvulsant activity. Suboptimal concentrations were present in 60% and potentially toxic levels were achieved in 11.8%. Other drugs with enormous variations in the dose–concentration relationship include digoxin, hydralazine, phenothiazines, perhexiline, propranolol, and tricyclic antidepressants. These variations may be caused by inadequate formulation of the drug leading to variable absorption, or may be inherent in the metabolic capacity of the individual. For drugs with high therapeutic ratios, the variation may not be important, but it can be seen that a dose which is safe in one individual may be highly toxic in another. Although it is not always true that the blood or plasma concentration of a drug is a relevant indicator of drug potential for activity, it is generally accepted that this is at least a step nearer the truth than the consideration of dose levels alone. Consequently, drug assays are very useful in titrating the drug dose for individuals in cases where the desired therapeutic plasma level is close to the toxic plasma level. This is the case for the heart drug digoxin where, because of the fact that clinical factors can alter the myocardial sensitivity to cardiac glycoside, there is a considerable overlap of the therapeutic and toxic concentrations. Thus it was reported that when plasma monitoring, and subsequent adjustment of the dose, was performed, the reported incidence of toxic side effects from this drug was considerably reduced.[9]

Where the purpose of the drug assay is to monitor the plasma levels with a view to altering the dose for optimal effect, it is obviously important to obtain results as quickly as possible. There are numerous marketed kits for performing such rapid assays, a field in which such techniques as radioimmunassay have become pre-eminent. Such techniques need to be specific, at least for the active constituents in the plasma (i.e., it is permissible to include active metabolites in the analytical response), they need to be quantitative rather than purely qualitative, and their precision and accuracy need to be compatible with their use in making decisions regarding plasma levels vis-a-vis previously experienced therapeutic and toxic levels. It is perhaps important to state that one of the most rapid methods of monitoring is by observing the clinical effect. Proponents of such drug monitoring should not be so committed to their thesis that they are blind to this fact. Table 1.2 lists a number of drugs for which therapeutic drug monitoring is considered useful in their proper management in the clinic.

TABLE 1.2
Drugs for Which Therapeutic Drug Monitoring
is Considered Useful

Drug	Method	Limit of detection (μg ml^{-1})	Therapeutic range (μg ml^{-1})	Toxic threshold (μg ml^{-1})
Amikacin	HPLC	1	10	
Acetaminophen	HPLC	0.2	—	250
Acetylsalicylic acid	HPLC	2	20–100	300
Carbamazepine	HPLC	2	8–12	12
Chloramphenicol	HPLC	0.7	10–20	25
Digitoxin	HPLC		0.01–0.03	
Digoxin	HPLC		0.001–0.025	
Disopyramide	HPLC		2–5	
Ethosuximide	HPLC	14	40–100	
Gentamicin	HPLC		2	
Imipramine	GC		>0.1	
Kanamycin	Bioassay		4–25	
Lignocaine	HPLC		2–5	
Phenobarbital	HPLC	2	15–40	40
Phenytoin	HPLC, GC	3	10–20	20
Procainamide	HPLC	1	4–10	10
Quinidine	HPLC	0.3	2–5	5
Theophylline	HPLC, GC	1	10–20	20
Tobramycin	HPLC		4–10	

1.2.6   PHARMACOGENETICS

One of the reasons for measuring drugs in biological fluids is to monitor concentrations of drugs which have a narrow therapeutic window and for which there is no consistent inter-individual relation between dose and blood levels. There may be reasons for this inconsistency including environmental factors, but in the last 15 years there has been considerable effort into disentangling a genetic basis for some of these inter-individual differences.

Sjöqvist et al.,[19] as long ago as 1967, postulated that the 36-fold differences in plasma levels achieved in different subjects on the same dose of the tricyclic antidepressant desipramine could be due to the varying activities of drug-hydroxylating enzymes. However, the critical period for this new discipline, which has been termed pharmacogenetics was during the late 1970s in two separate developments. In England, routine clinical pharmacology trials with debrisoquine, an antihypertensive agent, showed that one of the volunteers had a severe hypotensive response to the drug. This clinical reaction was linked to an inability of the volunteer to metabolize the compound by the usual route to 4-hydroxydebrisoquine.[20] At about the same time, in Germany, increased side effects were observed in some patients receiving an alkaloid drug with antiarrhythmic actions[21] and these side effects were linked to an inability to metabolize sparteine by N-oxidation. Further studies revealed that both oxidative metabolic reactions are under monogenic control and that simple assays could be developed to distinguish poor metabolizers from extensive metabolizers. In a typical test, the subject is dosed with the test drug (debrisoquine, sparteine, dextromethorphan) and urinary unchanged drug and the major metabolite assayed over a given time. The ratio of the two values is an index of metabolizing capacity. Such a test was applied to several thousand subjects in clinical studies and revealed that about 10 to 15% of a Caucasian population were poor metabolizers, while only 1 to 2% of Asian or Oriental populations were poor metabolizers.[22] Thus, it can be seen that the presence of poor metabolizers can have an effect on the outcome of trials for drugs which are metabolized by these enzymes. The problem is whether to deliberately exclude such subjects from trials or to include them as in a normal population, and then try to identify patients at risk, when the drug is used in practice. Poor metabolizers can, of course, be identified using the simple tests described; this is termed phenotyping.

By the 1980s, evidence had accumulated that the enzyme responsible for the metabolic oxidation of debrisoquine was a cytochrome P450,[23-25] called CYP2D6 according to the evolved nomenclature of pharmacogenetics. This enzyme was found to be absent in the livers of poor metabolizers.[26] The human cDNA for this enzyme has been cloned and its gene characterised,[27] so that putative poor metabolizers can be identified by genotyping rather than phenotyping. Nevertheless, because other environmental factors may be involved in determining the metabolic capacity, irrespective of the genotype, the analysis of drugs and metabolites in body fluids is likely to remain the mainstay of classification. Table 1.3 shows some of the drugs shown to be subject to this type of polymorphism. An interesting inclusion in this list is codeine which depends for its analgesic action on conversion to morphine; patients deficient in this P450 enzyme will not benefit from this action of codeine, thus explaining the often puzzling claim by some people that certain analgesics are better for them than others.[45] Certainly if a person is aware of his P450 status, he may be able to treat the odd headache more economically!

Another example is the metabolism of the recreational drug ecstasy or methylenedioxymethamphetamine (Figure 1.2). This compound is demethylenated to a neurotoxic catechol metabolite by a specific P450 enzyme, CYP2D6.[46] The possible consequence of this for ecstasy abusers who are poor metabolizers needs to be established. While such subjects may be less susceptible to chronic neurological effects of the drug, since they should produce less of the catechol metabolite, they may be at increased risk of acute toxicity because of impaired metabolism of the parent drug.[47]

There are several other cases of metabolic polymorphism, distinct from the debrisoquine/sparteine type described here. Kupfer et al.[48] noted unusual sedation in a volunteer following normal doses of mephenytoin. In this case, which was confirmed in the volunteer's blood relatives, only one of the two major metabolic pathways of mephenytoin, stereoselective 4′-hydroxylation of S-mephenytoin, was deficient, while N-demethylation to nirvanol was unaffected. The poor metabolizer can again be classified using simple analytical procedures to determine a hydroxylation index, or by measuring the ratios of the S- and R-isomers. Several other drugs, including hexobarbital, diazepam, nordiazepam, omeprazole and proguanil, have been shown to be substrates for the same polymorphic enzyme.[49]

Inter-individual differences in acetylation capacity have been known and studied for over 40 years, and the distribution differences in different populations is quite

TABLE 1.3
Debrisoquine-Sparteine Type Polymorphic Metabolism

Drug	Type	Pathway	Clinical effect	Reference
Amitriptyline	Antidepressant	*N*-dealkylation	Possible CNS	28
Amitryptiline	Antidepressant	Hydroxylation		29
Bufuralol	β-Antagonist	Aromatic and ring hydroxylation	Overdose of β-blocker	30
Clomipramine	Antidepressant	Hydroxylation		
Codeine	Narcotic analgesic	*O*-demethylation	Ineffective analgesia	
Debrisoquine	Antihypertensive	Ring hydroxylation	Hypotension	28, 31, 32
Desipramine	Antidepressant	Aromatic hydroxylation		33
Dextromethorphan	Antitussive	*O*-demethylation		
Eucainide	Anesthetic	*O*-demethylation	Ineffective	
Guanoxan	Antihypertensive	Aromatic hydroxylation	Hypotension	
Haloperidol	Tranquilizer	Oxidation-reduction	Increased side-effects	34
Imipramine	Antidepressant	Aromatic hydroxylation		28, 35
Indoramine	Antihypertensive	Aromatic hydroxylation	CNS toxicity	36
Metoprolol	β-Antagonist	Ring hydroxylation, *O*-demethylation	Overdose of β-blocker	37
Nortriptyline	Antidepressant	Ring hydroxylation		38
Paroxetine	Serotonin uptake inhibitor			39
Perhexiline	Antianginal	Aliphatic hydroxylation	Peripheral neuropathy, hepatoxicity	40
Phenacetin	Analgesic	*O*-demethylation	Methemoglobinemia	
Phenformin	Antidiabetic	*O*-demethylation	Lactic acidosis	41
Propranolol	β-Antagonist	Aromatic hydroxylation	Overdose of β-blocker	42
Sparteine	Oxytocic	*N*-Oxidation	Uterine contraction. CNS toxicity	21
Thioridazine	Tranquillizer	Ring sulfoxidation	Over-sedation	43
Timolol	β-Antagonist	*O*-Demethylation	Overdose of β-blocker	44

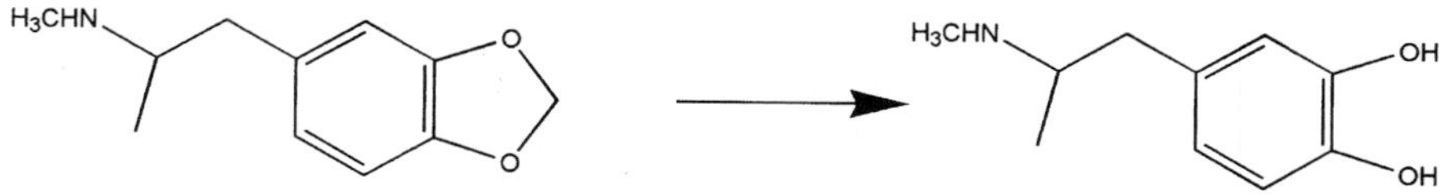

FIGURE 1.2 Metabolism of methylenedioxymethamphetamine ('Ecstasy') to a neurotoxic catechol metabolite.

remarkable, ranging from 5% in Canadian Eskimos to 90% in Moroccans.[50] Because of the role of acetylation in biotransformation or in the activation to carcinogens,[51] the acetylation polymorphism has been extensively studied in recent years.[52]

The test drug for the acetylator phenotype is usually isoniazid. In one common routine, isoniazide is given to the subject and the blood concentration is measured 4 hours later; populations will show a bimodal distribution into the fast and slow acetylators. Because acetylation can activate or deactivate drugs, then the correlation of phenotype with propensity to unwanted drug reaction is not so straightforward. A lupus erythrematosus-like syndrome is associated with poor metabolizers receiving hydralazine or procainamide, whereas there is a poor response to antituberculosis therapy with isoniazid in rapid acetylators.[53] Some examples of drugs metabolized by acetylation and therefore subject to a variable response in different populations are shown in Table 1.4. Recently, it has been suggested that regulatory authorities may require phenotyping of subjects in bioequivalence studies on new chemical entities.[63]

1.3 DRUGS IN RESEARCH AND DEVELOPMENT

In a recent book[64] on the development of drugs in the pharmaceutical industry, only 23 lines (out of nearly 300 pages) were devoted to the role of analytical services. This can make the analyst feel either very humble or very indignant. Selby[65] would have taken the latter view, maintaining that the analytical department was the hub in the wheel of all pharmaceutical research and development, having a crucial role, at least in all laboratory-based work. Even for the relatively narrow subject of the

TABLE 1.4
Drugs Subject to Variable Acetylation in
Population Groups

Drug		Reference
Aminoglutethimide	Antineoplastic	54, 55
Caffeine	Stimulant	56
Dipyrone	Analgesic	57
Hydralazine	Antihypertensive	58
Isoniazid	Antituberculous agent	59
Sulfapyradine	Sulfonamide	60
Sulfamerazine	Sulfonamide	61
Sulfamethazine	Sulfonamide	62

analysis of drugs in biological fluids in this book, the importance of the analyst is wide ranging as is indicated in Figure 1.1.

The topics discussed so far have referred to drugs already in use. However, the most extensive application of drug analysis in body fluids is in the development of a drug from its status as a new chemical entity to its final marketed form. The analytical method is applied at almost every stage of the development and, as in the various situations already described, the required characteristics of the method will depend on the particular stage of development. The following is a discussion of these various stages taken in the order in which they are encountered in the research and development process.

1.3.1   PHARMACOLOGY

Analysis of drugs in biological fluids can begin as early in the research and development process as in the initial pharmacology tests (i.e., the drug discovery phase). This depends on the philosophy of the particular research group, which could depend on the type of drug being researched. Some argue that the search is for a new chemical entity which has the required effect in the whole body, and once this is found, then the drug can be more extensively researched; the use of plasma and blood analyses prior to this decision is irrelevant and an unnecessary expense. Others argue that a knowledge of absorption, distribution, and elimination characteristics (see the later discussion on pharmacokinetics) is useful for interpreting pharmacological findings and could lead to better drug design. While both approaches have their points, the author believes that in recent years with more and more attention being paid to drug design — often by computer and long before the drug is synthesized, let alone administered to an animal — then the monitoring of the drug's progress in biological systems becomes vital to interpreting events and making development decisions.

One of the most seductive reasons given for the measurement of drugs in biological fluids in the research phase is based on the proposition that the concentration in plasma is directly related to the intensity of the effect; the assumption being that the concentration is proportional to the concentration at the site of action. It is usual in *in vitro* and in *in vivo* pharmacological tests to quote activity in terms of concentration, or in terms of a known standard. For example, we may say that the new compound is active in a particular *in vitro* test at a concentration of 10 ng ml^{-1}, or 0.5 mM, or only one-half as active as diazepam in the same test; we might also find that for the same compound one-twentieth of the dose is required to obtain the same effect as diazepam in an *in vivo* test. The difference in activity relative to diazepam in the two tests would be attributed to the difference in the ability of the compound and the test standard to reach the site of action in the *in vivo* test. Then, a more valid comparison for *in vivo* activity would be obtained in assessment of plasma concentrations of the two drugs at equieffective doses, with the expectation that the findings would be closer to the *in vitro* comparisons. It is generally accepted that the main factors determining the levels of drug achieved in plasma following oral dosing, are absorption, distribution, elimination, and the various physiological and pathological conditions which in turn affect these parameters. Thus the drug researcher, using the results of dose-plasma concentration-effect studies, will have a clearer idea of the intrinsic activity of the drug, but will, in addition, have some insight into the factors affecting drug delivery. The researcher can then decide if any conditions can be altered to obtain optimal *in vivo* effects, rather than accept or reject the drug candidate on the basis of single, whole-body, pharmacological observations.

An important consideration at this stage may be the extent of absorption of the drug. A low percentage absorption may not appear too great a problem if the drug has a demonstrable useful effect; one could merely increase the applied dose. However, low absorption (if one thinks of absorption as the amount of the drug molecule migrating from the gastrointestinal tract to the bloodstream) inevitably means variable absorption. Thus, a well-absorbed drug may vary between 90 and 95% total absorption, which would give variable plasma levels of mean values ±2.7% (2.5 divided by 92.5); a poorly absorbed drug with a smaller absolute range of absorption, say 1 to 2% would give plasma levels of mean values ±33% (0.5 divided by 1.5). The arithmetic is inescapable and in modern drug research it is unwise to settle on a poorly absorbed drug for development on the grounds that it is the only one that works. Sooner or later, the drug will fail on this variable consequence of poor absorption. The availability of a sound analytical method is a prime factor in this phase of decision making.

If the relation between plasma level and effect is to be developed rationally, then the nature of the drug is important. The most straightforward class of drugs in this respect is antibiotics, where the concentration in biological fluids is by definition equal to the effect, as the analytical method is almost always a microbiological one. Other classes of drugs, unfortunately, do not show the same obliging relationship. Antidepressants, for example, may take weeks of continuous dosing before their effects are seen,[9] even though the plasma levels at this time are no higher than the levels achieved after the first few doses. Another example is pentoxifylline, whose effect on improvement of walking distance in patients is seen only after 3 months of treatment.[66] Pentoxifylline itself has a very short half-life and the effect is not caused by drug accumulation.[67] While such phenomena can only spur the researcher to investigate the true relationship between plasma level and effect, in the absence of this knowledge, it may be useful to correlate plasma levels with biochemical measurements (such as catecholamine uptake for antidepressants and red cell deformability for vasodilators) but it is still necessary to demonstrate that these biochemical changes are related to the eventual desired effect.

Although classical bioassays are not extensively used in modern drug research as assay methods for drugs, being largely superseded by more reliable and consistent chemical assays, there has been a resurgence of assay methods depending on specific biochemical phenomena, exemplified by immunoassays and receptor assays (see Chapter 8). Such biochemical events may be controlled by single molecular interactions, so that, in theory, such assay methods could be extremely sensitive. However, they may be less specific than chemical assays, since the biochemical event may also be triggered by closely related compounds such as metabolites. In the case of receptor assays, one could argue that this is not such a bad thing because the assay would then measure total potential activity of the sample being measured. However, this argument is flawed for at least two reasons: (1) the typical receptor assay measures molecules that bind to the receptor and does not distinguish between agonists and antagonists, and (2) not all molecules in the plasma may be able to penetrate to the site of action to the same extent as the active compound. The receptor activity, then, is no more valid than using total immunoactivity of plasma samples — the immunoactivity of a compound being quite incidental to its pharmacological effect.

1.3.2   TOXICOLOGY

Once a new chemical entity has been shown to have a useful pharmacological effect the next step is generally to establish its safety by performing toxicity testing in animals. Initially, toxicology involves the administration of increasing single doses of drug to groups of animals (acute toxicology) to establish the toxic doses of single administrations. If enough animals are used and there is a clearly apparent lethal effect of large doses then it is possible to establish the LD_{50} (the dose which will kill half the animals in the test) for the drug in a particular animal using the particular route of administration. Such tests are administered by experienced technicians, using single doses to single animals, usually as solutions, and there is generally no doubt that the animals have indeed been dosed according to the experimental protocol. These well-controlled conditions allow the LD_{50} to be determined with reasonable accuracy, and there is usually no need to monitor plasma levels to ensure the validity of the test. However, if the increasing dose does not result in proportionately increased absorption (i.e., a constant fraction should be absorbed over the whole range of dosing), then the LD_{50} calculation may be subject to distortion, usually as a result of poor absorption at the higher doses. Alternatively, the larger doses may saturate the drug-metabolizing enzymes, thereby giving a proportionately greater dose to the systemic circulation at the higher doses. Thus it may be useful to determine plasma concentrations of drugs in acute toxicity tests to establish a better relation between toxicity and the exposure of the animal to the test drug. Unfortunately, this is rarely done and most acute toxicity testing is based on the pragmatic approach whereby the LD_{50} is established without the refinement of plasma level determination.

The next stage of toxicity testing is to dose animals with amounts of drug much lower than the established LD_{50}, but on a repetitive schedule over long periods (sometimes even for years). This stage is called the chronic toxicity testing phase. In this phase, it is impractical to dose large numbers of animals on an individual basis every day for such long periods. Instead, the drug is introduced into the animals' feed or drinking water. This method may introduce a considerable error in the amount of drug actually dosed. The error can be caused by a variety of factors: (1) the individual animals may not eat and drink to the same extent; (2) if the drug is not thoroughly mixed into the diet, then some animals may receive more than others; (3) and some animals may even be selective in picking out morsels of food without (or with!) the drug. A large error can be introduced if the feed needs to be mixed a long time in advance, so the animal is receiving food with no drug due to deterioration, or even worse, the animal may receive a toxic degradation product. The fact that the drug is given in the diet and only dosed with food may impart quite different absorption characteristics compared with the acute dosing regimen, where the drug would be given without food.

Thus, the likelihood of a false negative can be high in such testing if one depends entirely on the stated dose for interpretation of toxicity. In some toxicological protocols, the dose may even be stated in terms of mg per kg feed, rather than mg per kg animal, indicating the imprecision of the dose given. The only sure way to normalize values in this type of toxicity testing is to relate the toxicity to the plasma levels of drug or metabolites, usually in the form of steady-state concentrations achieved over long-term dosing. These sort of steps are now being demanded by regulatory authorities.

Despite this theoretically desirable approach to chronic toxicity testing, however, full implementation according to these recommendations could raise as many problems as it solves. For animals to be bled on a regular basis would raise problems for the observation of toxicological manifestations, particularly for small animals where a large fraction of the animals blood may be required for the assay. In such instances a parallel group of animals may be set up for assay purposes. If the whole animal is needed for the assay then a large parallel group is needed at the start of the study; such refinements will make the already high cost of toxicity testing even higher, when one considers the number of animals required.

Another considerable difficulty is when small animals have free access to food and water containing drug, and there is no measure of the time between the last dose and the time of death. This is especially important with drugs of short half-life (and drugs tend to have shorter half-lives in these animals than in larger animals or in humans). The problem is exacerbated in chronic testing where the animal becomes too ill to drink or feed, and the dose may be lower during this phase than stated in the protocol. For larger animals, such as dogs, which tend to eat all available food as soon as it is given, the timing between dose and sampling may be better controlled. Thus, although plasma monitoring in toxicity trials is a desirable goal, one should not underestimate the pitfalls in implementation and interpretation.

The question of specificity in such monitoring should be considered. If one remembers that the prime reason for analysis is to show that the animal has received the drug, then it is only necessary to establish a response which is related to the presence of the drug in the appropriate biological fluids. Hence, it will be unnecessary to distinguish metabolites from the parent, assumed active, drug. It is only if one wishes to relate concentration to effect that a degree of specificity is required, and until 10 years ago this was not generally considered in chronic toxicity experiments. Thus, it may be acceptable to apply a relatively nonspecific method for plasma assay, such as radioimmunoassay (see Chapter 8) for such monitoring, and this approach of including metabolites in the measurement will also help overcome the problem of drugs with very short half-lives.

Recently, toxicology has moved toward a more quantitative footing. Toxicologists have become concerned with the effects of dose levels, effects which can be critically dependent on dose, when one considers that large doses, capable of saturating metabolizing enzymes, are usually being administered. The time scale of toxicity has also been studied in increasing detail. These new areas of awareness have given rise to a fascinating new discipline, termed toxicokinetics, in which the quantitative assay of drug and metabolites in biological fluids and tissues occupies a pivotal position. These studies, however, are directed at furthering our understanding of toxicological mechanisms, rather than being an essential part in the development of a specific new chemical entity.

Toxicokinetics has been defined as the rates of absorption, tissue distribution and redistribution, enzymic and nonenzymic biotransformation, and excretion as related to toxicologic end points.[68] Toxicokinetics may be distinguished from pharmacokinetics in that it sets out to examine the fate of foreign compounds when given to organisms in relatively large quantities; transport systems as well as metabolizing enzymes may become saturated, protein binding of the drug and endogenous compounds may change, and the drug itself may have a different effect at the high concentrations reaching the biophase. The role of metabolites in the biological effect may be more important and the actual mechanisms for producing metabolites or active intermediates need to be elucidated. The requirements of analytical methodology for application in toxicokinetic studies have been reviewed.[69]

1.3.3   PHASE I CLINICAL TESTING

Assuming that the drug survives the toxicity testing then the next stage is to begin cautious testing in humans. Although termed tolerance testing, it is the nearest thing to a toxicology study in the target species. A very low dose is first given to a small number of closely monitored healthy volunteers. This dose would be considerably below the expected therapeutic dose and the volunteers are observed for any signs of undesired side effects. The dose is then increased to a preset limit, often in excess of the proposed therapeutic dose and the top dose in this series may even be the largest dose of the drug that is ever administered. For many drugs, the clinical effect may be only manifest itself in patients, that is, as alleviation of symptoms, with no direct effect in volunteers. Some drugs, however, may have direct effects on volunteers (e.g., diuretics, which cause increased elimination of body water into the urine) and in such cases the drug action would be expected to be seen in the volunteer tolerance studies. If such an effect is not seen, or if it not seen until unexpectedly high doses, it is important to establish whether this is from poor activity of the drug, or whether some other reason can be postulated for the lack of effect, such as poor absorption in the test subjects, or increased metabolism compared with the pharmacological test animals. Measurement of plasma levels is extremely useful at this point. If the levels are reasonably high, then it may be assumed that the drug itself is inactive in humans, and the analytical method will have been crucial in the termination of development.

If, on the other hand, the levels are lower than anticipated would be necessary for a pharmacological effect, then the consideration of absorption and metabolic factors needs to be addressed. In the case of poor absorption, in the sense of poor transport across the gastrointestinal membranes, then the formulation chemist may be able to improve the situation; if extensive metabolism has occurred, then there may be reason to terminate the project, or the drug may be used only as a parenteral formulation, which may limit its appeal. Many of the newer drugs which are peptides are expected to be very vulnerable to metabolic attack, even at the gut membrane, and their advantages may make it worthwhile to devise very sophisticated systems to deliver them.

Unfortunately, the measurement of plasma concentrations following an oral dose of drug does not enable the differentiation between poor absorption and extensive metabolism of drug by the liver prior to its delivery to the general circulation (the so-called first-pass effect). To obtain further information on this aspect, then, the analysis of plasma following intravenous dosing is necessary. However, at this stage of development, an intravenous form of the drug may not be available, either because such a delivery route is not anticipated, or the relevant toxicity testing has not been performed. Although it is feasible to develop an intravenous formulation for this specific study, it may involve unwanted, expensive intravenous toxicity testing and the development decision may need to be made on the basis of animal metabolism studies. Such considerations also apply to drugs without a demonstrable effect in normal volunteers, but without the extra guidance afforded by the effect.

1.3.4   METABOLISM

The initiation of animal metabolism studies will be made at the same time as the tolerance testing in humans. Drug metabolism is a fascinating subject in itself and is discussed more fully in Chapter 2. In this chapter, the author will only mention some of the reasons for such studies in animals. One of the more popular reasons for such studies is to establish that the pathways of metabolism in the species chosen for toxicity testing are the same as those in humans. The philosophy is that if the metabolism pathways are different in the two species (humans vs. experimental animal), then it is possible that a metabolite not formed in humans may be the toxic principle in the animal, hence giving rise to misleading extrapolations to a toxic effect in humans. Conversely, if a toxic metabolite is formed in humans but not in the animal, then the toxic potential would not be predicted from the animal studies and there would be an unwelcome surprise during the clinical testing. If a test species, or a combination of test species, cannot be found which is subject to the same chemical exposure as humans, then it may be necessary to synthesize the metabolites formed in humans and test them separately for toxicological effects. The study of drug metabolism is therefore justified as being a science which can validate the toxicity species used. In the best ordered research organizations, the drug metabolism studies would actually be used to select the species to be used. Because the toxicity testing can be the most expensive phase of preclinical development, such a rational approach can make a significant difference to the cost, efficiency, and success of the research program; cost, because it will be unnecessary to repeat expensive toxicity testing in an alternative species or with synthesized metabolites; efficiency, because it minimizes the time lost in such repeat studies, just at the time when the drug program should be being accelerated; and success, because the correct decision is made when the toxicity report arrives. However, in reality, custom and practice still dictate that the principal species used in toxicology are rat and dog, and metabolism studies tend to concentrate on these two species, until there are specific indications to perform studies in other species.

Metabolism studies also play a role in the selection of the species used for pharmacological testing. Recently, this aspect has been observed with the emphasis being placed on safety in the regulation of new drugs. Drug metabolism experts have spent much of their efforts validating the toxicity species as described above. However, a knowledge of the metabolism of the drug in the species used for pharmacology will alert the researcher to the possibility of the formation of active metabolites. For example, a drug which is extensively and rapidly metabolized but is apparently extremely active may well act through one of its metabolites. An early recognition of this fact, possibly coupled with findings of little metabolism in a species in which the drug is inactive, enables the researcher to consider the possible development of the metabolite as a drug in its own right.

Full metabolism studies, as indicated in Chapter 3, involve the description of absorption, distribution, chemical transformation, and elimination of drug and comparison of such events in humans and experimental animals. An extremely useful aid in such studies is a radiolabelled form of the drug, usually ^{14}C rather than ^{3}H, to minimize the loss of label either by metabolic exchange or metabolic elimination. For a full description of metabolic pathways, the metabolites can be isolated by following the label through various purification steps, using standard radioactive monitoring procedures, and subsequent final characterization of partially purified material by accepted physicochemical methods. The full elucidation of structures of all metabolites is often extremely complex, and an open-ended commitment to such a program may not be appropriate in the early stages of drug development. Nevertheless, as separation techniques become more rapid and can be used in direct conjunction with such powerful identification tools as mass spectrometry and nuclear magnetic resonance, the complete production of metabolic pathways of candidate drugs is more feasible than a decade ago.

In the meantime, the use of radiolabel in initial studies can be supported by using the specific analytical method which will have been developed at this stage, to analyze samples containing radiolabel. The discrepancy between total radioactivity and specific identification of parent drug will give an early and useful indication of the extent of metabolism, as illustrated in Figures 1.3 and 1.4 for the comparison of profiles of drugs which are extensively or poorly metabolized.

The advantage of using animals for metabolism studies is that they can be used to investigate the changes in metabolism when different dosage routes and schedules are used without the extensive toxicology back-up that would be necessary for such tests in volunteers. Thus, the comparison of oral and intravenous testing to obtain data on the first-pass effect can be performed on animals, preferably a large animal such as a dog, where serial samples can be taken from the same animal, with crossover designs also being used for a group of animals. Thus, in a typical experiment, plasma would be analyzed for parent drug in a group of 5 to 6 dogs, dosed on two separate occasions (once intravenously, once orally) with the test drug and the resulting profiles compared (Figure 1.5). Although comparison of total radioactivity in such an experiment could also indicate first-pass metabolism, the effect would be masked if the metabolites had the same half-life as the parent drug. By using a specific assay method, however, the extent of first-pass metabolism in animals can be clearly characterized, and on the basis of experience with similar drugs the transference of the animal findings to humans can be considered. The decision of

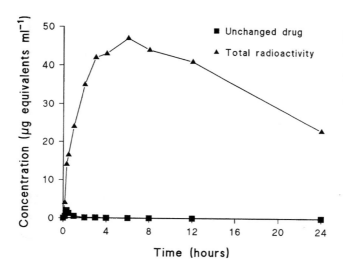

FIGURE 1.3   Plasma concentration–time profile of total radioactivity and unchanged drug illustrating the case where the orally administered drug is extensively metabolized.

whether to expect poor bioavailability in humans may then be possible without resorting to an intravenous study in humans. Obviously, such decisions will be of high quality if the corresponding comparative metabolism is known, which returns us to the argument of selecting test species which are as close to humans in their metabolism as possible.

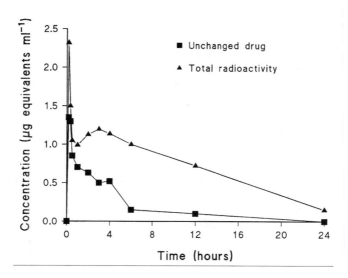

FIGURE 1.4   Plasma concentration–time profile of total radioactivity and unchanged drug illustrating the case where the drug is metabolized to a limited extent.

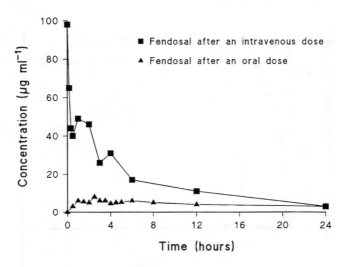

FIGURE 1.5 Comparison of plasma profiles of the same dose of a drug administered orally and intravenously to a dog, illustrating an extensive first-pass effect in this species.

1.3.5 PHARMACOKINETICS

We have now reached the stage where the typical candidate drug has been shown to be active in animal tests, safe in toxicity tests, tolerated in quite high doses in humans, absorbed in humans without too rapid metabolism to inactive species, and, in certain cases, to be active in humans. The next stage is to develop the pharmaco-kinetics of the compound. This term is given various definitions in the scientific literature and there is some confusion as to whether the term includes all aspects of what the animal does to the drug, as opposed to what the drug does to the animal (pharmacodynamics) or whether the term should be applied only to the specific mathematical description of the disposition of the active drug in the body. The former definition would include all aspects of absorption, distribution, transformation, and elimination. For purposes of this discussion, and during most of this book, the author shall use the narrower use of the term where rates of various transfers and transfor-mations are considered but not full descriptions of factors affecting absorption distribution and elimination, nor will it be used to cover the actual chemical trans-formation for which the term drug metabolism is usually applied.

The mathematics of pharmacokinetics has been the subject of numerous reviews and treatises. It is possible to discuss the subject in great depth and a clear under-standing and application of the principles involved will make it possible to reach conclusions not immediately apparent from the raw data. However, it is just as important to appreciate the limitations of analytical results before applying the principles of pharmacokinetics to such results; the pharmacokinetic conclusions may be justified by the mathematics, but the quality of the data may make such conclu-sions invalid. In this book, the author does not intend to give a mathematical

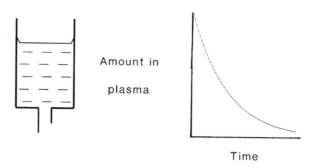

FIGURE 1.6 Simple hydraulic model model illustrating the elimination of a drug from plasma by a first-order process. The concentration decreases exponentially with time.

exposition of pharmacokinetic principles, but rather to give an extended discussion of a simple model which helps to appreciate the principles and their repercussions in any pharmacokinetic application.

The model the author has found to be extremely useful in understanding the broader aspects of pharmacokinetic theory is the hydraulic model first proposed by E. R. Garrett[70] but now somewhat modified. In this model, the concentration of drug in, for example, the plasma is represented by the height of the water in a cylindrical vessel (Figure 1.6). The rate at which drug is removed from plasma by natural processes is, at any particular time, proportional to the concentration and this is represented in the model by an outlet at the base of the cylinder, with the rate of flow through the outlet being directly proportional to the height of the fluid (concentration of drug). Mathematically, this is a simple mono-exponential decay, which quite accurately describes the situation where the drug introduced into the body by a single intravenous injection is eliminated from the body. Being a single mono-exponential decay, the rate of elimination can be described, from the observed curve, with a characteristic half-life.

To represent an oral dose, a second vessel is added to the model, representing a supplier of drug to the plasma or central compartment (Figure 1.7). The initial level in this supplying vessel (stomach or gut representing the site of absorption) corresponds to the total amount of drug available for absorption. As for the central compartment this loss from the site of absorption and into the central compartment is represented by an outlet at the base of the vessel; the absorption process is therefore also represented in the model as an exponential one. In this model, at any one time, the actual height of the fluid in the central vessel is determined by the two first-order processes (i.e., absorption from the gut, and elimination from the plasma) and can be described mathematically by the equation familiar to basic pharmacokinetics — the Bateman function:

$$y = Ae^{k_1 x} - Be^{k_2 x} \tag{1}$$

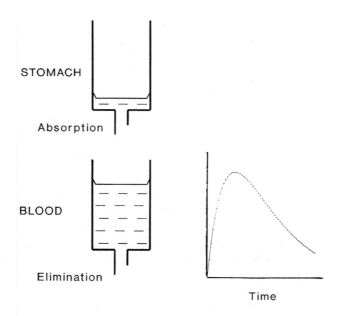

FIGURE 1.7   Hydraulic model illustrating the absorption of drug from the absorption site into the central compartment and subsequent elimination. Both processes are first-order and the plasma concentration–time profile is the result of two exponential processes.

where y is the concentration in the central compartment, k_1 and k_2 are the two controlling rate constants, A and B are constants depending on the sizes of the vessels, etc., and x is the time variable.

For describing the concentration time profile for a drug, y is the concentration in the plasma at time x and the two rate constants k_1 and k_2 are the absorption and elimination rate constants, respectively. Even after several decades of discussion, the term absorption in pharmacokinetics still is apt to be confusing. As described here, it is simply the action of being transported across a membrane into the blood. In real pharmacokinetic experiments, however, the researcher is usually measuring the appearance of drug in the plasma rather than disappearance from the site of absorption; the pharmacokinetic analysis of the plasma curve, assuming the simple Bateman function, will then produce a value for the absorption constant which is distorted by any loss of drug between the two sites, for example by metabolism or degradation of the drug. A useful alternative term to circumvent this confusion is invasion (used in the German literature), which avoids the necessity to explain that absorption includes such processes as transfer across the gut membrane and metabolism by the liver before the drug reaches the systemic circulation.

The outlet from the central compartment represents all processes that contribute to the elimination of drug from this compartment. The two major processes are metabolism and elimination via the excretory organs such as kidney. Thus, the model can be further refined by designating two separate outlets. If the "metabolism" outlet is large in comparison with the "kidney" outlet, then obviously metabolism will be

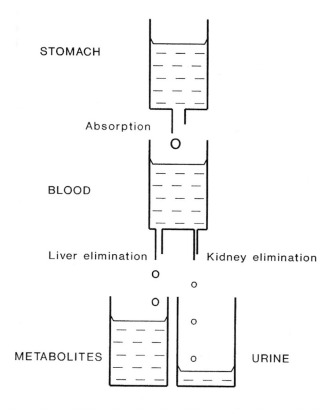

FIGURE 1.8 Hydraulic model illustrating absorption of drug into the plasma and subsequent elimination by two separate processes with different rates.

the principal method of elimination; if the converse is the case, then the drug will tend to be eliminated unchanged into the urine (Figure 1.8). This simple extension of the model demonstrates that in the case of renal failure, the drug may still be eliminated at a reasonable rate, but the elimination will be by metabolism rather than excretion; the patient is at risk, not through build-up of drug (which can be avoided by lower doses, but by production and build-up of toxic metabolites, which would not be seen in the healthy subject receiving the same drug). The simple model can be developed further to include other physiological processes. A mathematical treatment of the model leads to extremely complex-looking equations, but the use of the hydraulic model will continue to provide a clear understanding and appreciation of the effect of these processes on pharmacokinetics without recourse to the mathematics.

An important concept in pharmacokinetics is the idea of distribution of drug in the body tissues other than the plasma, including, of course, the target tissue. This distribution can be represented by a vessel connected to the central compartment at the base of the two vessels. The flow of fluid from the central compartment to this new vessel (the body compartment) will depend, as does the elimination, on the level of the fluid in the central compartment; however, there is also a reverse flow depending on the height of fluid in the new vessel, so that a dynamic equilibrium between the two compartments will be set up. At equilibrium, the level in the two vessels will be the same. This does not say that the concentrations in plasma and tissue will be identical, but it provides a useful description of the concept of volume of distribution, where the rest of the body is described by the volume it would have to be, if the concentration was the same as in the plasma. In our model, the volume of distribution is represented by the total volume of the various vessels; the amount of drug in any one compartment is still represented by the level of the fluid, but the true concentration needs to account for the volume, which in the model can be varied by varying the diameter of the cylindrical vessels. A body compartment represented by a cylinder with a large diameter much larger than the central vessel will be one with a large volume of distribution, that is, it will sequester a large proportion of the total drug and will have an actual concentration higher than that of plasma. As the level in the central vessel falls by other processes, the dynamic equilibrium ensures that the fall is partly compensated by the store in the body compartment and the observed decay in the plasma-time curve will not be the pure elimination curve. In pharmacokinetic nomenclature, this is usually described as a "deep" compartment, although in this model, the term "wide" compartment would be more appropriate; similarly a "narrow" compartment would more appropriately describe an organ with little affinity for the drug, rather than the more usual "shallow" compartment.

Another factor in the redistribution process is the rate of transfer between the vessels, which will depend on the bore of the interconnecting tube; for a wide bore, there will be almost instantaneous equilibrium (a rapidly equilibrating tissue), for a very restricted connection, the equilibration will be slow, but such tissues will retain drug, and at higher levels than the plasma levels, long after it has all but disappeared from the plasma.

The model so far has assumed all processes are first-order processes, that is, the rate of transfer depends at all times on the various concentrations of drug at the particular time. Where the first-order rate no longer applies, it is simple to introduce physical changes to the model, which can readily show the effect of phenomena which are not first-order processes. For example, drug may be introduced to the central compartment, not by a single instantaneously distributing injection, but by a constant infusion over a period of time. This can be represented in the model by a constant head reservoir so that the supply of new drug to this compartment is a zero-order process. To represent other types of delivery, the supplying vessel can be designed with specific shapes so that the rate of release is a particular function of the amount remaining.

A more complex picture occurs when a process is episodic, such as in repeat dosing, or when drug is reabsorbed following biliary excretion. In the former case, the model is simply created by refilling the supply vessel at appropriate intervals and the resulting rise and fall of the levels in the other compartments will simulate the concentrations of drug expected on repeat dosing regimens. For biliary recirculation, the bile can be represented by a vessel similar to an elimination compartment, with a pump that periodically transfers part or all of its contents back to the absorption vessel.

Despite the often surprising findings in following the concentration–time profile of drugs in biological fluids, the simple hydraulic model can almost always be

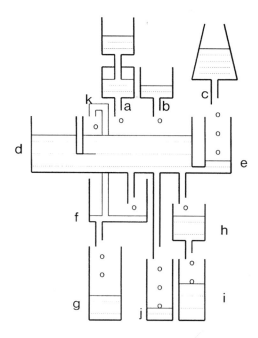

FIGURE 1.9 An extended hydraulic model illustrating a number of physiological processes. (a) absorption by a zero-order process; (b) absorption by a first-order process; (c) absorption by a non-linear process; (d) concentration in plasma and in rapidly equilibrating tissues; (e) concentration in a slowly equilibrating tissue; (f) concentration in the bile; (g) fecal excretion; (h) metabolites in the plasma; (i) metabolites in the urine; (j) unchanged drug in the urine; and (k) biliary recirculation.

adapted to picture the processes involved. An extended model which can represent most of the kinetics of drugs in the body is shown in Figure 1.9.

It can now be seen how the understanding of pharmacokinetics and interpretation of changing plasma concentrations can be used to understand further the disposition of drugs in the rest of the body. Pharmacokinetics consists largely of fitting mathematical functions to the observed concentrations of drug in biological fluids; if the mathematical functions represent physiological functions, then the analysis can be used to relate changing plasma levels to levels of drug at the sites of action. Thus, if the plasma concentrations arising from an oral dose of drug can be loosely fitted with a biexponential function of the type mentioned previously, then the drug can be said to be rapidly distributed to equilibrating tissues after absorption, and is eliminated by a process depending only on the plasma concentration. If a triexponential equation is necessary for an acceptable fit, then either the distribution phase is less rapid, and affects the overall shape of the plasma curve, or there is a distribution into a slowly equilibrating compartment, so that the observed, longer terminal half-life is determined by this slow rate of equilibration, rather than by the true elimination rate. It is not a true half-life, however; if the analytical method is sensitive enough, very low concentrations will be observed and the longer the drug can be measured, the longer this half-life will appear. This is a very good example why an understanding

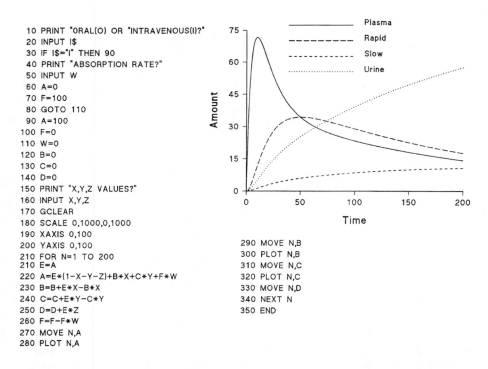

```
 10 PRINT "ORAL(O) OR "INTRAVENOUS(I)?"
 20 INPUT I$
 30 IF I$="I" THEN 90
 40 PRINT "ABSORPTION RATE?"
 50 INPUT W
 60 A=0
 70 F=100
 80 GOTO 110
 90 A=100
100 F=0
110 W=0
120 B=0
130 C=0
140 D=0
150 PRINT "X,Y,Z VALUES?"
160 INPUT X,Y,Z
170 GCLEAR
180 SCALE 0,1000,0,1000
190 XAXIS 0,100
200 YAXIS 0,100
210 FOR N=1 TO 200
210 E=A
220 A=E*(1-X-Y-Z)+B*X+C*Y+F*W
230 B=B+E*X-B*X
240 C=C+E*Y-C*Y
250 D=D+E*Z
260 F=F-F*W
270 MOVE N,A
280 PLOT N,A
290 MOVE N,B
300 PLOT N,B
310 MOVE N,C
320 PLOT N,C
330 MOVE N,D
340 NEXT N
350 END
```

FIGURE 1.10   BASIC program and printout to illustrate drug profiles generated by the hydraulic model. In the example shown, the relative transfer rates were input as 0.04 (absorption), 0.005 and 0.0005 (the rapidly and slowly equilibrating tissues, respectively), and 0.002 (urinary elimination). The amounts of drug in the plasma, rapidly equilibrating tissue, slowly equilibrating tissue, and urine are shown with time.

of distribution must be coupled with an understanding of stated limits of detection in order to correctly interpret pharmacokinetics.

Distribution of drug into the body compartment is generally rapid compared with absorption and elimination and the distribution term is generally too small to affect the goodness of fit following an oral dose. For an intravenous dose, however, this distribution phase has a marked effect on the initial plasma levels and this term would need to be taken into account.

In the first edition of this book,[71] a very simple BASIC program was described, which enabled quite complex pharmacokinetic models to be constructed. This method of representation is shown in Figure 1.10. An alternate, and equally simple method is to use a spreadsheet, such as LOTUS 1-2-3, SUPERCALC, EXCEL, or QUATTRO. In the spreadsheet method, the rate constants to be used in the model are first placed in cells near the top of the sheet, say: cell B2 for the fraction of the drug absorbed in unit time; cell D2 for the fraction of drug removed from the blood to the urine. For the main part of the spreadsheet, a line of cells is set up with the initial values of drug amounts in the various compartments: gut, blood, and urine. The cell below each of these initial values contains the calculation for the new amount in that compartment after the unit time interval. Thus, if the amount in the gut at zero time is set at 100 (cell B4) then the new value in cell B5 is calculated as +B4–B4*B$2. (The term B$2 rather than B2 is used for convenience when formulas are copied to other cells using the usual spreadsheet methods.) Similarly, the amount in blood after the same interval is +C4+B4*B\$2–C4*D\$2 and in urine is D4+C4*C\$2. When these formulas are copied for the desired number of time steps and the values resulting in the columns are plotted as a time sequence then the classical pharmacokinetic curve shown in Figure 1.7 is generated. More complex models can be simulated merely by extending the spreadsheet rightward and placing the appropriate formulas in the cells for as many compartments as the experimenter desires. Using this method, the identical curves can be generated as shown in Figure 1.10 for a drug absorbed into the bloodstream and distributed to a slowly equilibrating tissue and a rapidly equilibrating tissue, as well as being excreted in the urine. The effect of changing any of the rate constants listed can be rapidly tested by changing the values at the top of the spreadsheet. This technique is very powerful for illustrating many of the principles of pharmacokinetics and perhaps, more importantly, for illustrating the consequences of physiological changes. For the purposes of this example, however, the author will limit it to pointing out that a very small amount of drug distributed to a slowly equilibrating compartment will result in a very long residual half-life, if the analytical method is sufficiently sensitive.

At this point, we must consider the importance of drug determinations in obtaining high-quality pharmacokinetic interpretations. If the analytical method does not have high sensitivity, then these terminal phases will not be found, and a very important aspect of the kinetics of a drug will consequently be overlooked. If the drug is extensively metabolized and the metabolites have different kinetic characteristics, as they almost certainly will, then an analytical method that does not distinguish metabolites will lead to completely erroneous pharmacokinetic descriptors for the drug. In fact, the true kinetic analysis of a plasma profile which measures more than one species will be the sum of a number of exponentials and would be impossible to fit to a unique equation enabling a physiological interpretation to be placed on the data.

Even when the method is sensitive and specific, it is apparent that for treatment which involves mathematical manipulations, the numerical values assigned must be as accurate and precise as possible. Comprehensive pharmacokinetic analysis of drug concentrations in biological fluids may also depend on precise timing of sampling, especially in the case of rapidly changing concentrations as, for example, immediately following a bolus injection.[72] Our concern, however, is on the importance of the laboratory values. Ironically, the full application of pharmacokinetics demands the most precise determinations at the lowest concentrations (i.e., the accurate determination of terminal half-life can be very susceptible to errors in determination of low plasma concentrations). The importance of this can be seen in the consideration of the only other pharmacokinetic concept introduced in this book. This is the concept of the area under the concentration-time curve (AUC). This concept is intended to convey a sense of the total exposure of the body to the drug over a given period — the units of AUC, concentration multiplied by time, indicate this. A drug with a long, flat profile could have a similar AUC to one with a short sharp profile, and the two would have the same total exposure. Often the drug researcher will wish to determine the AUC on extrapolation to infinite time to obtain the true total exposure, the AUC_{∞}. To calculate AUC_{∞}, it is necessary to calculate the AUC from the available data, usually by a trapezoidal approximation and then add the terminal portion derived from the product of the last measured point and the terminal half-life. Because the half-life can only be calculated from these later, lower concentrations, then both the concentration and the half-life can be subject to large errors and hence the added portion of the AUC, which may be a significant part of the total, will be subject to large errors. It is important for the analyst to understand this before dismissing values measured toward the limit of sensitivity of the method as being of little importance, when these values are to be used in bioequivalence studies.

It is usual to apply analytical methods of proven quality to definitive pharmaco-kinetic studies in the early stages of administration of drug to humans. This should establish the true elimination half-lives of the drug and also its volume of distribution if an intravenous experiment is performed. The kinetic parameters can then be used to predict the course of plasma or tissue concentrations for various dosing regimens. It is then standard to test these predictions by suitable sampling and assaying in appropriate experiments. Thus, in these early phases of testing in humans, the analytical method is critical and is extensively used. Important decisions may be made on the basis of these experiments and the importance of high-quality data cannot be overestimated. In a typical series of experiments in humans, it might be thought necessary to establish linear kinetics (i.e., an increase in dose gives a proportional increase in plasma concentrations at the equivalent times after dosing), bioavailability (comparison of AUC of orally and intravenously administered drug), and the extent of accumulation following multiple dosing. This phase of drug testing could well involve 10,000 plasma and urine analyses, a considerable expense, even within the enormous expenses of modern drug research and development.

1.3.6 FORMULATION DEVELOPMENT

After establishing the basic pharmacokinetics of the drug, the analytical method is then used to monitor the performance of the marketed dosage form. Initially, the dosage form is likely to be designed as the most bioavailable oral formulation. Such studies will involve the administration of oral solutions and the developed dose to volunteers in a cross-over design with the measurements of appropriate plasma profiles for comparison with resulting AUC, peak concentration, and time to peak and hopefully these will be identical (Figure 1.11). For further development it may be necessary to extend the peak plasma levels (i.e., to slow down the absorption, by changes in formulation), while still ensuring the total exposure of the body to the same amount of drug. In such formulations the time to peak will be later, the actual value of the peak concentration will be lower, but the AUC_∞ should be unchanged in the ideal case. In practice the AUC_∞ for sustained release formulations will always be less than that for the best bioavailable form, since the sustained release formulation cannot stay at the site of absorption indefinitely and the true biexponential curve will be truncated at the point at which the formulation leaves the site of absorption; at this point, the observed curve will revert to the normal decay curve for the elimination of drug from the plasma (Figure 1.12). The requirements for the analytical method during formulation develop-ment are similar to those for establishing basic pharmacokinetics except that specificity for the unchanged drug compared with metabolites is not so critical, provided metabo-lism proceeds by first-order rate processes.

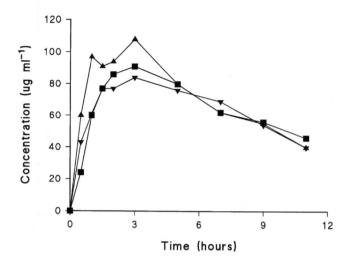

FIGURE 1.11 Comparison of plasma concentration–time profiles of tolbutamide in subjects dosed with three different formulations of the drug, illustrating the bioequivalence of the formulations.

1.3.7 PHASE II CLINICAL TESTING

Development now proceeds to early clinical trials. Analysis of drugs in biological fluids has two distinct roles in this phase. First, it may be necessary to confirm the kinetic picture of the drug, already seen in volunteers, in patients, as a prelude to

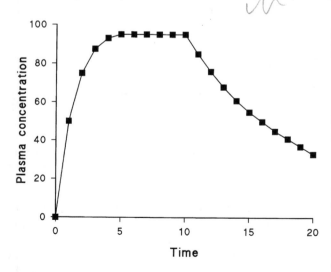

FIGURE 1.12 Effect of removal of sustained-release drug from the absorption site on the plasma concentration–time profile. There is a sustained plasma concentration as the drug is delivered to the systemic circulation until the formulation leaves the absorption site and then the plasma level decays according to the inherent half-life of the drug.

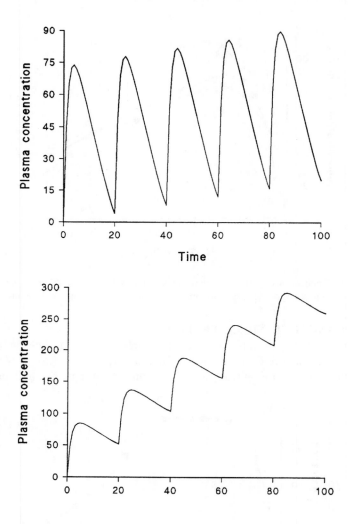

FIGURE 1.13   Effect in change of rate of metabolism of a drug on the accumulation of drug in the tissues following repeat administration. In the first case, accumulation is relatively modest; a decrease of Vmax to one-half of its normal value results in accummulation of drug in the tissues.

determining the biological effect in patients. The kinetics may be influenced by the disease state itself, particularly in cases of renal or hepatic failure, or in hyper- or hypotension, and this could lead to a reconsideration of the dose necessary in a patient compared with doses in volunteers. An example of this is when the drug is intended for use, or is likely to be used in aged populations. Metabolism is often slower in such populations and more variable, as exemplified by fentiazac,[73] which leads to unanticipated accumulation if the kinetics are assumed to be the same as for healthy, young volunteers. Figure 1.13 illustrates the effect on accumulation of a drug

TABLE 1.5
Special Populations for Which
Pharmacokinetic Considerations Should Be
Applied in Clinical Phase II Studies

Elderly
Renal insufficiency
Hepatic disease
Obesity

which is metabolized at different rates in different populations. Kinetic studies in aged populations were included in early clinical trials by many pharmaceutical companies as soon as this problem was recognized, and is recommended by the World Health Organization.[74] Several more categories of special populations are now recommended for special pharmacokinetic consideration (Table 1.5)

The second role in analysis in Phase II clinical trials is testing for compliance (i.e., providing assurance that patients in the trial are actually taking the drug). In some ways, this may be considered analogous to the situation already described for animal toxicity trials where there may be concern that the animal may not have been exposed to the test drug. However, in the clinical trial situation, the concern is whether the patient has actually taken the drug as instructed. This can be a great problem in medicine. Patients have been demonstrated, on the whole, to be very noncompliant with doctor's instructions, whether in change of lifestyle or in taking medicines, unless highly motivated to do so.[75] Often, the advantage of one drug over placebo may be real, but if large numbers of patients in the trial are not taking the drug, or are taking drugs other than those intended in the trial, the differences will not be observed. Clinical trials are very expensive to perform and obviously the quality of the decisions arising from a clinical trial is considerably enhanced if the trialist knows absolutely that all the members of each group in the protocol are receiving the prescribed drug, comparator, or placebo. At this point in the drug's development, the extent of the absorption of the drug has been established and the compliance test is not designed to confirm absorption, as in the exercise of monitoring animal toxicity studies, but is intended to confirm that the drug has actually been administered. In its simplest form, this may consist of asking the subject to bring the bottle back to the clinic on each visit and counting the unused tablets, or even to use intelligent medicine bottles complete with microchip that counts the number of times the bottle has been opened and reclosed between visits to the clinic.

A more direct test is to take blood samples and analyze them for the drug. However, this is obviously inconvenient, and a more usual approach is to ask the patient to bring or supply a urine specimen at each visit — a common procedure not likely to cause the patient any apprehension. In fact, analysis of the sample does not have to be for unchanged drug but could even be for an excipient of the tablet, or other marker. The essential point is to establish that the patient has ingested the drug. However, analysis for drug-related products (i.e., metabolites, particularly in urine) does produce more information, particularly in patients who may, legitimately or otherwise, be taking other drugs.

TABLE 1.6
Results of Compliance Studies During
Phase II Clinical Trials of Isoxepac

Trial 1: Isoxepac vs. placebo

| | Number of subjects detected with | | |
	Isoxepac	No isoxepac	Totals
Aspirin	53	69	122
No aspirin	107	134	241
Totals	160	203	363

Trial 2: Isoxepac vs. aspirin

Aspirin	64	173	237
No aspirin	144	38	182
Totals	208	211	419

Tests developed for compliance purposes obviously need to be specific for a chemical entity related to the dosage form whether it be an excipient, a marker, the drug, or its metabolites. Because the timing of the sampling following the supposed dose is likely to be unclear, quantitative information will be of little value. The sensitivity, however, needs to be such that the drug can be detected up to 24 h after dosing. Because of the conditions under which the compliance tests are made (often in doctors' surgeries) then the test should be as simple as possible (a dipstick type of assay is ideal). In one of my laboratories we developed a thin-layer chromatography method for an anti-inflammatory drug, isoxepac, which, though rather complex for routine use in the average doctor's surgery, nevertheless allowed analysis of large numbers of samples during the clinical testing phase. This test had the advantage that the urine could also be tested for aspirin metabolites. The test was applied to patients in several trials of isoxepac and some of the results are summarized in Table 1.6. In the first trial, subjects were ostensibly being tested against placebo although they were allowed to take aspirin if they felt the need. The compliance test showed that half of the patients could be identified as isoxepac takers, a very good figure indeed, assuming the other half belonged to the placebo group. However, significant numbers in each group were also shown to be taking aspirin. In interpretation of the clinical results of the trial, a comparison of the group identified as isoxepac-non-aspirin against the group identified as non-isoxepac non-aspirin would be expected to indicate more accurately whether isoxepac was better than placebo than merely comparing groups according to the protocol.

In the second trial, aspirin was used as the comparator and it can be seen that more than half of the patients were aspirin takers including a significant number of isoxepac takers. Again, the most valid comparison of clinical efficacy would be obtained by comparing the aspirin-non-isoxepac group and the isoxepac-non-aspirin group. Unfortunately, in the trials described here, the development of the drug was terminated before the clinical analysis was completed.

TABLE 1.7
Requirements for Methods for the Analysis of Drugs in Biological Fluids for Different Purposes

Purpose	Requirement				
	Specificity	Sensitivity	Precision	Accuracy	Speed
Forensic toxicology	Low	Low	Low	Low	Medium
Overdosage	Medium	Low	Low	Low	High
Drug monitoring	Medium	Medium	Low	Low	High
Pharmacology	High	Medium	Medium	Medium	Low
Toxicology	Low	Medium	Low	Low	Medium
Phase I	Medium	High	Medium	Medium	Medium
Metabolism	High	Medium	Medium	Medium	Low
Pharmacokinetics	High	High	High	Medium	High
Formulation development	Medium	High	Medium	Medium	High
Phase II	High	High	High	High	High
Compliance	Low	Medium	Low	Low	Medium
Pharmacodynamics	High	High	High	High	Medium
Pharmacogenetics	High	Low	Medium	Medium	Low

1.3.8 PHARMACODYNAMICS

The question of correlating plasma drug levels with pharmacological response was referred to at the beginning of this discussion on the use of analysis of drugs in biological fluids in research and development. When a successful drug has been developed to the point of full clinical trials the wheel has turned full circle and there is a renewed intellectual drive to attempt to relate plasma levels with effect in human volunteers or in patients. There is greater justification in pursuing this research when the drug has proven efficacy, and a thorough understanding of the levels necessary to obtain the desired effect makes formulation design and dosage regimen more sensible. This could be extremely useful, for instance, in designing a sustained release formulation to provide exactly the right plasma concentration and to avoid the near toxic effects of large doses which may be thought necessary to achieve effective levels for a reasonable time. Apart from avoiding the short-term high concentrations, a suitably designed sustained release form will result in an overall smaller exposure of the body to the drug and therefore, presumably, a safer medication. Thus, the exercise of analysis of drugs in biological fluids is a key factor in the establishment of a concentration–effect relationship and the monitoring of dosage forms to achieve this optimal performance.

In this opening chapter, the author attempted to describe the many reasons for analyzing drugs in biological fluids during the research, development, and use of drugs. The discussion has brought in many different disciplines, full discussions of which are beyond the scope of this book, but the author hopes he has given enough of a thumbnail sketch to enable the analyst to appreciate the importance of various aspects of such work. The final table in this chapter (Table 1.7) summarizes the various topics which have been discussed and indicates the relative importance of the features of the appropriate methods to use. A more detailed discussion of the terms, precision, accuracy, sensitivity, and specificity is found in Chapter 10 which focuses on development and evaluation of the analytical method.

2 Special Problems with Biological Fluids

CONTENTS

2.1 INTRODUCTION

In its simplest form, the analysis of drugs consists of presenting the sample to a measuring device and receiving the answer to the questions "What?" and "How

35

TABLE 2.1
Biological Samples Listed in Descending
Order of Fluidity Corresponding to
Degree of Difficulty of Analysis

Liquids	Cerebrospinal fluid
	Tears
	Sweat
	Saliva
	Urine
	Bile
Mixed	Plasma
	Serum
	Blood
	Feces
Solids	Brain
	Heart, kidney, liver
	Lung, muscle
	Bone

much?" However, a biological fluid is not a simple mixture, but a very complex one containing many different components which may subtly react with one another and hence contribute to some sort of interference with the end point, perhaps by elevating the response, or by masking it, or by altering the actual values by degradation (pH factors, enzymes). This chapter will describe some of the special problems which arise from the nature of the fluid and its constituents.

The biological fluids that are most commonly analyzed are whole blood, plasma (or serum), and urine. Because preparation of plasma or serum is so simple and gives a dramatic decrease in interfering components for most types of assay, whole blood is less often analyzed, although for small animals and forensic toxicology the direct analysis of whole blood may be unavoidable. Less common fluids requiring analysis include bile, sweat, milk, and saliva. Saliva may be considered particularly useful where it can be shown to reflect accurately the concentration in blood as it is a noninvasive technique. Spinal fluid may also be analyzed, particularly for compounds acting on the central nervous system as the fluid may be considered to be in close proximity, in physiological terms to the target organ — the brain.

2.2   PROPERTIES OF THE MEDIA

2.2.1   GENERAL COMMENTS

As few analytical end points are specific enough to be able to assay the drug directly in the fluid, the first problem is to separate the drug from as much of the endogenous material as possible.

Maickel[76] has produced an excellent practical review of the application of separation science to the analysis of biological samples. The ease with which samples can be analyzed increases with the degree of fluidity, cerebrospinal fluid generally being the easiest fluid to handle, while whole blood is the most difficult (Table 2.1). To increase the fluidity, solids, and semisolids submitted for analysis are often subjected to mechanical procedures. As shown in Table 2.2, these procedures may affect the

TABLE 2.2
Procedures Used for Disrupting Biological Samples Prior to Solvent Extraction for Analysis[76]

Characteristics	Mortar-Pestle		Blades		Other	
	Potter-Elvejehen	Teflon-Glass	Waring Blender	Virtis-Sorvall	Sonication	Chemical
Mechanism	Shear	Shear	Cut/shear	Cut/shear	Vibration	Hydrolysis
Cooling possible?	Easily	Easily	No	Easily	Perhaps	Not normally
Speed variation	Continuous	Continuous	Stepped/continuous	Continuous	—	—
Metal contamination	None	Nominal	Severe	Severe	Minimal	None
Foaming	Slight	Slight	Moderate	Moderate	Moderate	Severe
Consistency	Poor	Poor	Good	Good	Good	Good
Limitations	Hard tissues	Hard tissues	Container size	Minimal	Type of sample	Harsh conditions
Safety	Poor	Fair	Good	Excellent	Good	Fair

TABLE 2.3
Characteristics of Aqueous Solvents Used for Preparing Biological
Samples Prior to Solvent Extraction[76]

Solvent	Advantages	Disadvantages
Distilled water	Relatively good solvent	Degree of ionisation may vary
	Does not destroy tissue constituents	Does not denature enzymes consistently
	Final pH near 7.0	Final pH may vary with tissue
Dilute acid (<0.5 N)	Relatively good solvent	Considerable protein denaturation
	Denatures many enzymes	Compounds may be acid sensitive
	Final pH < 7.0	
	Foaming is minimal	
Strong acid (>0.5 N)	Good solvent	Clumping and aggregation may occur
	Denatures all enzymes	
	Precipitates proteins	Compounds may be acid sensitive
	Final pH < 4.0	Tissue constituents may break down
Dilute alkali (<0.5 N)	Relatively good solvent	Considerable protein denaturation
	Denatures many enzymes	Compounds may be alkali sensitive
	Final pH > 7.0	May cause foaming (soaps)
Strong alkali (>0.5 N)	Relatively good solvent	Clumping and aggregation may occur
	Denatures all enzymes	Compounds may be alkali sensitive
	Precipitates proteins	Tissue constituents may break down
	Final pH > 10.0	Foaming generally serious

sample in several undesirable ways, some leading to actual changes in concentration of the drug in the sample (temperature effects, metal chelation, and conjugate hydrolysis) and some actually making subsequent handling more difficult (foaming, emulsions, rupture of red cells). The solvent used as the medium for such procedures is critical and Maickel has reviewed the general advantages and disadvantages of the most often used aqueous systems (Table 2.3). Subsequent ease of extraction of the drug in the aqueous fluid into an organic solvent will depend on the solvent used. In general, the aim is to extract as much of the drug as possible, but leaving behind the undesired interfering material. This is most commonly done by a combination of steps where the polarity of the extracting solvent (Table 2.4) or the pH of the aqueous phase is adjusted.

2.2.2   BLOOD

Blood is the most complex of the biological fluids mentioned. As collected from a subject or animal, the blood consists of a buffered clear fluid containing solubilized proteins, dissolved fats and solids, and suspended cells, Luckily, the major constituent, the red blood cells, or erythrocytes, can be separated from the clear fluid, or plasma, by simple centrifugation. However, if the blood is not treated carefully, the cells can burst and separation of undesirable components becomes more difficult. For example, ferric ions released from erythrocytes may chelate with some analytes causing poor extraction from the aqueous phase.[77] The cells can be caused to burst by heating or by freezing, or by mechanical means such as stirring, but the most common cause is by changing the ionic strength of the surrounding fluid by the addition of water; the resulting osmosis causes the cells to swell and rupture. It is for

TABLE 2.4
Order of Polarity and Other Relevant Properties of Solvents Used for the Extraction of Drugs from Biological Fluids

Solvent	UV limit (nm)	Boiling point (°C)		Dielectric constant
n-Hexane	210	69	Least polar	1.890
Cyclohexane	210	81		2.023
Carbon tetrachloride	265	77		2.238
Benzene	280	80		2.284
Toluene	285	111		2.438
Di-*iso*propyl ether	220	68		3.88
Diethyl ether	220	35		4.335
Amyl acetate	285	149		
Chloroform	245	61		4.806
Dichloromethane	235	40		9.08
1,2-Dichloroethane	230	83		10.65
Methyl-*iso*butylketone	330	116		
Ethyl acetate	260	77		6.02
n-Butanol	215	118	Most polar	17.8

this reason that any manipulation to change the volume of a whole blood sample should be performed using isotonic saline.

If blood is allowed to stand without the addition of anticoagulating agents, then the red cells will eventually clot and the resultant fluid (serum) can be decanted. Serum is, in most respects, similar to plasma except that it does not contain the soluble factors that lead to the clotting phenomenon. On the other hand, if anticoagulants are added and plasma subsequently prepared, then these factors remain in the plasma and may give rise to subtle differences when serum and plasma are analyzed. Blood is generally not directly extracted, instead plasma or serum is always prepared first and subsequent procedures are based on these fluids. If blood is extracted directly, it should be handled with care to minimize rupture of the red cells as described above.

2.2.3   PLASMA AND SERUM

The chief feature of plasma and serum is the presence of large amounts of protein. Obviously, the protein itself is chemically and physically different from the small-molecule drugs normally being measured. However, there is often a strong affinity between such proteins and drugs, and straight removal of protein such as by ultrafiltration or dialysis could also remove a large fraction of the drug. On the other hand, any direct measurement of drug could possibly miss the "total" drug present and measure only "free" drug. In microbiological assays, only the free drug is measured but because standards are also run in the same fluid as that being measured, it is assumed that the quoted result would be the total amount of antimicrobial agent present. Although one may argue that the free drug is the physiologically important entity, as most drugs are predominantly protein bound, the free concentrations are extremely low and it is usual to measure the total drug in plasma or serum samples. Thus, the problem becomes one of physically destroying the binding of drug to protein and then extracting the total drug for analysis. It should be pointed out that as the fraction bound is usually a constant percentage of the total, then the analyst is only exploiting an amplification factor to make the analysis simpler. At present, this is a lucky break for the analyst, as he or she does not need to develop extremely sensitive methods for the unbound fraction; however, there have been recent claims that meta-stable conformers of some drugs may be the active forms[78] and if these claims are true than the elucidation of their pharmacokinetics would require the analysis of the specific conformer in biological fluids — a daunting task by any standards.

The first step in preparing a plasma or serum sample for analysis is to obtain a protein-free aqueous solution suitable for extraction with an organic solvent. The simplest and oldest method is to precipitate the protein and isolate the filtrate. The protein is denatured by the precipitation and its drug-binding ability is destroyed thus releasing total drug into the filtrate. Although caustic solutions are equally effective in denaturing and precipitating proteins in plasma and serum, they are generally not favored because of their propensity to form soaps with subsequent difficulty in extraction. Popular acidic reagents for protein precipitation include trichloroacetic acid, perchloric acid, and tungstic acid. However, such strong acids could have detrimental effects on the drug to be extracted and any such procedure should be tested on standard compounds rather than be used as convenient general reagents. Stevens and Bunker[79] performed a survey on the use of these acids, as well as the use of 5 M HCl at 80°C, aluminum chloride and ammonium sulfate, and concluded that although no method was generally applicable, aluminum chloride was the "safest" reagent for precipitation of proteins in whole blood; the apparent low recoveries of several drugs being due to precipitation of the drug with the protein, and degradation.

The instability of drugs at low pH can be circumvented by using organic solvents to denature and precipitate proteins. Dell recommends methanol or ethanol, with at least two volumes of ethanol being necessary to precipitate all plasma proteins.[80] Another popular solvent, particularly as a prelude to HPLC, is acetonitrile. Mathies and Austin[81] described a procedure for the analysis of anticonvulsants and analgesics whereby plasma is mixed with an equal volume of acetonitrile, the solution is saturated with sodium-bisulfate/sodium chloride, and the upper phase is directly injected onto the chromatographic column.

Protein can also be denatured using proteolytic enzymes, a procedure that should avoid the possibility of damage to the analyte using chemical-type denaturation. Such procedures are generally found in the preparation of tissue for drug analysis, but the enzyme subtilisin has been successfully used for the digestion of plasma proteins. Osselton[82] showed that the better recoveries of drug from tissue using enzyme hydrolysis compared with direct extraction was also obtained in the analysis of whole blood and plasma. In Osselton's procedure, 200 μl plasma is buffered to pH 10.5 with 50 μl Tris buffer, enzyme is added, and the mixture incubated at 55°C for 1 h before extraction with 50 μl butyl acetate. The organic phase is then analyzed by an appropriate method. Osselton points out that the enzyme hydrolysis is most useful for general screening procedures where it is not usually possible to optimize the normal extraction procedures for a specific compound. A number of other proteolytic enzymes have been proposed (Table 2.5), which presumably could be adapted for biological fluid assays with detriment to the cited analyte. The various methods of separating protein from solutions containing analytes are reviewed in Table 2.6.

TABLE 2.5
The Use of Proteolytic Enzymes for the Preparation of Biological Samples Prior to Drug Analysis

Drug	Enzyme	Reference
Amitriptyline	Subtilisin	82
	Trypsin, proteinases	83
Chlorpromazine	Subtilisin, papain, ketodase	84
	Trypsin, proteinases	84
Diazepam	Subtilisin	82
	Trypsin, proteinases	83
	Ketodase	84
Diphenhydramine	Subtilisin	82, 84
	Ketodase, papain	84
Methaqualone	Ketodase, subtilisin, papain	84
Methylmercury	Subtilisin	85
Phenobarbitone	Trypsin, proteinases	83
Phenylbutazone	Subtilisin	82
Phenytoin	Subtilisin	82
	Trypsin, proteinases	83
Salicylic acid	Subtilisin	82
Trimipramine	Subtilisin	82

TABLE 2.6
Denaturation and Precipitation of Whole Blood, Plasma, or Serum Proteins Prior to Drug Analysis

Method	Comment	Reference
Heat at 90°C for 5–15 min	Not very efficient, may decompose analyte	76
Freeze-thaw cycles	Not very efficient, time consuming	76
Saturation with ammonium sulfate	Moderately efficient, high salt concentration in supernatant, final pH about 7	76, 86
$ZnSO_4$-sulfosalicylic acid	Clear solution, mild enough for RIA	87
$ZnSO_4$-sodium hydroxide	Excellent efficiency, fine precipitate final pH about 7, suitable at low temperature	76
Metaphosphoric acid	Excellent efficiency, reagent needs to be kept cold, acidity (pH < 3) may decompose analyte	76
Perchloric acid	Excellent efficiency, reagent needs to be kept cold, explosion hazard. Final pH (<3) may be detrimental to analyte. Most basic compounds can be safely extracted	76, 86
Trichloroacetic acid	Good efficiency, reagent needs to be kept cold and may be difficult to remove from analyte	76
Tungstic acid		88
Ethanol	Two volumes of ethanol are required for complete denaturation, suitable if drug is unstable at low pH	80
Acetonitrile	1.5 volumes required for complete denaturation. Loose floc minimizes coprecipitation of analyte. Particularly suitable for subsequent HPLC	89–91
Aluminum chloride	Better than ammonium sulfate or tungstic acid for basic compounds	86

TABLE 2.7
Protein Binding of Some Basic Drugs

Drug	pKa	%	Bound to	References
Aminoglutethimide	4.2, 11.9	25–30	α_1-acid glycoprotein	93
Dipyridamole	6.4	98	α_1-acid glycoprotein, albumin	94
Fluvastatin	—	67, 99	α_1-acid glycoprotein, albumin	95
Furosemide	3.8	99	Albumin	93
Phenytoin	8.3	90	Albumin	93
Tacrolimus (FK506)	—	75–80	Plasma protein, albumin	96

Protein binding is a phenomenon that is important for the transport of drugs in the blood, and sometimes for the solubility of the drug also. As detailed above, the usual methods of extraction tend to break down such binding so that the total drug in the fluid is usually measured. There may be occasions when it is desirable to measure the so-called free drug as well to be aware of the extent of protein binding normally encountered. Acidic drugs such as barbiturates are generally considered to be strongly bound, whereas basic drugs tend not to be. However, there are exceptions to this rule (Table 2.7). Protein binding of phenytoin is affected by valproic acid.[92]

Although plasma is a very complex fluid, its composition is remarkably stable and even in patients there are rarely gross changes in the composition that the analyst will not have accommodated before evaluating a method. Thus the pH is never outside the range 7.30 to 7.50 and the total protein and salt concentrations are also well controlled. The lipid content, however, can vary considerably, often being associated with the timing and nature of meals. If a plasma concentration–time curve is being constructed, lipid content can be a factor affecting analysis and especially fatty samples which may require a specific partitioning step to remove lipids. Indeed, in such situations a partition step in the standard procedure may be necessary, even though development using low-lipid control plasma may not have indicated that such a step was necessary.

It is important to recognize that some changes in plasma constituents can have an impact on the actual levels of drug, as distinct from factors which interfere with the analysis. An example of this is the potential for some drugs to displace others from their protein-binding sites. In such cases, the physiological effect of the drug may depend on only the unbound, or free, concentration, yet because there are fewer binding sites available, the total drug measured will appear inappropriately low for the observed effect. Thus, many researchers advocate the analysis of the free drug as the true indicator, notwithstanding the increasingly difficult problem imposed on the analyst. For such determinations, ultrafiltration methods have become popular to separate free drug from bound drug using commercially available filters.[97-99] However, such low concentrations as would be found for the free drug may be outside the limit of detection; an alternative strategy would be to assess the percentage bound by equilibration with a radiolabeled drug in the same plasma sample prior to the usual analysis for total drug.

Not all analysts subscribe to the philosophy of protein denaturation prior to extraction. Although drugs may be strongly bound to plasma proteins, the binding is

TABLE 2.8
The Use of C-18 Bonded Phases for Prepurification of
Biological Samples Prior to Drug Analysis

Drug	Reference
Amitriptyline, imipramine, nortriptyline, desipramine	103, 104
Diazepam, desmethyldiazepam	105, 106
Temazepam, oxazepam	105
Demoxepam, *N*-desmethylchlorodiazepoxide, chlordiazepoxide	106
Terbutaline, salbutamol, fenoterol	107
Cyclosporin A	108
Zeranol, stilboestrol	109
Protryptiline	104
Probenecid	110
Ethynylsteroids	111
Chloramphenicol	112
Doxorubicin	113
Mebendazole	114
Phenobarbitone, phenytoin, primidone, ethosuximide, carbamazepine	115

reversible. Hence at an appropriate pH, the sample can be extracted with an organic solvent. If the partition into the organic solvent for the un-ionized form of the drug is sufficiently high, then the binding equilibrium can be shifted sufficiently to allow efficient extraction into the organic solvent. The strategy would be to adjust the pH with the minimum amount of buffer and then extract with a relatively large volume of solvent. The advantage of this procedure is that one avoids a filtration or decantation step which makes possible adaptation as a routine procedure for large numbers of samples, a real problem for laboratories supporting toxicology and clinical pharmacology studies.

This pressure on the laboratory to perform large numbers of assays has resulted in the search for simplifying this important first step in the analysis, possibly by reducing the number of operations to be performed. Thus, the initial extraction may be replaced by adsorption of the analyte in solution onto a solid matrix and subsequent elution with a much smaller volume of solvent. Several methods have been proposed which allow the raw sample to be added to a column of adsorbent, which may be charcoal,[100-102] modified silicas (Table 2.8), or ion-exchange resins (Table 2.9). This subject is discussed more fully in Chapter 12.

A final point that needs to be made regarding the constituents of plasma or serum is the possible presence of enzymes which may continue the degradation of the drug after the sample has been collected. Particularly important in this respect are the non-specific esterases which will convert esters to the free acid and alcohols. The nature and activity of these esterases may vary considerably from one species to another.

2.2.4 URINE

Urine, unlike plasma or serum, is generally free of protein and lipids and therefore can usually be extracted directly with an organic solvent. Urine does, however, have a wide variation in its composition, this most obviously being seen in the dark amber

TABLE 2.9
The Use of Exchange Resin XAD-2 for Pre-Extraction of
Drugs from Biological Samples

Drug	References
Amphetamine, diphenhydramine, haloperidol, flurazepam, chlordiazepoxide	116
Phenformin	117
Morphine	118
Phenobarbitone	83
Amitriptyline	83, 116, 119
Diazepam, chlorpromazine	83, 116
Codeine	116, 118
Barbiturates	120, 118, 119
Oxycodone	121
Meperidine, phencyclidine, lidocaine, methapyrilene, propoxyphane, imipramine, doxepin, cocaine	116,119

of an overnight specimen compared with the pale urine collected during the day. The gross composition of urine depends very largely on the diet and this can account for a quite startling range of colors. Fortunately the normal type of compound found in urine is water soluble, whereas most drugs are lipid soluble and can be extracted with an appropriate solvent. One of the greatest difficulties that arises, however, is the volume of urine that may be produced over fixed time intervals. For most analytical tests it is usual to perform the analysis on a fixed volume of sample; with urinary excretion, however, it is the amount of substance excreted in a certain interval which is of interest, not its concentration, and the amount is obtained by multiplying the volume by the concentration. The problem that arises is twofold; if the urine volume is large, then the sample is dilute and the method may be operating near its limit of detection, and the inevitable errors at the low concentrations will be multiplied on adjustment for the volume. The large relative error becomes a large absolute error in assessing the amount of urinary excretion. Thus in a series of urines collected from a single subject, the individual urine samples will be subject to different errors and there will be considerable uncertainty in the total drug excreted. It is wise to be aware of this factor when drawing conclusions over total drug excretion; the error in one urine collection could well exceed the total in another collection in the same series.

A related problem has been noted in the assay of the urine of athletes, when random test samples, particularly for abnormal levels of caffeine, have been taken following a race. The sport has set a concentration limit for caffeine in urine; however, the urine can become very concentrated particularly when the athlete is dehydrated and a normal level of caffeine can easily turn into a prohibited level. Sports authorities have been castigated for taking an unscientific view of the problem and its interpretation.[122] On the other hand, the nature of the urine sample can often cause false negatives. Lafolie[123] prefers to assess the possibility of false negatives by reference to creatinine content rather than the color differences in urines as suggested by Simpson et al.[124]

Urine has a wide range of pH values, predicated to a great extent on the diet, or on medication. Antacids, for example, if absorbed, may cause the urine to be alkaline;

strongly acid urine is less likely, the normal pH range being 5.5 to 7. This applies to the pH of the urine as excreted. The nature of the excreted drug could depend on the pH at this time. Weakly basic drugs such as amphetamine are more efficiently excreted in acid urine, whereas weakly acidic drugs are more excreted in alkaline urine.[125] Urine which is left standing will slowly lose carbon dioxide and become more alkaline, resulting in the precipitation of inorganic phosphates. In addition to the pH change, which can cause changes in degradation of the analyte, the analyte may coprecipitate with these phosphates and be lost for the analysis. Thus, stored urine will have a different composition from fresh urine and analytical methodology and assessment of stability of drug in this matrix must account for the properties of both types.

One aspect of urine that may be overlooked when used in quantitative studies is the method of sampling and the amount sampled. Does the subject provide a complete emptying of the bladder so that a total production of drug or metabolite into the urine over a given period can be calculated? Subjects asked "to provide a sample" may give the minimum and it may be of a first catch rather than a midstream sample, which may or not be representative. Cone et al.[126] studied this in connection with marijuana metabolites and concluded the concentrations were similar for consecutively collected samples of the same urine voided.

2.2.5 MILK

Milk is not a very usual fluid for the analysis of drugs, but it is occasionally of interest when trying to establish whether drugs may be transferred from mother to infant by this route. Most authors seem to be able to adapt existing methods for urine[127] or serum[128-130] without much difficulty. The main problem would seem to be the presence of fats in milk (approximately 4.5% in mature human breast milk), and a method for phylloquinone uses lipase to hydrolyze fats in the sample prior to HPLC.[131] Alternatively, the defatting step that is recommended above for processing plasma samples may be sufficient, as applied by Heintz et al.[132] for the analysis of tenoxicam in human breast milk. In this procedure, fats were removed by washing 0.5 ml breast milk, buffered to pH 3 to 4, with 10 ml *n*-hexane.

2.2.6 CEREBROSPINAL FLUID

Cerebrospinal fluid is another biological fluid which is thought to be closer to the site of action of drugs and natural biological agents than the usually assayed plasma or serum. Not surprisingly, attempts have been made to analyze drugs with pain-killing activity in this fluid. Samples are, however, not so easily obtained, although when they are available, techniques that have been developed for plasma or serum seem to be readily applicable. Such unmodified assays have been reported for β-endorphins and other peptides,[133] enkephalin,[134] morphine,[135] and nimodipine.[136]

2.2.7 BILE

Bile is a complex and variable fluid which is not frequently used for the assay of drugs, partly because of the difficulty of collection and also because any systems of removing bile will interfere with the process being investigated. Nevertheless, liver perfusion studies *in vivo* will often involve such analyses, particularly where conjugat-

TABLE 2.10
Comparison of Compositions of Saliva, Blood, Plasma, and Serum

	Saliva	Blood	Plasma	Serum
Water	99.5%	—	90–95%	90–95%
Specific gravity	1.002–1.008	1.05–1.06	1.025–1.029	1.024–1.028
pH	5.17–6.77	7.36	7.32	7.32
Total protein	262 mg/100 ml	—	6550 mg/100 ml	—
Total lipid	2.8 mg/100 ml	—	500 mg/100 ml	—
Fibrinogen	—	—	330 mg/100 ml	—
Cholesterol	<1 mg/100 ml	—	—	c200 mg/100 ml
Chloride	102.8 mg/100 ml	300 mg/100 ml	350 mg/100 ml	350 mg/100 ml
Hemoglobin	0	15 g/100/ml	—	—

ing enzymes are being investigated. Because of its character, methods developed for fluids with well-established consistency are not immediately adapted for bile. For example, blood, urine, vitreous humor, and cerebrospinal fluid could be analyzed directly for morphine using commercial RIA kits, but bile needs to be made alkaline and extracted with dichloromethane:propanol and the residue reconstituted in serum.[136]

2.2.8   SALIVA

Saliva is a colorless, transparent or translucent, somewhat viscid material of low viscosity. Its attraction for bioanalysts is that it is relatively free of interfering substances, is easily extracted by organic solvents, and is thought to reflect the levels of nonprotein bound drug in the blood. For nonionized drugs such as steroids, the drug should readily cross from plasma to saliva and this would represent a noninvasive method of monitoring.

Normal subjects may produce up to 2 l of saliva in 24 h, with a more or less continuous flow of 15 ml h^{-1} between meals. Collection of a reasonable sample in a reasonable time is tedious for normal subjects but larger samples can be readily obtained by stimulation of the glands by chewing an inert material such as parafilm or by using citric acid. Schramm et al.[137,138] have described and evaluated a device consisting of sucrose granules enclosed in polyurethane-bonded cellulose semipermeable membranes, which, when placed in the mouth, collect an ultrafiltrate of saliva. This ingenious device bypasses several of the problems of collection and handling of the fluid. Stimulated saliva is also reported to be more consistent in its pH value (pH 7.0 to 7.8).[139,140] The composition of saliva compares well with those of blood, plasma, and serum (Table 2.10).

Authors reporting on the analytical methods for determining drugs in saliva do not seem to have experienced any particular problem in adapting existing methods for other fluids such as plasma or serum; minor problems have been noted with absorption of phenytoin to the mucoid proteins of saliva[141] and occasional adsorption of drugs to the stimulation material. The principal problem with analysis of drugs in saliva is not a technological problem but is to establish the relevance of results of such analyses. Assuming the transfer of drug from blood to saliva is by passive diffusion, then the relationship between saliva and blood concentrations can be expressed from theoretical

TABLE 2.11
Theoretical Distribution of Acid and
Basic Drugs Between Saliva and Plasma

Acid drugs			pH of saliva			
pKa		6	6.5	7	7.5	8
	2	0.04	0.13	0.40	1.26	3.98
	3	0.04	0.13	0.40	1.26	3.98
	4	0.04	0.13	0.40	1.26	3.98
	5	0.04	0.13	0.40	1.26	3.98
	6	0.04	0.13	0.40	1.26	3.98
	7	0.04	0.13	0.40	1.26	3.97
	8	0.08	0.16	0.42	1.25	3.87
	9	0.31	0.37	0.57	1.19	3.13
	10	0.81	0.82	0.88	1.05	1.60
	11	0.98	0.98	0.99	1.01	1.07
	12	1.00	1.00	1.00	1.00	1.01

Basic drugs			pH of saliva			
pKa		6	6.5	7	7.5	8
	2	1.00	1.00	1.00	1.00	1.00
	3	1.00	1.00	1.00	1.00	1.00
	4	1.00	1.00	1.00	1.00	1.00
	5	1.00	1.00	1.00	1.00	1.00
	6	1.01	1.00	1.00	1.00	1.00
	7	1.10	1.03	1.01	1.00	1.00
	8	1.92	1.27	1.06	0.99	0.97
	9	7.87	2.98	1.43	0.94	0.79
	10	20.28	6.55	2.21	0.84	0.40
	11	24.53	7.77	2.47	0.80	0.27
	12	25.06	7.93	2.51	0.79	0.25

considerations. Researchers can then decide whether the levels to be expected are within their analytical capabilities and once analytical values have been obtained, they can make the appropriate extrapolations to determine blood concentrations. By applying the usual Henderson-Hasselbach equation, the ratio of drug concentrations for a weak acid (pKa) in saliva and plasma can be expressed as:

$$\text{ratio} = \frac{1 + 10^{(\text{pH of saliva-pKa})}}{1 + 10^{(\text{pH of plasma-pKa})}} \times \frac{\text{fraction unbound in plasma}}{\text{fraction unbound in saliva}}$$

with a corresponding equation for weak bases of

$$\text{ratio} = \frac{1 + 10^{(\text{pKa-pH of saliva})}}{1 + 10^{(\text{pKa-pH of plasma})}} \times \frac{\text{fraction unbound in plasma}}{\text{fraction unbound in saliva}}$$

Table 2.11 shows the calculated ratios for acid and basic drugs assuming a plasma pH of 7.4 and minimal protein binding in saliva. The pH of the saliva can have a

TABLE 2.12
Saliva and Plasma Distribution of Acid
and Basic Drugs

Drug	pKa	Saliva/Plasma ratio	References
		Acidic drugs	
Carbamazepine	—	0.37	142
Diazepam	3.3	0.03	143
Ethosuximide	9.5	1.0	144
Phenobarbital	7.2	0.41	144–148
Phenytoin	8.3	0.13	144–148
Primidone	—	1.0	144, 147, 148
Tolbutamide	5.4	0.01	145, 149
		Basic drugs	
Antipyrine	1.4	0.9	144, 145, 150
Pethidine	8.7	—	145
Procainamide	9.4	1–9	145, 151
Propranolol	9.5	0.2–2.7	145

dramatic effect on the distribution and at low pH values and for acidic drugs that are 99% bound to plasma proteins, the concentration in saliva would be a 0.0004 that of plasma. Thus, for any meaningful interpretation, it is essential to measure the pH of saliva on collection. Table 2.12 lists several drugs for which correlations between unbound plasma concentrations and salivary concentrations have been investigated.

2.2.9   BLISTER FLUID

Blister fluid obtained from suction-induced skin blisters is similar to interstitial fluid and exudate in a mild inflammatory reaction and contains proteins and lipids.[152,153] It has been used for studying the pharmacokinetics of ibuprofen.[154]

2.2.10   SYNOVIAL FLUID

The analysis used was the same used for plasma for anti-inflammatories by HPLC,[155] and similarly for the analysis of enantiomers of flurbiprofen in synovial fluid and plasma of patients with rheumatoid arthritis; the limitation of the synovial fluid was only on the small amount of sample available.[156] Also, the same method of extraction was used for synovial fluid as for plasma in the assay of diclofenac.[157]

2.2.11   AQUEOUS HUMOR

Analysis was the same as for plasma for imipenem by HPLC.[158]

2.3   DRUG METABOLITES

All biological fluids have the potential presence of metabolites of the drug being investigated. This phenomenon has been particularly exposed since the advent of chromatographic methods of separation. Before these procedures, the classical methods of analysis included metabolites because their properties were naturally similar to the properties of the parent drug. The analyst needs to be aware of the type of metabolite that will be produced from a particular drug. This is a highly complex subject and the complete metabolism of any drug cannot yet be predicted from its structure alone and the analyst must be guided by experiments performed specifically to elucidate the metabolic pattern for the drug in various biological fluids. It should not be forgotten, either, that as metabolism is an ongoing process and metabolites themselves have their own kinetics, that the amounts of metabolite relative to parent drug will also change with time. The following is a brief account of the more common metabolic pathways that have been described for drugs and other small foreign compounds. Descriptions of details on the assay of specific drugs should always include a description of the metabolites to be expected and the effect they may have on analysis.

Drug metabolism has been described as a detoxication mechanism. This, however, could be a misleading term. Dutton[159] has pointed out that the actual mechanisms that act on foreign compounds have evolved to protect the animal against certain compounds that would otherwise be toxic and that this process of evolution has ensured that these mechanisms are successful against naturally occurring compounds. However, when the organism is challenged with a new compound, then the evolution time is not sufficient to evolve the specific detoxication required and the new compound is simply a good or bad substrate for the enzymes it encounters. The resulting products can just as easily be more rather than less toxic and the term "detoxication," when one means metabolism is to be avoided. As explained below, most transformations do result in more water-soluble compounds which would be more readily excreted.

Williams[160] has classified the stages of metabolism into two phase. In the first phase, the drug is attacked by a single enzyme to effect a simple change in its structure. Such reactions are usually oxidations, reductions, or hydrolyses which usually result in a structure with a convenient functional group or "handle" which is utilized in the second phase. The second phase involves a synthetic step where a water-soluble function is added to a functional group of the drug or of its phase I metabolite to form a water-soluble conjugate which is readily excreted by the body usually via the urine. The main site of both phases of metabolism is the liver. Although other tissues may also be capable of drug transformations, they generally have lower capacity and the metabolism by liver overwhelms the metabolism by other routes. Some drugs may exist in a form which is directly amenable to second phase metabolism, such as carboxylic acids which may be conjugated with a variety of sugars or sugar acids or with amino acids.[161]

2.3.1 PHASE I METABOLITES
2.3.1.1 Oxidation

Aromatic compounds such as benzene are metabolized by direct hydroxylation, benzene itself being metabolized to phenol, and naphthalene being metabolized to β-naphthol. Further hydroxylation to catechols may also take place. Aliphatic compounds are metabolized by ω-hydroxylation to the primary alcohol. Above a certain chain length, however, ω-1 hydroxylation predominates as exemplified by the xanthine, pentifylline which is metabolized by oxidation of the hexyl side chain (Figure 2.1).[162] Alcohols are further oxidized to aldehydes and ketones and aldehydes subsequently

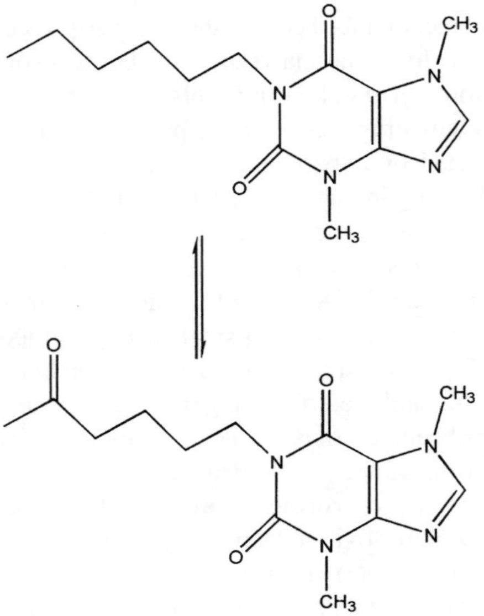

FIGURE 2.1 Oxidation of the hexyl side-chain of pentifylline, an example of ω-1 (omega) hydroxylation.

oxidized to acids. Thus, the metabolism of an aliphatic side chain such as that for pentifylline will proceed via this series of oxidations to yield a complex mixture of acids, ketones, and alcohols — all eliminated in the urine as shown in Figure 2.2.

In fact, the oxidation of alcohols to the corresponding aldehydes or ketones is a reversible reaction and the relative amounts of the two forms depend on the structure of the drug and on the species. When the xanthine oxpentifylline is administered, the reduced metabolite appears in plasma in a constant ratio to the amount of oxpentifylline[163] although the ratio is different for different species (Figure 2.1). Aromatic ethers are converted to phenol by *o*-dealkylation, the alkyl group being lost as the corresponding aldehyde (Figure 2.3) as described by Brodie et al.[164] Similarly, thiol ethers such as 6-methylthiopurine (Figure 2.4) are oxidized by *S*-dealkylation.[165]

Aromatic amines are hydroxylated to the corresponding hydroxylamino compounds; sulfanilamide is hydroxylated at the N^4-amino group to give *p*-hydroxylaminobenzenesulfonamide (Figure 2.5).[166] Tertiary amines may be oxidized to the *N*-oxide, the main route of metabolism of the benzodiazepine, loprazolam (Figure 2.6).[167] Secondary amines are dealkylated to primary amines and the corresponding aldehyde (Figure 2.7). Thioethers are converted by oxidation to sulfoxides, an example being the metabolism of the phenothiazine, quinuclidinyl-3-methyl-10-phenothiazine (Figure 2.8).[168,169] Thiones may be oxidized to the corresponding oxo compound as in the metabolism of thiobarbital to barbital (Figure 2.9).[170] Metabolism is sometimes reversible as for example in the interconversion of 4-amino-5-chloro-2-[(methylsulfinyl)ethoxy]-*N*[2-(diethylamino)ethyl]benzamide and its sulfide and sulfone metabolites in rats.[171]

Primary amines such as mescaline may be oxidatively deaminated, being converted to the corresponding phenylacetic acid (Figure 2.10).[172]

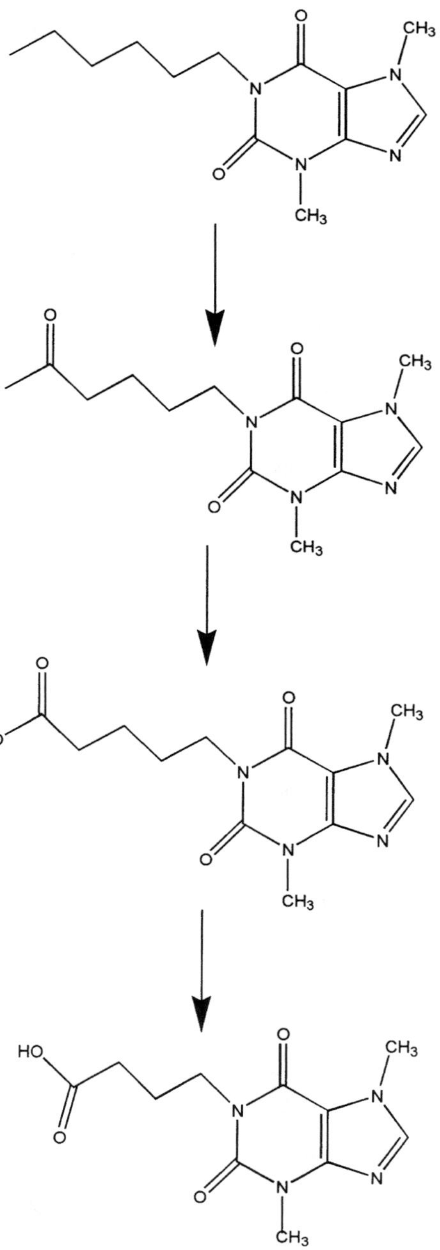

FIGURE 2.2 Metabolism of pentifylline by successive oxidation of the hexyl side-chain.

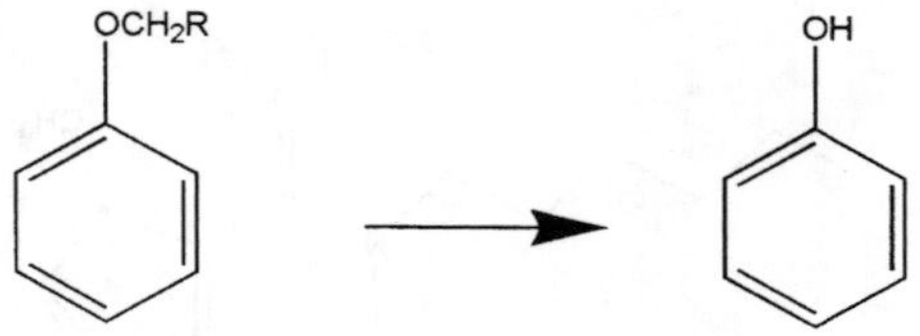

FIGURE 2.3 *O*-Dealkylation of aromatic ethers.

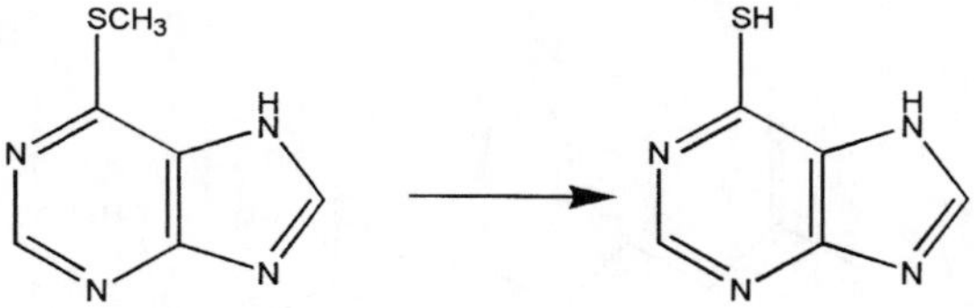

FIGURE 2.4 *S*-Dealkylation of 6-methylpurine.

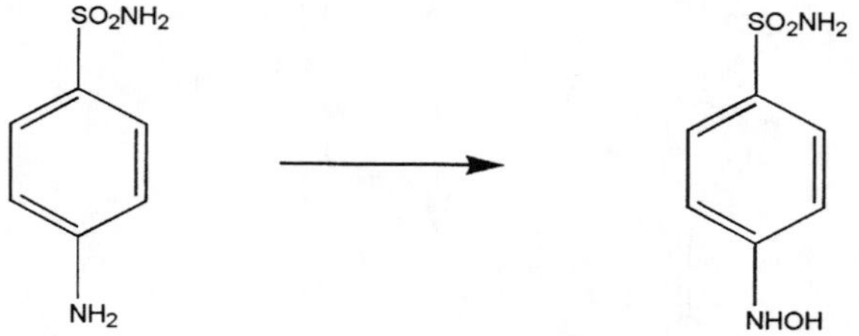

FIGURE 2.5 *N*-Hydroxylation of sulfanilamide.

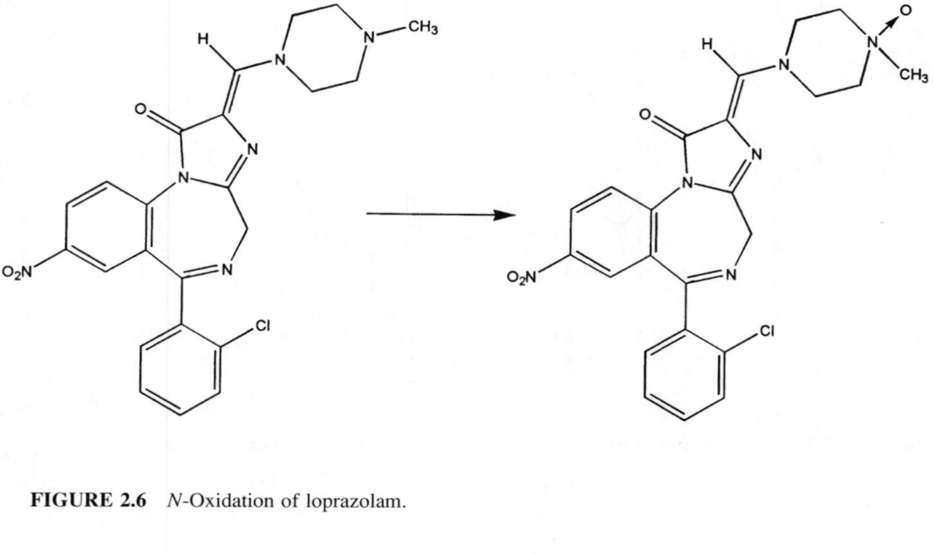

FIGURE 2.6 *N*-Oxidation of loprazolam.

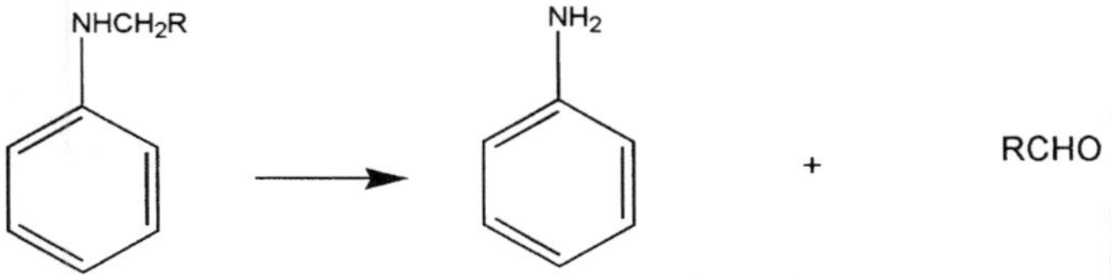

FIGURE 2.7 Dealkylation of secondary amines.

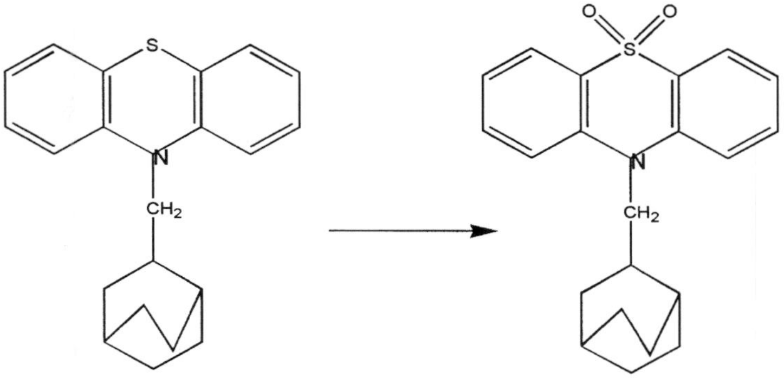

FIGURE 2.8 *S*-Oxidation of a phenothiazine.

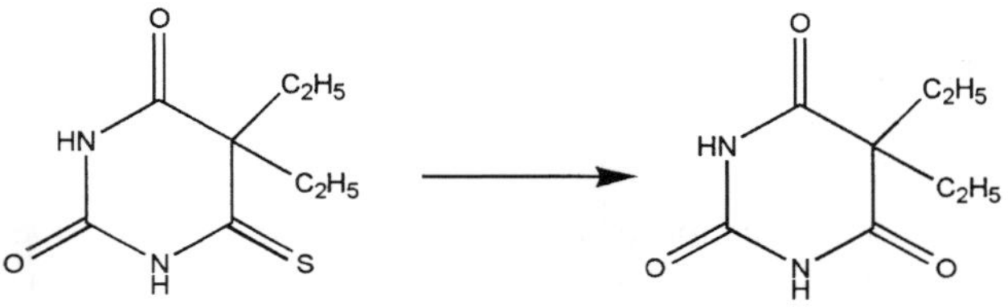

FIGURE 2.9 Conversion of thiobarbital to barbital.

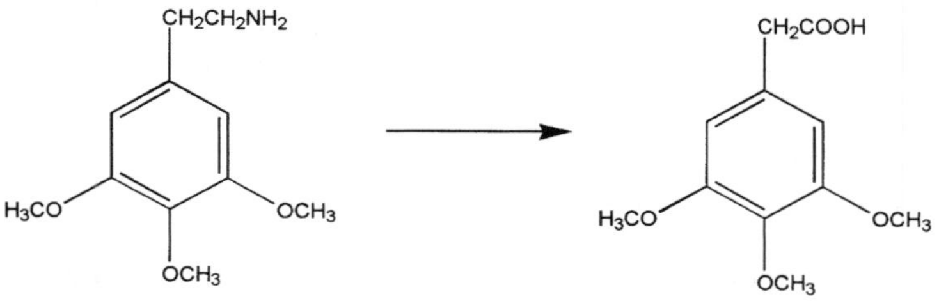

FIGURE 2.10 Oxidative deamination of mescaline.

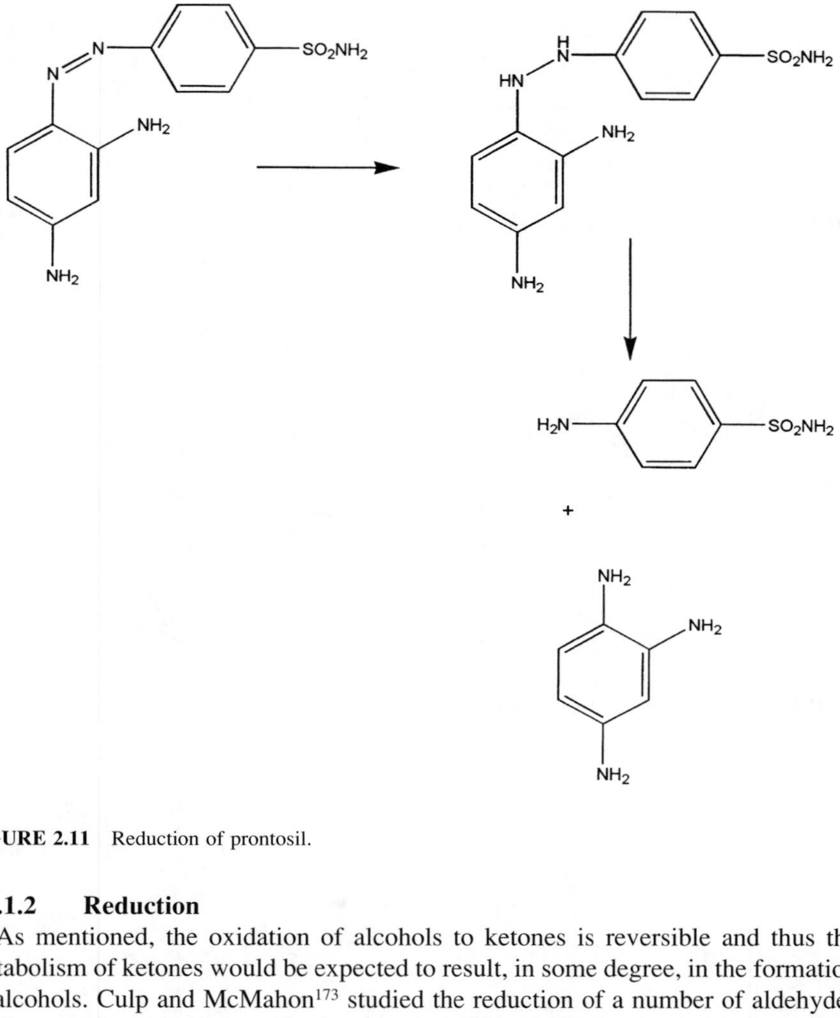

FIGURE 2.11   Reduction of prontosil.

2.3.1.2   Reduction

As mentioned, the oxidation of alcohols to ketones is reversible and thus the metabolism of ketones would be expected to result, in some degree, in the formation of alcohols. Culp and McMahon[173] studied the reduction of a number of aldehydes and ketones in an *in vitro* study showing the effect of substituents on the reduction.

Carbon-carbon double bonds are also reducible to the saturated equivalent, although this appears to be relatively rare. Nitrogen-nitrogen double bonds, as in prontosil, are reduced to primary amines, probably via the hydrazine (Figure 2.11).[174] Nitro groups, particularly aromatic nitro groups are reduced to the primary amine, an example being the metabolism of nitrazepam (Figure 2.12).[175]

2.3.1.3   Hydrolysis

All types of esters are readily hydrolyzed by a variety of specific and nonspecific esterases.[176,177] However, these esterases may be specific for certain species and tissues and the hydrolysis is not easy to predict. Amides are also hydrolyzed but not as readily as esters. For example, the amide analog of procaine, procainamide, unlike procaine itself, is only slowly hydrolyzed in the body (Figure 2.13).[178]

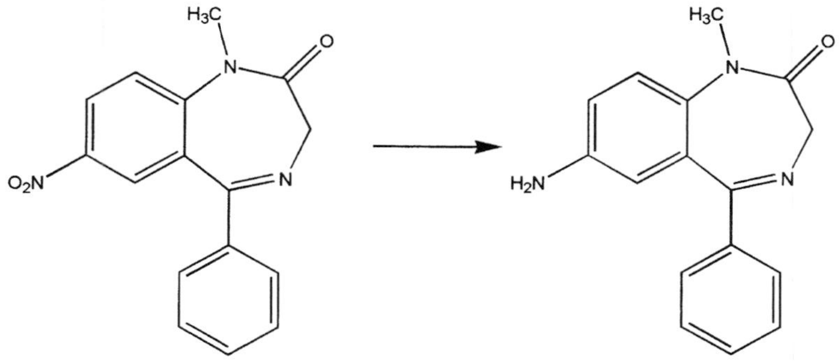

FIGURE 2.12   Reduction of nitrazepam.

2.3.2   PHASE II METABOLITES
2.3.2.1   Glucuronidation

Conjugation with glucuronic acid (Figure 2.14) is the most common, and most extensively studied mode of phase II metabolism. The conjugation may occur through an ether-type linkage with alcohols, phenols, or through an ester-type linkage with carboxylic acids. The resulting highly water-soluble derivatives are then excreted in the urine. A feature of such conjugates, however, is that they are readily hydrolyzed by dilute alkali and therefore may often be the cause of anomalously high urine levels of unchanged drug, especially if the urine is allowed to become alkaline on standing.[179]

Hydroxylamines are readily conjugated with glucuronic acid, an example being the conjugation of *N*-hydroxy-2-acetylaminofluorene (Figure 2.15).[180] Another unstable glucuronide is that formed by the enol form of a ketone (Figure 2.16).[181] *S*-Glucuronides are formed with thiols, being analogous to *O*-glucuronides formed with alcohols.[182] *N*-Glucuronides, such as that formed with nomifensine[183] are extremely unstable at pHs other than neutral, so much so that there is doubt as to whether nomifensine itself circulates in plasma as the free compound. The conjugate appears to be the major form circulating in the plasma and if extreme care is taken when preparing the samples, only ng ml^{-1} is found in the plasma of subjects given therapeutic doses[184] rather than µg ml^{-1} levels initially reported. Glucuronic acid may also conjugate with the barbiturates through *N*-glucosylation as described in human urine for phenobarbital[185-188] and amobarbital.[189]

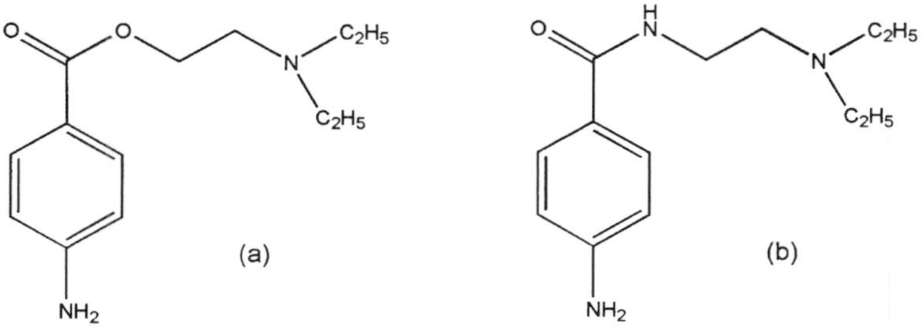

FIGURE 2.13   Structures of (a) procaine; (b) procainamide.

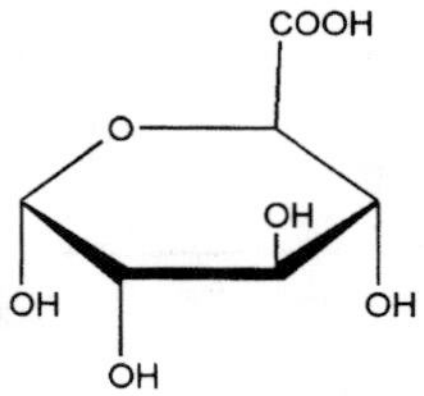

FIGURE 2.14   Structure of glucuronic acid.

2.3.2.2   Sulfation

Conjugation with sulfuric acid is the second most important method of conjugation for drugs and their metabolites, particularly for phenols and alcohols. Aromatic amines may be sulfated to form sulfamates (Figure 2.17).[190]

2.3.2.3   Acetylation

Conjugation with acetic acid is often classified as a phase II metabolic step, but is often associated with direct acetylation of the parent drug rather than as a second step in the metabolic pathway. Acetylation of aromatic amines and sulfonamides is common (Figure 2.18).[191] Acetylation is of interest, not only as a method of eliminating foreign compounds, but in characterizing the metabolic capacity of individuals or populations — a subject discussed more extensively in Chapter 1.

Glycine may conjugate with carboxylic acids to form hippuric acids; the classic example being the conversion of benzoic acid to hippuric acid itself (Figure 2.19).[192] Methylations occur when the drug substrate has a resemblance to endogenous substrates for this reaction such as for the methylation of catecholamines and histamines.

As mentioned, the metabolic transformation of drugs does not guarantee deactivation of the molecule. All of the above reactions are capable of transforming an administered molecule into one which is either pharmacologically more active, or which is toxic. A number of examples are given in Table 2.13. All of the phase I and phase II reactions are of course, not mutually exclusive. It can be seen that one reaction could produce a product which could be a suitable substrate for another and a sequence of reactions could be postulated incorporating a number of the steps outlined above. For example, the structure *N*-methylphenylethylamine could suffer

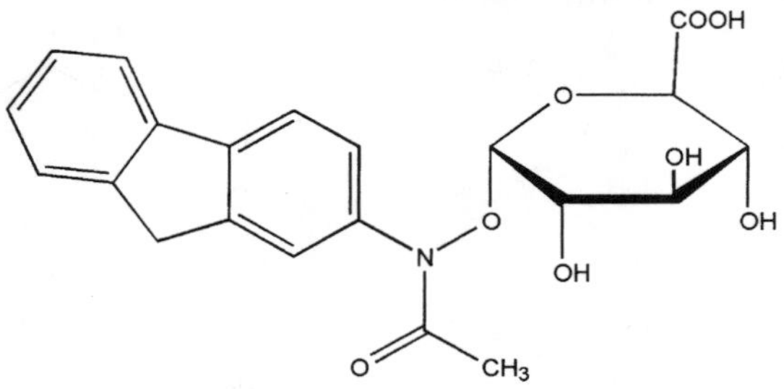

FIGURE 2.15   Glucuronic acid conjugate of a hydroxylamine, *N*-hydroxy-2-acetylaminofluorene.

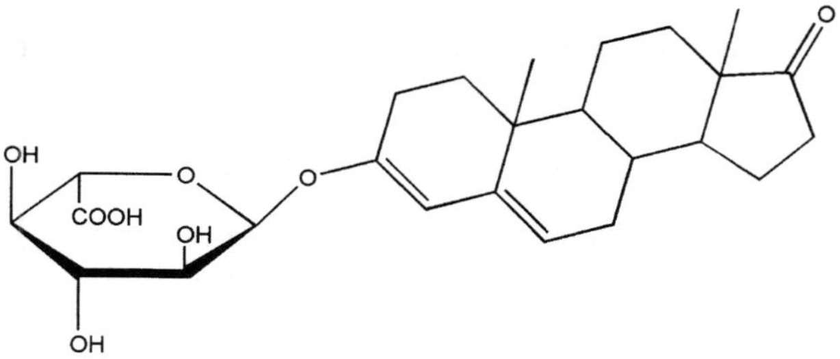

FIGURE 2.16 Unstable glucuronic acid conjugate of the enol form of androstenedione.

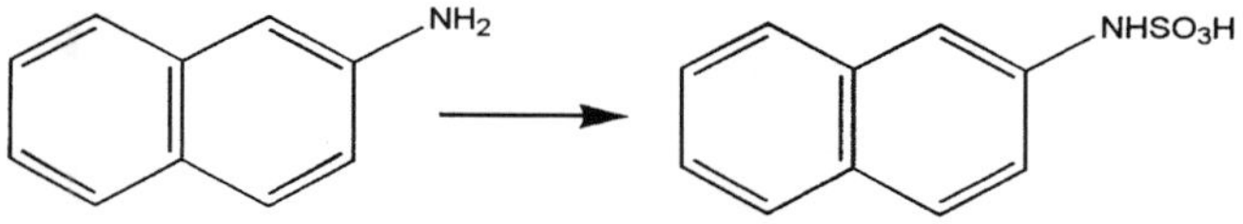

FIGURE 2.17 Sulfation of aromatic amines.

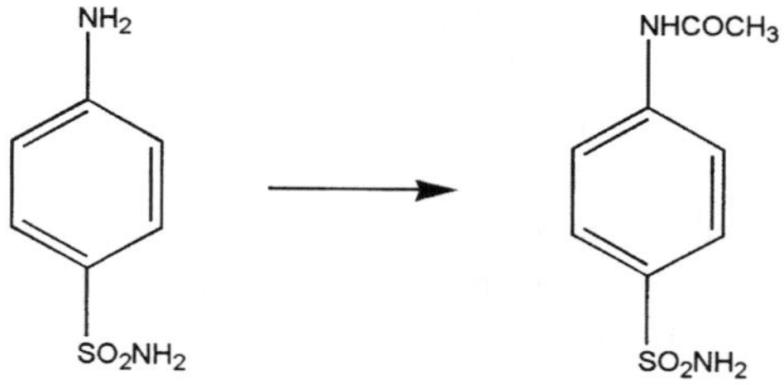

FIGURE 2.18 Acetylation of sulfonamide.

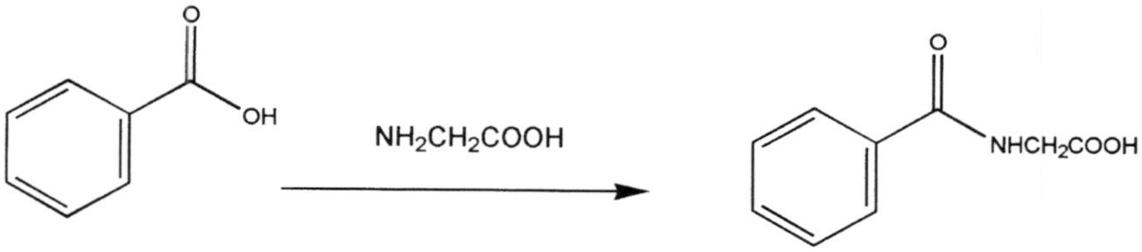

FIGURE 2.19 Conjugation of benzoic acid with glycine to form hippuric acid.

TABLE 2.13
Examples of Metabolism Producing More Active or More Toxic Compounds

Compound	Metabolic pathway	Effect	Reference
Acetanilide	Deacetylation	Aniline causing methemoglobinemia	193
Acetanilide	Oxidation	Active metabolite acetaminophen	193, 194
Acetohexamide	Hydrolysis	Hypoglycemic metabolite	195
Allopurinol	Oxidation	Active metabolite oxipurinol	196
Bambuterol	Esterase	Terbutaline prodrug	197
Carbamazepine	Epoxidation	Active metabolite	198
Clobazam	N-demethylation	Longer lived norclobazam	199
Codeine	Demethylation	Active metabolite morphine	200
Diazepam	N-desmethylation	Active in autonomic nervous system	201
Diphenoxylate	Hydrolysis	Active metabolite diphenoxylic acid	202
Glutethimide	4-hydroxylation	Metabolite causes coma, ataxia	203, 204
Isoniazid	Hydrolysis	Acetylhydrazine promotes hepatitis	205
Lignocaine	N-dethylation	Toxic glycinexylidine	206
Primidone	Oxidation	Active metabolite phenobarbitone	207

successively demethylation, oxidative deamination, and conjugation with glycine. On the other hand, the drug molecule may be subject to several different concurrent conversions as exemplified by chlorpromazine (Figure 2.20) which is metabolized by 7-hydroxylation, S-oxidation, and N-demethylation. These three conversions alone lead to eight possible metabolites, and at least 20 urinary metabolites of chlorpromazine have been characterized[208] with as many as 200 being postulated.[209]

All of these reactions may occur with any particular drug. However, the extent of any particular reaction or chain of reactions depends on the ability of drug and intermediate metabolites to act as substrates for the particular enzyme and the rapidity with which the substrate is removed from the metabolizing sites. Thus, a drug such as isoxepac which is rapidly converted to a water-soluble, readily excreted glucuronide is not exposed to other metabolizing enzymes to any great extent and is largely excreted in the urine as this conjugate and only small amounts are excreted as hydroxymetabolites.[210]

This short description should give one an idea of the range of metabolic reactions that can occur for drug molecules but by no means attempts to show how one could predict the metabolism of any particular compound. This could well be a goal for those who study the mechanisms of drug metabolism, but most researchers agree that

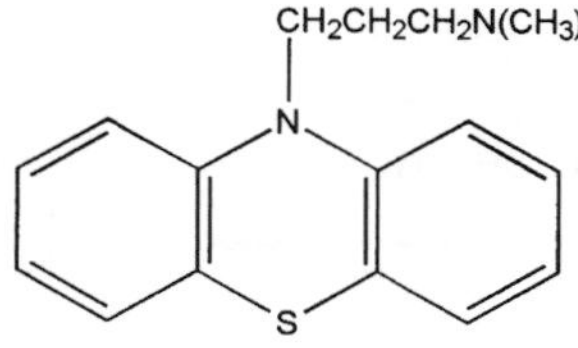

FIGURE 2.20   Structure of chlorpromazine.

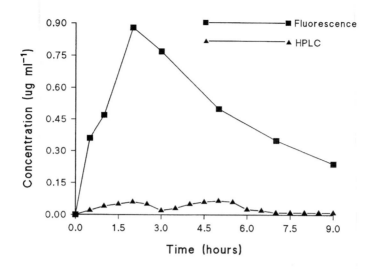

FIGURE 2.21   Plasma concentrations of triamterene as measured by a non-specific fluorescence method and a specific HPLC method following an oral dose of 50 mg triamterene to a human subject.

the state of the art is such that each new chemical entity needs to be studied separately to ascertain the metabolites which are likely to be in biological fluids and which may affect analytical procedures.

2.3.3   SIGNIFICANCE OF METABOLITES

Older methods of analysis of drugs in biological fluids, such as fluorimetry or ultraviolet spectroscopy, tended to include metabolites in the measurement, because these compounds have similar structures to the drug under study. Triamterene, a diuretic, is extremely fluorescent and this property was used to devise a very sensitive fluorescence assay.[211] However, with the advent of chromatographic methods of separation of closely related molecules (Chapters 5, 6, and 7), it was realized that much of the fluorescence in plasma and urine was due to two metabolites, the hydroxy metabolite and possibly its sulfate conjugate.[212] Figure 2.21 shows the difference in the plasma concentration-time profiles for this drug using the direct fluorescence method and a very specific chromatographic method.

Even using chromatographic methods, the analyst needs to be aware of the presence of labile conjugates that could revert to parent drug if over enthusiastic methods are used for preliminary extraction. Thus, at least two separate chromatographic methods for the determination of nomifensine in human plasma were published[213,214] before it was pointed out that the circulating form of the drug was the labile conjugate and the true levels of unchanged nomifensine were only one-hundredth, if that, of the previously reported concentrations following therapeutic doses.[184] As elaborated in later chapters, it is important to consider the possibility of the interference by metabolites in particular methods of analysis.

It may also be important in drug analysis, if the metabolite itself has activity, to measure the concentration of metabolite, in addition to measuring the concentration

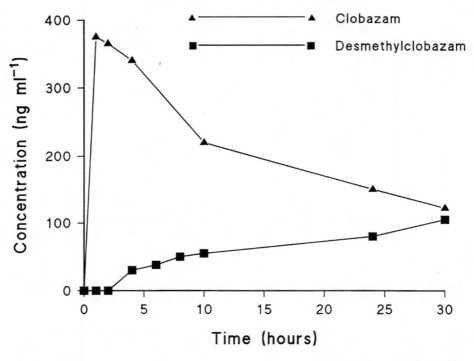

FIGURE 2.22   Plasma concentrations of clobazam and its metabolite desmethylclobazam following an oral dose of 30 mg clobazam to a human subject.

TABLE 2.14
Interference in Drug Analysis by Coadministration of Other Drugs

Drug	Method	Interference	Reference
Acetaminophen	Colorimetric	Salicylates	220
N-Acetylprocainamide	HPLC	Codeine, caffeine	221
Chloramphenicol	HPLC	Methicillin, theophylline	222
Demoxepam	HPLC	Phenytoin	223
Imipramine	HPLC	Thioridazine	224
	EMIT	Thioridazine	224
Maprotiline	HPLC	Benzodiazepines	225
Meprobamate	TLC	Barbiturates	226
Methotrexate	HPLC	Triamterene	227
Mexilitine	HPLC	Disopyramide	228
Opiates	RIA	Pholcodine	229
	EMIT	Pholcodine	229
Phenylbutazone	Colorimetric	Barbiturates	230
Procainamide	HPLC	Acetaminophen	221
Theophylline	RIA	Dimenhydrinate	231
	HPLC	Salicylic acid	232

of the parent drug. If the compounds are equiactive, and have identical pharmacokinetics, then it may be permissible to tolerate an assay which measures the sum of the two; for example, pentoxifylline does not separate from its reduced active metabolite on nonpolar gas chromatography columns.[215] However, if the compounds are not equiactive, then the significance of the pharmacokinetics derived from such measurements will be difficult to interpret. A clearer picture emerges when the drug and metabolite can be determined separately, but preferably, as far as the analyst is concerned, in a single assay. Thus the measurement of clobazam,[216] a short-lived benzodiazepine, and its desmethyl metabolite, a long-lived but less active benzodiazepine, is a very useful exercise (Figure 2.22). It is now well recognized that pharmacokinetic studies must now only be performed when a specific analytical method is used to determine drug concentrations in biological fluids, even for enantiomers as described in more detail in Chapters 6 and 7. Smoking may alter the pharmacokinetics and metabolism of many therapeutically important substances, mainly due to induction of drug-metabolizing enzymes.[217-219]

2.4   PRESENCE OF OTHER DRUGS

In the development of drugs, when animals and normal volunteers are used to investigate the pharmacokinetics of novel compounds, the presence of other drugs is not a problem. However, when the drug is being used in practice, either in patients or in other investigations, then it is important to consider the impact of other drugs. In forensic toxicology and related investigations, the problem is that the drug is unknown and the presence of other drugs is part of the problem being investigated. However, once the first drug has been identified, the investigation is not complete, as in an overdose case the selected corrective measures may not be appropriate for other drugs present in the patient.

Once a development compound begins clinical studies, it is important to establish that the same kinetics apply as those found in normal volunteers; an example of this is the fact that many antiarthritic drugs have a longer half-life in elderly patients than in young volunteers and the recommended dosage may lead to toxic symptoms that could otherwise be avoided. It is also likely that patients may be receiving other drugs and the potential for interference in the analytical method must be considered. Several such interferences have been described (Table 2.14). Although the analyst who develops the original method will attempt to make the method reasonably specific and can guard against a limited selection of coprescribed drugs, the potential for interference from such sources is enormous, particularly when one considers the range of metabolites that is also possible from any one particular drug. Consequently, this is a problem that the analyst needs to keep continually in mind and deal with when it arises.

A more controllable situation arises, however, when a drug is being developed in combination with another. On the one hand, a fixed combination of drugs commonly coprescribed allows the safety of a combination to be assessed more accurately; on the other hand, a fixed combination may not be right for a particular patient. Although there was a rise of fixed combination drugs during the 1980s, such innovations have lessened recently as regulatory authorities have begun to treat these preparations more cautiously.

Common combinations have included painkillers, cough mixtures, antibiotics (to provide a broad spectrum of activity), and oral contraceptives (progestin and estrogen combined). Table 2.15 lists a number of these combinations that are currently marketed in the United Kingdom, excluding antibiotic combinations and the numerous preparations containing codeine, paracetamol (acetaminophen), or acetylsalicylic acid (aspirin). Naturally, in the development of such combinations, it is well recognized that the potential effect of one component on the other's kinetics (formulation or biological) must be checked. The ideal solution is when the same analytical system can be used to assay both drugs simultaneously. In these situations, chromatographic systems are invaluable. Unfortunately, this does mean that extraction and prepurification needs to be optimized for both compounds, at the same time minimizing the potential interference from both sets of metabolites.

2.5   DEGRADATION BEFORE ANALYSIS

Degradation before analysis is often a result of straightforward chemical decomposition due to the instability of the drug itself, particularly in fluids such as urine which may have extremes of pH, or which lack the protective effect of being bound to plasma proteins. Acidification is often used as a preservative in urine, and although it may stabilize some labile conjugates such as N-glucosides of barbiturates,[233] it hydrolyze others more rapidly. Thus, it is particularly important in drug metabolism studies to be able to distinguish true metabolism (or lack of it) from such false results.

Degradation may also be due to enzymic activity which continues in the sample after collection. This is particularly true of plasma or serum esterases. The enzymic activity of such esterases depends on the conditions of collection and storage of the sample. For example, the ibuprofen prodrug, ibuprofen piconol, was hydrolyzed by esterases in blood, plasma, and serum, but the rate of hydrolysis was slowed to varying extents by citrate, heparin, or ethylene diamine tetracetic acid which were used as anticoagulants.[234]

In the section on metabolism, it was noted that many drugs and metabolites are conjugated to organic and inorganic acids. For a long time in drug metabolism studies, such conjugation was an end step in metabolism prior to immediate excretion and it was usual to hydrolyze the sample to measure total drug, particularly in urine samples. Thus several methods have been proposed for releasing the aglycone for further analysis. Dilute acids or bases are acceptable methods. However, the drug itself may be unstable to such procedures, or the hydrolysis may not be complete. Enzymic hydrolysis is generally proposed because it has the advantage of being a chemically mild procedure and also, if the enzyme preparation is

TABLE 2.15
Some Combination Drugs

Name	Supplier	Components	Use
Co-Phenotrope	Gold Cross	Diphenoxalate	Antidiarrheal
		Atropine	Anticholinergic
Pameton	Sterling Winthrop	Paracetamol	Analgesic
		Methionine	Antidote to paracetamol
Moduretic	Du Pont	Amiloride	K+-sparing diuretic
		Hydrochlorothiazide	Thiazide diuretic
Navisapre	Ciba	Amiloride	K+-sparing diuretic
		Cyclopenthiazide	Thiazide diuretic
Burinex A	Leo	Amiloride	K+-sparing diuretic
		Bumetanide	Loop diuretic
Frumil	Rhone-Poulenc	Amiloride	K+-sparing diuretic
		Frusemide	Loop diuretic
Dyazide	SK&F	Triamterene	K+-sparing diuretic
		Hydrochlorothiazide	Thiazide diuretic
Dytide	SK&F	Triamterene	K+-sparing diuretic
		Benzthiazide	Thiazide diuretic
Kalspare	Cusi	Triamterene	K+-sparing diuretic
		Chlorthalidone	Thiazide diuretic
Aldactide	Gold Cross	Hydroflumethiazide	Thiazide diuretic
		Spironolactone	K+-sparing diuretic
Lasilactone	Hoechst	Frusemide	Loop diuretic
		Spironolactone	K+-sparing diuretic
Tenoretic	Stuart	Atenolol	β-Antagonist
		Chlorthalidone	Thiazide diuretic
Beta-Adalat	Bayer	Atenolol	β-Antagonist
		Nifedipine	Vasodilator
Co-Betaloc	Astra	Metoprolol	β-Antagonist
		Hydrochlorothiazide	Thiazide diuretic
Lopresoretic	Geigy	Metoprolol	β-Antagonist
		Chlorthalidone	Thiazide diuretic
Corgaretic	Squibb	Nadolol	β-Antagonist
		Bendrofluazide	Thiazide diuretic
Trasidrex	Ciba	Oxprenolol	β-Antagonist
		Cyclopenthiazide	Thiazide diuretic
Lasipressin	Hoechst	Penbutolol	β-Antagonist
		Frusemide	Loop diuretic
Viskaldix	Sandoz	Pindolol	β-Antagonist
		Clopamide	Thiazide diuretic
Sotazide	Bristol-Myers	Sotalol	β-Antagonist
		Hydrochlorothiazide	Thiazide diuretic
Moducren	Morson	Timolol	β-Antagonist
		Amiloride	K+-sparing diuretic
		Hydrochlorothiazide	Thiazide diuretic
Prestim	Leo	Timolol	β-Antagonist
		Bendrofluazide	Thiazide diuretic
Hydromet	MSD	Methyldopa	Antihypertensive
		Hydrochlorothiazide	Thiazide diuretic
Capozide	Squibb	Captopril	Antihypertensive
		Hydrochlorothiazide	Thiazide diuretic
Innozide	MSD	Enalapril	ACE inhibitor
		Hydrochlorothiazide	Thiazide diuretic

TABLE 2.15 *(continued)*
Some Combination Drugs

Name	Supplier	Components	Use
Carace	Morson	Lisinopril	ACE inhibitor
		Hydrochlorothiazide	Thiazide diuretic
Duovent	Boehringer Ingelheim	Fenoterol	Sympathomimetic
		Ipratropium	Bronchodilator
Franol	Sanofi Winthrop	Ephedrine	Sympathomimetic
		Theophylline	Bronchodilator
Intal Compound	Fisons	Cromoglycate	Anti-allergic
		Isoprenaline	Sympathomimetic
Actifed	Wellcome	Dextromethorphan	Cough suppressant
		Pseudoephedrine	Sympathomimetic
		Triprolidine	Antihistamine
Benylin	Warner-Lambert	Diphenhydramine	Antihistamine
		Dextromethorphan	Cough suppressant
		Codeine	Narcotic analgesic
Davenol	Whitehall	Carbinoxamine	Antihistamine
		Ephedrine	Sympathomimetic
		Pholcodine	Cough suppressant
Dimotane	Whitehall	Brompheniramine	Antihistamine
		Codeine	Narcotic analgesic
		Pseudoephedrine	Sympathomimetic
Phensedyl	Rhone-Poulenc	Codeine	Narcotic analgesic
		Promethazine	Antihistamine
Tixylix	Intercare	Pholcodine	Cough suppressant
		Promethazine	Antihistamine
Dimotapp	Whitehall	Brompheniramine	Antihistamine
		Phenylephrine	Sympathomimetic
		Phenylpropanolamine	Sympathomimetic
Eskornade	SK&F	Diphenylpyraline	Antihistamine
		Phenylpropanolamine	Sympathomimetic
		Phenylephrine	Sympathomimetic
Expurhin	Galen	Chlorpheniramine	Antihistamine
		Ephedrine	Sympathomimetic
Tuinal	Lilly	Amylobarbitone	Barbiturate
		Quinalbarbitone	Barbiturate
Pamergan	Martindale	Pethidine	Narcotic analgesic
		Promethazine	Antihistamine
Bactrim	Roche	Trimethoprim	Antimicrobial
		Sulfamethoxazole	Sulfonamide
Flagyl Compak	Rhone-Poulenc	Metronidazole	Antibacterial
		Nystatin	Antifungal
Maloprim	Wellcome	Pyrimethamine	Antimalarial
		Dapsone	Antileprotic
Loestrin	Parke-Davis	Norethisterone acetate	Progestin
		Ethinyloestradiol	Estrogen
Mercilon	Organon	Desogestrel	Progestin
		Ethinyloestradiol	Estrogen
Conova	Gold Cross	Ethynodiol diacetate	Progestin
		Ethinyloestradiol	Estrogen
Eugynon	Schering Health	Levonorgestrel	Progestin
		Ethinyloestradiol	Estrogen
Femodene	Schering Health	Gestodene	Progestin
		Ethinyloestradiol	Estrogen

sufficiently specific, it will give information on the type of conjugate which is hydrolyzed. The study of conjugates by their hydrolysis reactions has been reviewed by Caldwell and Hutt.[161]

Recently, however, the importance of conjugates in their own right has been recognized. The advent of high-pressure liquid chromatography with its ability to separate compounds of widely differing polarities, and new stationary phases that will accept biological fluids directly has enabled analysis of conjugates and parent compounds to be performed with relative ease. Thus, the necessity for hydrolysis and the consequent loss of information has decreased.

2.6   SAFETY CONSIDERATIONS

Most pharmaceutical research laboratories are well aware and take all necessary precautions against the untoward effects of the drugs they are handling and this becomes more important as drugs become more and more active. The activity of the drug itself is not often such a problem for the analytical laboratory specializing in determining drugs in biological fluids, since the amount of drug being handled is very small. However, because the samples are of biological origin, there may be biological hazards, particularly in samples from patients with infectious or contagious diseases.

Recently, it has been recognized that even samples from healthy subjects in phase I and phase II clinical trials may pose a health hazard, with the possibility of the human immunodeficiency virus (HIV) being present in a significant fraction of a seemingly normal population. Additionally, blood bank plasma, often used for development work, or for preparing standards and quality control samples may be suspect, as has been shown where hemophiliac patients have been infected with HIV after treatment with blood products from such a source.[235,236] However remote the possibility of this infection being transmitted through analytical procedures, safety experts advise that protective measures (i.e., use of surgical gloves or even surgical masks to guard against aerosols) must be taken because the consequences of infection are so devastating.

Chemical or thermal treatment of biological samples[237,238] may seem the obvious method of decontamination, but of course any harsh treatment could very well degrade the drugs to be measured. Thus, if strong disinfectants are used on samples prior to analysis, the effect on the analyte must first be checked. Most methods of analysis begin with an extraction step using organic solvents and this is often sufficient to deactivate any biological hazards. Good et al.,[239] recognizing that the HIV virus could be deactivated by heat treatment, investigated the effect of those conditions on the anti-AIDS compound zidovudine and its glucuronide as part of a study of plasma levels of the drug and conjugate during treatment; no degradation of the compounds was observed after a 1-h incubation at 58°C.

Irradiation using X-rays is used routinely in the food industry and in the medical industry to sterilize surgical instruments and de Bree and van Berkel have suggested this approach be used for deactivation of HIV in biological samples.[240] It was shown that even relatively high doses of irradiation (up to 10 Mrads), which completely inactivated HIV had no effect on a number of drugs of diverse structure (i.e., fluvoxamine, clovoxamine, flesinoxan, idaverine, *N*-desmethylidaverine, and eltoprazine) in serum samples. It was concluded that such treatment could be used routinely on biological samples to make them harmless as far as the risk of HIV infection was concerned, provided it was shown that the analyte in question was unaffected by the irradiation.

3 Special Problems in Structures of Drugs

CONTENTS

3.1 SMALL ORGANIC MOLECULES

The vast majority of chemical compounds used as drugs are small molecules of molecular weight 200 to 300 Da. Most of these compounds are weak acids or weak bases and the drugs themselves are usually in the form of salts, the particular salt being chosen for its physicochemical properties in the formulation stages rather than its pharmacology; once in the body, buffering in the blood and other fluids makes the original salt form mainly irrelevant to its subsequent biological activity. A few of these small molecular weight drugs may be uncharged, relatively lipophilic molecules, such as the steroids. Other uncharged drugs include esters, which may be active or may act as prodrugs releasing either an acid drug or an alcohol, and quaternary amines. Some examples of these different types of structures are given in Table 3.1.

Because most modern drugs have been carefully chosen from large numbers of similar synthetic compounds, rather than as synthesized versions of naturally occurring compounds, they are more likely to have favorable physicochemical properties, including good stability under adverse conditions of heat, exposure to light, and humidity. Nevertheless, in biological fluids, the drug may be present in very small concentrations and amounts, and in the development of an analytical method, the stability of the small amounts of analyte encountered needs to be checked. Structural features that may become more important at high dilutions include, for example, ketones (isomerization), olefines (epoxidation), alcohols (oxidation), and amines (oxidative deamination).

TABLE 3.1
Examples of Organic Molecules Used as Drugs

Uncharged
    Steroids
    Acetylsalicylates
    Phenacetin
    Meprobamate
Weak acids
    Ethynylestradiol
    Phenol
    Valproic acid
    Barbiturates
Weak bases
    Benzodiazepines
    Propanolamines
    Amitryptiline
    Amphetamines
    Morphine
    Chlorpromazine

The neutral character of steroid drugs, esters, and ethers means that the partition of such compounds into organic solvents is independent of the pH. This property gives good scope for partial purification of such compounds by simply removing the acids and bases from an organic extract by successive washes with dilute acid and dilute alkalis; final washing with distilled water and drying by simple filtration through sodium sulfate or cellulose gives a relatively clean extract for urine samples, but the presence of lipids in plasma or serum is more problematic and requires careful choice of organic solvent for extraction. On the other hand, aqueous samples containing weak bases and weak acids as analytes can be readily defatted with a hexane wash. Thus, the charged nature of the small-molecule drug is an important consideration in the analytical determination.

3.1.1 CONJUGATES

Few drugs are administered as conjugates, but many drugs form conjugates *in vivo*. It was originally thought that formation of conjugates was a final step in making xenobiotics water soluble for elimination in the urine and thus little attention was given to the analysis of the conjugates. However, the presence of enzymes responsible for xenobiotic metabolism, including the formation of conjugates, is now recognized as an important parameter in determining the individual's genetic control of his own metabolism. As a result, the assay of intact conjugates has become an important consideration for the analyst. The various enzymes under such control are discussed in detail in Chapter 1.

The classical procedure in determining conjugates was to assay the sample for unchanged drug and then after a suitable hydrolytic procedure determine the "total" drug; the difference between the two procedures would be considered the amount of conjugated drug. By using selective hydrolysis, either by carefully controlled pH conditions, or by using enzymes considered specific for particular conjugates, such

TABLE 3.2
Analysis of Intact Conjugates of Drugs Using HPLC

Drug	Conjugate	Fluid	Detection	Reference
Coumarin	Hydroxy glucuronides	—	—	242
Equilin	Sulfate	Serum	HPLC	243
Estrone	Sulfate	Serum	HPLC	243
7-Hydroxycoumarin	Glucuronide	—	—	242
Ibuprofen	Glucuronide	—	—	244
Indomethacin	Glucuronides	Plasma, urine	—	245
Morphine	3- and 6- glucuronides	Plasma	Fluorescence	246
Morphine	3-glucuronide	Human urine	CIMS	247
Naproxen	Glucuronide	Plasma, urine	—	248
Naproxen	Glucuronide	Plasma, urine	—	249, 250
Oxazepam	Glucuronide	Urine	CIMS	251
Zidovudine	Glucuronide	Serum	HPLC	239

as β-glucuronidase for morphine glucuronide[241] and sulfatases, attempts could be made to determine the individual conjugates. However, chemical hydrolysis could also degrade the drug itself, and enzyme hydrolyses were less specific than hoped, or else would not always be complete. Furthermore, commercially available enzymes, such as those from *Helix pomatia* tended to introduce other interfering substances thus making subsequent analysis more difficult.

The recent trend has been to analyze samples for unchanged material, rather than submit to the risk of incorrect interpretations. There have been many successful approaches to direct profiling of drugs and drug metabolites in biological fluids. The ability of liquid chromatography to work with a wide variety of structurally diverse compounds has made it very useful in this area and the powerful structural tools of physical chemistry are now also being used to assess the content of untreated biological fluids, sometimes with no chromatographic purification.

Table 3.2 lists a few of the compounds and their conjugates that have been analyzed directly using chromatographic methods.

3.1.2 CHIRALITY

The first edition of this work was like similar works of its period in that it gave only scant attention to the chiral nature of some drugs, or to the fact that almost all synthesized drugs which had an asymmetric carbon were marketed as racemates. Workers in the field must have been aware of such things; after all, molecular asymmetry had been studied widely in the middle of the nineteenth century. The whole basis of organic chemical structure is based on the tetrahedral carbon atom and the concept of mirror-image isomers is an inevitable consequence. Many of the synthetic drugs initially produced by the pharmaceutical industry were semisynthetic and based on naturally occurring precursors such as steroids and β-lactams. Fortuitously, the marketed compounds were optically pure, and any consideration of enantiomeric forms of the drug did not normally arise.

With the introduction of new drugs which were obtained by total synthesis from readily available precursors, the development and marketing of chemically indistinguishable racemates became much more likely for several reasons: the racemate was prepared because the precursors were achiral and because chirally directed synthetic methods were uncommon; the racemates were tested because preparative methods of separation were also uncommon. The developed form would then naturally follow from the tested form, rather than introduce apparently unnecessary and expensive further separation procedures. In addition, all analytical methodology at the time appeared to be achiral (or used standards which masked the chiral nature of an analytical response). Likewise, many development scientists may not have been aware that the racemate was being developed. The complacency was fuelled (if that is a word that can be applied to complacency) by the lack of any indication that the use of racemates as drugs was in any way undesirable. Two of the most successful drugs recently, the β-adrenoceptor blocking agent propranolol, and the analgesic ibuprofen are both marketed as racemates, but there has been no indication of untoward clinical effects arising from the racemic nature of these preparations. Similarly, the hypnotic thalidomide was used as its racemate, and the suggestion has often been made that if the optically pure form, which was responsible for its hypnotic effect, had been separated and used then the tragedy of the early 1960s would not have happened. However, animal evidence that the teratogenic and hypnotic effects can be separated in this way is inconclusive,[252-254] and of course the hypothesis is untestable in humans.

Propranolol and ibuprofen probably represent, respectively, cases where one isomer is pharmacologically inactive at the doses used,[255] or the inactive enantiomer is converted *in vivo* to the active isomer.[256] As such, any clinical implications may not be expected, and there would be no harm in using the cheaper produced version of the drug. However, ignoring the presence of the second isomer may lead the unsuspecting pharmacokineticist or clinical pharmacologist to assume that the presence of enantiomers can also be ignored. If indeed the enantiomers are handled by metabolizing enzymes and transport proteins in an identical manner, there would be no problem, but considering such processes can be highly stereochemically selective, then it is not surprising that pharmacokinetic differences in pairs of enantiomers have been described (Table 3.3).

Because of the unavailability of chirally selective analytical methods, and perhaps more importantly because of the general unawareness that a problem may exist, the separate pharmacokinetics of enantiomers were not reported for many drugs in the literature until about 10 years ago. Therefore, there is an extensive pharmacokinetic literature which reports many drugs as having biphasic elimination from the plasma, which could be erroneous. To illustrate how this error can occur, consider the hypothetical case of a drug which is eliminated from the plasma by a single process, but with different rates for the two enantiomers. The observed decay using a nonchiral analytical method to monitor blood levels would be as shown in Figure 3.1, where the simulation is for enantiomers having elimination constants of 0.005 and 0.04 min^{-1} (corresponding to half-lives of 140 and 17.5 min). The usual methods of pharmacokinetic analysis (using stripping methods) on the data shown in this figure yield a biexponential decay with elimination constants of .006 and .033. Simple pharmacokinetic theory would suggest that multiple dose treatment with the drug would result in

TABLE 3.3
Differing Pharmacokinetics in Humans for Pairs of Enantiomers of Drugs Marketed as Racemates

Drug	Comments	Reference
Alprenolol	S peak of 8 ng ml^{-1}, R peak 1.8 ng ml^{-1}	257
Bufuralol	(+) peak of 18 ng ml^{-1}, (−) peak of 35 ng ml^{-1}	258
Disopyramide		259
Disopyramide	(+) clearance of 131 ml min^{-1}, (−) clearance of 76 ml min^{-1}	260
Fenfluramine		261
Flecainide		262
Flurbiprofen	(+) clearance of 57 ml min^{-1}, (−) clearance of 75 ml min^{-1}	156
Flurbiprofen	(+) clearance of 20 ml min^{-1}, (−) clearance of 16 ml min^{-1}	263
Glutethimide	1.1 mg (+) in urine, 2.1 mg (−) in urine	264
Halofantrine	1.4–2.4 of R compared with S in plasma	265
Ibuprofen	(+) clearance of 74 ml min^{-1}, (−) clearance of 68 ml min^{-1}	266
Metoprolol	(+) clearance of 56 ml min^{-1}, (−) clearance of 62 ml min^{-1}	267
Mexilitine	Little difference	268
Misonidazole	(+) clearance of 48 ml min^{-1}, (−) clearance of 41 ml min^{-1}	269
Nicoumalone (acenocoumarol)	R-(+) clearance of 25 ml min^{-1}, S-(−) clearance of 250 ml min^{-1}	270
Nicoumalone (acenocoumarol)	R clearance of 35 ml min^{-1}, S clearance of 500 ml min^{-1}	271
Nitrendipine	R clearance of 1.62 ml min^{-1}, S clearance of 1.51 ml min^{-1}	272, 273
Pentobarbital	(+) clearance of 40 ml min^{-1}, (−) clearance of 32 ml min^{-1}	274
Phenprocoumon	(+) clearance of 0.85 ml min^{-1}, (−) clearance of 0.70 ml min^{-1}	275
Propranolol	l half-life of 52 min, d half-life of 24 min	276, 277
Propranolol	No first-pass difference	278
Propranolol	(+) clearance of 5000 ml min^{-1}, (−) clearance of 2000 ml min^{-1}	279
Propranolol	2:1 ratio of (+)/(−) clearances	280
Propranolol	(+) clearance of 1200 ml min^{-1}, (−) clearance of 1000 ml min^{-1}	281
Quinidine/quinine	Quinidine clearance of 99 ml min^{-1}, quinine clearance of 25 ml min^{-1}	282
Tocainide	S-(+) peak of 8 µg ml^{-1}, R-(−) peak of 4 µg ml^{-1}	283
Tocainide	(+) clearance of 106 ml min^{-1}, (−) clearance of 197 ml min^{-1}	284
Verapamil	(+) clearance of 1720 ml min^{-1}, (−) clearance of 7460 ml min^{-1}	285
Verapamil	(+) clearance of 800 ml min^{-1}, (−) clearance of 1400 ml min^{-1}	286
Vigabatrin	(+) clearance of 82 ml min^{-1}, (−) clearance of 90 ml min^{-1}	287
Warfarin	(+) half-life of 45.5 h, (−) half-life of 33 h	289
Warfarin	(+) clearance of 1.5 ml min^{-1}, (−) clearance of 2.45 ml min^{-1}	290
Warfarin	(+) peak plasma level of 58 µg ml^{-1}, (−) peak plasma level of 33 µg ml^{-1}	288

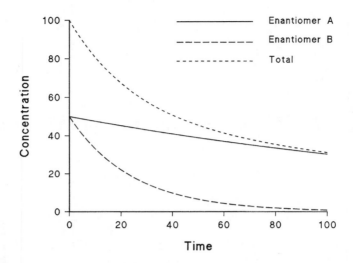

FIGURE 3.1 Simulation of enantiomers eliminated from plasma at different rates. The upper trace is the curve generated if a nonchiral analytical method was used to monitor plasma levels, and is the sum of the two lower curves representing the separate enanatiomers.

a reasonable steady-state level, with a concomitant continuous drug activity of the drug. This conclusion would, of course, be mistaken if the active compound was the enantiomer with the shorter half-life — the value that was probably written off as the "disposition phase." It is a matter of debate, however, whether the demands of simpler pharmacokinetic analysis would be sufficient for deciding whether the drug should be developed as a single isomer.

Approximately 10 years ago, Ariëns[291] caused some comment, but perhaps not enough outrage by describing the nonactive enantiomer in racemic preparations as unnecessary ballast, akin to levels of impurity of 50% in the true drug. Ariëns was acknowledged to be correct, but because there was no particular problem, little seemed to be done. However, as more and more reports appeared of separation of pharmacological effect associated with different enantiomers, the wisdom (indeed the advantages) of separately developing pure isomers was recognized. This new realism coincided with the development of techniques in synthetic and analytical chemistry of optically active compounds. The process development chemist in the pharmaceutical industry is responsible for creating an efficient and economic method of preparing commercially useful amounts of new drugs. It is now recognized that producing optically pure entities is only one of the routine problems to be solved. Concurrent with all this activity in research and development, the regulatory authorities have taken an interest, particularly in light of the potential, specific toxicity of the therapeutically active partner. Under the current U.S. regulatory guidelines, when a new drug substance is asymmetric, the various stereoisomers should also be separated or synthesized independently and their physical, chemical, and pharmacological properties evaluated.[292,293] Keeping in mind that cholesterol has 256 possible isomers, these guidelines, perhaps, should not be taken too literally. It is evident, however, that the chiral nature of a new drug substance is an important consideration.

A recent paper on aspects of chiral high-performance liquid chromatography in pharmaceutical analysis[294] includes the current European regulatory viewpoint on racemates vs. single isomers. Chiral problems are not limited to the parent drug; in some cases a nonchiral center can be introduced by metabolic processes, such as by reducing ketones,[295,296] including nafimidone[297] and oxcarbazepine,[298] and other metabolic routes.[299-301]

The success of the bioanalyst in devising chiral separations for following separate enantiomers following dosing of racemates has undoubtedly influenced the trend toward single-isomer products. This success has, paradoxically, diminished the very reason for the development of such methods; if the drug dosed is a pure enantiomer, then there will be no need to follow the isomer that was not dosed! Whelpton foresaw this situation several years ago and made the ingenious suggestion that the developed chiral analysis could be applied, with the unused enantiomer serving as a perfect internal standard or carrier, being considerably cheaper in equipment than the equivalent mass-spectrometry isotope dilution methods.[302]

3.2   PEPTIDE AND PROTEIN DRUGS

Although almost all of the drugs mentioned so far have been small organic molecules, the increasing pace of medicinal research has led to other types of molecules, possibly more closely related to the macromolecule exerting biological effects. Some 200 protein drugs are now in active clinical trials. It is, however, a matter of debate as to whether the increased knowledge on the mechanism of action of these drugs will lead to designs of new drugs based on these proteins or whether the understanding will lead to better design of the more traditional small-molecule drugs. Nevertheless, it is apparent that many of the newer specific drugs will be created by modification of peptides, which will pose a different set of challenges to the analyst.

The development of small molecules has been based on varying the chemical structure of known bioactive molecules, usually having extraordinarily simple structures given the range of functions they perform. These structures include steroids, histamines, adrenaline, and γ-aminobutyric acid. On the other hand the peptide and protein drugs are much more labile and need to be developed and manufactured under more delicate conditions to ensure that they retain their activity, which often will reside in a single conformation. In addition, large protein molecules are liable to induce antibody formation in the host animal which may limit their usefulness as drugs.

There are many early examples of natural proteins used as drugs, often as replacement therapy using the drug protein extracted from natural sources or even from animals. These would include insulin, luteinizing hormone, and follicle stimulating hormone. In some cases, only part of the peptide chain is necessary for biological activity and protein fragments may often be used in therapeutics. Other active molecules may be shorter chain peptides and often, too, fragments of these chains may have sufficient activity to be used as conventional drugs, particularly if they are more stable and less immunogenic than their long chain parents. Some examples of protein and peptide drugs are listed in Table 3.4.

TABLE 3.4
Proteins and Peptides Used as Drugs

Proteins
  Insulin
  FSH
  LH
  Growth factors
  Digestive enzymes
  Fibrinolytic enzymes
Peptides
  ACTH
  Peptidoglycans
  Neuropeptides
  Toxins from fungi
  Snake venoms
  Tachykinins
  β-Endorphin
  Vasopressin
  Oxytocin
  Somatomedins
  Calcitonin

Immunotoxins are molecular conjugates formed by monoclonal antibodies linked to toxic agents, which then target particular cells; in therapeutics the targeted cells are tumors. A typical and possibly most common toxin used in this way is ricin A.[303] The desired conjugates are prepared identically as conjugates for radioimmunoassay development and these large protein molecules will have characteristic problems for the development of analytical methods.

The major problem with hormones is how to ensure that they are not denatured. For analytical work, retention of activity may not be so critical and assays can be developed which are specific for the denatured protein and hence for the drug. Much of the analytical work in this area has focused on natural proteins and peptides and is rather outside the scope of this volume, which is concerned largely with small, relatively stable molecules. Most methods of analysis will utilize the biological effect, or the immunogenic potential so that radioimmunoassays can be developed, or will utilize liquid chromatography systems where the analyte may remain stable in the solvent system used. Alternatively, the protein may be degraded to a specific component which is assayed by conventional means, as for example, the peptidoglycans can be assayed by the muramic acid content. Analytical methods being developed for protein and peptide drugs are mostly derived from the chemistry of natural products and endogenous hormones. A survey of this enormous subject is beyond the scope of this book which limits itself to the analysis of traditional small-molecule compounds. Some key references for the analysis of proteins and peptides are listed in Table 3.5.

3.3   PRODRUGS

Prodrugs are chemical structures which are not pharmacologically active, but in the body are converted to an active molecule. In general, the prodrug has chemical

TABLE 3.5
Analysis of Proteins and Peptides

Type	Method	Reference
Enkephalin peptides	HPLC- mass spectrometry	134
Growth hormone	Gel filtration chromatography	304
Hypophysial hormones	HPLC	305
Neuropeptides	HPLC-radioimmunoassay	306
Pituitary proteins	Ion-exchange HPLC	307
Polypeptides and proteins	Reversed-phase HPLC	308
Proteins	Reversed-phase HPLC	309
β-Endorphin in plasma	Cation-exchange HPLC	310
Vasoactive peptides	Chromatography	311
C-Peptide	Radioimmunoassay	312
Neurokinins	HPLC-radioimmunoassay	313
γ-MSH	HPLC-Thermospray MS	314

characteristics, such as stability, lipophilicity, state of ionisation, that increase its chances of reaching its site of action, or which protect intervening biological structures from the effect of the active drug (taste, toxicity). Ideally, the prodrug should be converted to the active form at the site of action, particularly if the active form would be toxic at other sites. This is particularly true of anticancer drugs which are usually directed at killing cells. Table 3.6 shows a number of prodrugs and their active degradation products.

The use of prodrugs does not offer a particular problem to the analyst in chemical terms. However, the pharmacokineticist needs to distinguish between the needs of measuring the parent compound or the active compound, depending on the object of the study. For example, if the project is to investigate the absorption of the administered drug (i.e., to investigate formulations), then it is sensible to measure the appearance of unchanged compound in the blood and to apply classical bioavailability theory. If, on the other hand, the investigator wants to compare potential utility of the drug, then the levels of the active product, preferably at the site of action, are more relevant. A particular form of prodrug is found in the conjugation of the drug molecule to polymers as described for the polypeptide R-(a-acetyl)eglin c conjugated to poly(oxyethylene) to give a macromolecule of molecular weight >20,000.[321]

TABLE 3.6
Prodrugs and Their Active Products

Administered form	Active form	Advantage	References
Chloramphenicol palmitate	Chloramphenicol	Hides bitter taste	315
Clindamycin phosphate ester	Clindamycin	Reduces pain at injection	316
Acetylsalicylate esters	Acetylsalicylic acid	Reduces gastric irritability	317
Pivampicillin	Ampicillin	Improved absorption	318
Propranolol hemisuccinate	Propranolol	Reduces first-pass metabolism	319
L-Dopa	Dopamine	Site-specific delivery	320

3.4   FORMULATION

Formulation is not usually considered as a problem affecting the assay of drugs in biological fluids, other than when the drug may be administered with another drug which may interfere in the analysis. Recently, however, novel forms of drug delivery are being developed and in some of these the drug may be attached, or contained within particles or other vehicles while being transported in the blood or other biological fluid. Depending on how firmly the drugs are attached or encapsulated, the analyst may face the problem of extraction, or indeed he may be called on to distinguish between the free drug and the drug attached to the vehicle. 2-Hydroxypropyl-β-cyclodextrin complexes of methyltestosterone and esters of 1,4-dihydropyridines have been prepared as injectable forms of the drugs with improved tolerance characteristics.[322] In following the pharmacokinetics of these preparations, Muller and Albers[322] did not note any special problems in assaying serum by standard HPLC methods. Similarly, doxorubicin encapsulated in liposomes could be readily extracted from the liposomes in plasma and measured by HPLC in the study of the pharmacokinetics of this novel drug delivery system.[323]

4 Spectroscopy and Fluorimetry

CONTENTS

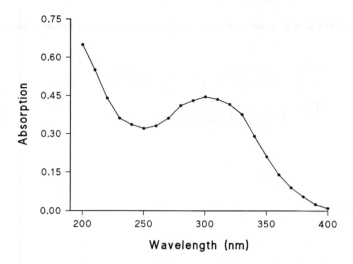

FIGURE 4.1 UV-absorption spectrum constructed by manual determination of optical absorption at specified wavelengths.

4.1   INTRODUCTION

The two topics of this chapter have a commonalty in that the amount of light emanating from the sample is measured and is assumed to be a function of the concentration of the compound under investigation in the sample. In spectrophotometry the sample is irradiated and the extent to which the incident light is absorbed is measured; in fluorimetry, the sample is irradiated and the resulting activated emission, at a different wavelength, is measured. It is therefore usual to treat the two phenomena sequentially and this is the procedure adopted here.

4.2   SPECTROPHOTOMETRY

4.2.1   DIRECT MEASUREMENT

Any sample of matter, be it gaseous, liquid, or even solid, will absorb all wavelengths to some extent when irradiated with light. If the proportion of light absorbed at a particular wavelength is plotted against the wavelength, a spectrum of absorption is produced. With older spectrometers, it was usual to produce the spectrum by such painstaking procedures, with the smoothness of the resulting spectrum depending on the number of points the analyst was prepared to plot (Figure 4.1). With the advent of continuous scanning and recording spectrophotometers, however, the more familiar continuous spectra are obtained (Figure 4.2).

Two types of information can be obtained from the absorption of a sample: quantitative and qualitative. Figure 4.2 shows the absorption spectrum of a solution of clobazam in ethanol. The range of wavelengths covered is that usually supplied by most commercial spectrophotometers and ranges from 200 to 800 nm — the so-called UV-visible region. The characteristic spectrum seen provides qualitative information on the identity or structure of the analyte. The quantitative information

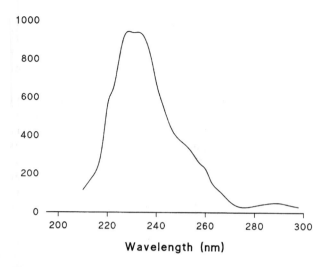

FIGURE 4.2 UV-visible absorption spectrum of clobazam in ethanol using a continuous scanning and recording spectrophotometer.

that can be obtained from a UV-visible spectrum depends on the extent of the absorption. This can be described in terms of the Bouguer-Lambert-Beer principles which state that:

1. The proportion of light absorbed is independent of the intensity of the incident light.
2. The intensity of radiation is reduced by the same fraction in each equal thickness of the sample.
3. Absorption is proportional to the concentration of the absorbing species.

Consequently, the rate of change of intensity of light with respect to the distance penetrated into the sample is directly proportional to the concentration multiplied by the intensity. For a sample placed into a cell of fixed width, the absorption will be proportional to the concentration of the material of interest. It is usual to express the capacity of the material to absorb UV and visible light by quoting the molar absorptivity ε,

$$\varepsilon = \frac{\log I_o - \log I}{c \cdot l} \times M \tag{1}$$

where I_o is the intensity of irradiation, I is the measured intensity, l is the path length (in cm), c is the concentration (in g l^{-1}), and M is the molecular weight of the absorbing species.

Although a useful notation used in the literature often quotes an E value for a 1% solution in a 1 cm cell, in practice the molar absorptivity should be calculated from a sample displaying an absorption in the range 0.3 to 0.6. When the absorption of the sample is to be used for quantification of the absorbing species, it is apparent that the best wavelength for measurement will be the absorption maximum, as this will give the optimal response. Thus for clobazam, the compound shown in Figure 4.2, the procedure would be to measure the absorption of the sample at 230 nm and relate this value to the absorption of standard solutions.

In using such a method for the analysis of drugs, the molar absorptivity has to be sufficiently high to give measurable absorption for the amounts of material available. Gifford[324] has provided a useful table which gives the concentrations of drug in the measuring cell which will give an absorption of 0.4 and compares this with the peak concentration in plasma following therapeutic doses. An extended version of Gifford's table is presented in Table 4.1 and indicates which drugs would bear consideration for quantitative analysis by direct absorption spectrophotometry. Thus, useful methods would be feasible for acetylsalicylic acid and tolbutamide, but not for amphetamine or haloperidol.

The molar absorptivity of the solute in spectrophotometry is determined by the chemical structure. A functional group on a molecule which is responsible for absorption is termed a chromophore. Table 4.2 lists the absorption characteristics of a number of functional groups found in drugs.

Some functional groups are lacking in useful absorption characteristics but when conjugated with a chromophore the wavelength of absorption and the intensity are both increased. Such groups are termed auxochromes and include hydroxyl functions, amines, and halogens. Increased intensity of absorptivity (hyperchromicity) and increased wavelength (bathochromicity) will also result if chromophores conjugate with one another.

UV-visible absorption spectra are characterized by broad bands with hills and valleys rather than by sharp peaks. Thus, compounds with the same functional groups will have similar absorption spectra, and the technique is not appropriate for identifying unknown components in a sample. Likewise, the technique is not especially specific when used as a quantitative method for drugs in biological fluids if metabolites with the same functional groups are present. Nevertheless, the technique can be extremely useful and uses relatively troublefree equipment. A typical procedure is described by McConnell et al.[325] for the assay of rifampicin. Plasma, buffered to pH 5 to 6, is extracted with a benzene:hexane mixture (which does not extract bilirubin, a potential interfering endogenous compound). The solvent is evaporated and the residue reconstituted in methanol prior to measurement of absorbance at 343 nm. The method is suitable for assay of rifampicin in plasma over the range of 1 to 45 μg ml^{-1}. Similar methods for other drugs can be developed by the judicious use of solvents and pH to minimize interference by endogenous compounds or metabolites.

The pH of samples used for measurement can dramatically affect the absorption spectrum, and this property should also be utilized in the development of a spectrophotometric method. The classic examples of this phenomenon are phenol and aniline. Phenol in its un-ionized form absorbs at 270 nm with a molar absorptivity of 1450. In alkaline solution, however, the phenate ion absorbs at 287 nm with an ε value of 2600. This increased absorbance is due to the presence of additional unshared electrons available for conjugation with the phenyl ring. Aniline, on the

TABLE 4.1
Comparison of Ultraviolet Absorbance of Commonly Prescribed Drugs and Their Therapeutic Plasma Concentrations

Drug	Concentration giving absorption of 0.4 absorption units (μg ml^{-1})	Approximate plasma concentration during therapy (μg ml^{-1})
Amitriptyline	8	0.05
Caffeine	8	2
Carbamazepine	15	10
Cephaloridine	10	10
Chloramphenicol	13	10
Chlordiazepoxide	4	2
Clorpromazine	3	0.45
Chlorpropamide	7	30
Cinnarizine	5	0.10
Codeine	80	0.03
Dapsone	4	1
Diazepam	3	0.20
Digoxin	15	0.002
Doxapram	285	0.3
Ethosuximide	470	150
Flupenthixol	5	0.01
Frusemide	4	1
Glutethimide	5	5
Haloperidol	10	10
Hydrochlorothiazide	6	0.10
Imipramine	13	0.10
Isoniazid	10	5
Lignocaine	300	5
Lorazepam	3	0.01
Medazepam	5	0.10
Methaqualone	3	2
Metronidazole	10	5
Neostigmine	200	0.10
Nitrazepam	3	0.03
Oxazepam	3	0.10
Oxprenolol	11	0.70
Perphenazine	4	0.01
Phenacetin	5	5
Phentolamine	9	0.03
Phenylbutazone	6	30
Phenytoin	130	20
Pimozide	26	0.01
Practolol	6	0.50
Prednisolone	9	0.20
Primidone	740	10
Procainamide	6	10
Propantheline	39	0.10
Propranolol	3	0.05
Quinidine	4	2

TABLE 4.1 (*continued*)
Comparison of Ultraviolet Absorbance of
Commonly Prescribed Drugs and Their
Therapeutic Plasma Concentrations

Drug	Concentration giving absorption of 0.4 absorption units ($\mu g \, ml^{-1}$)	Approximate plasma concentration during therapy ($\mu g \, ml^{-1}$)
Quinine	4	5
Salicylic acid	15	70
Sulfadiazine	5	30
Sulfapyridine	5	0.01
Temazepam	4	0.50
Tetracycline	8	1
Theophylline	7	10
Thioridazine	3	0.1
Tolbutamide	8	100
Triamterene	5	0.01
Trifluoperazine	5	0.002
Verapamil	12	0.10
Warfarin	9	8

other hand, already has a pair of unshared electrons available for conjugation in the nonprotonated form; the absorption maximum of aniline in neutral or alkaline solution is at 280 nm with an ε value of 1430. The addition of acid results in protonation of the unpaired electrons, conjugation is decreased, and the absorption maximum shifts to 254 nm with an ε value of 160.

The use of such shifts in spectral properties has been particularly useful in the analysis of barbiturates.[326] Barbiturates have the general formula shown in Figure 4.3, and in alkaline solution two absorbing species are normally formed as shown in Figure 4.4. Some barbiturates, such as methylphenobarbitone and narcobarbital have one of the nitrogen atoms already substituted and the full ionization is not possible. At neutral and acid pH values, the barbiturates are un-ionised and absorption is negligible. At about pH 9.5, the second species is formed and absorption at 240 nm is observed. At pH 13, the fully ionized species with extended conjugation is formed and the absorption maximum increases to 255 nm. For quantitative analysis of barbiturates the increase in absorption at 260 nm on changing the pH from 10 to 13.4 is the basis of Broughton's classical paper.[326] A recommended method for the differentiation and quantification of barbiturates in serum, plasma, or whole blood, is as follows. The sample is shaken with chloroform and 3 M NaH_2PO_4, and the mixture centrifuged. The separated chloroform is extracted with 0.45 M NaOH (pH 13) and the absorption spectrum from 230 to 300 nm of this alkaline phase recorded. Saturated NH_4Cl solution is then added to the cell to increase the pH to 9.45 and the spectrum is recorded again. The resulting spectra are superimposed and the concentration of barbiturate calculated from the difference of absorption at 260 nm. Additionally, a comparison of the absorptions at 240 and 270 nm of the pH 9.5 spectrum can enable the particular barbiturate to be identified.

TABLE 4.2
Absorption Characteristics of Functional Groups
Commonly Found in Drug Molecules

Group or Compound	Wavelength	Extinction
–OH	180–185	500
–SH	190–200	1,500
–Cl	170–175	300
–Br	200–210	400
–I	255–260	500
–O–	180–185	3,000
–S–	210–215	1,250
–S–S–	250	400
–N<	190–200	2,500–4,000
–CH=CH–	163 and 174	15,000 and 5,500
–CO_2	170–185	40–100
–CONH–	175	7,000
–C=N	<170	
–CO–NH–CO–	190–200	10,000–15,000
–NO_2	200–210	15,000
–N=N–	350–370	10–15
–C=O	180–190	2,000–10,000
>S=O	210–230	1,500–2,500
>SO_2	<190	
Benzene	204; 256	7,900; 200
Naphthalene	220; 286; 312	130,000; 9,300; 270
Azulene	236; 269; 357	22,000; 45,700; 400
Quinoline	228; 270; 315	40,000; 3,200; 2,500
Isoquinoline	218; 265; 313	63,000; 4,200; 1,800
Pyridine	195; 251; 270	7,600; 2,000; 450
Pyrimidine	243; 298	2,000; 300
Pyrazine	260, 327	6,300; 100
Pyridazine	246; 340	1,300; 320
Purine	220; 263	3,000; 8,000
Pyrrole	210; 240	5,000
Furan	205	6,5000
Thiophene	231	7,100
Imidazole	207	5,000
Pyrazole	210	3,200
Isoxazole	211	4,000
Thiazole	240	4,000

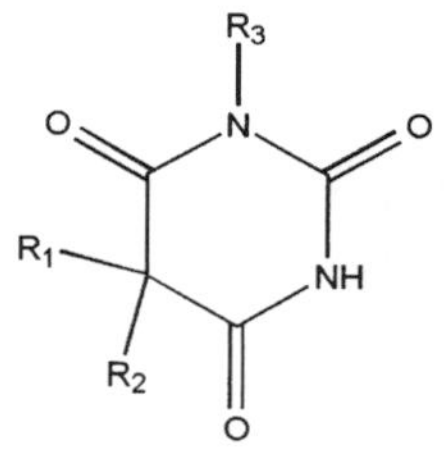

FIGURE 4.3 General formula for barbituric acids.

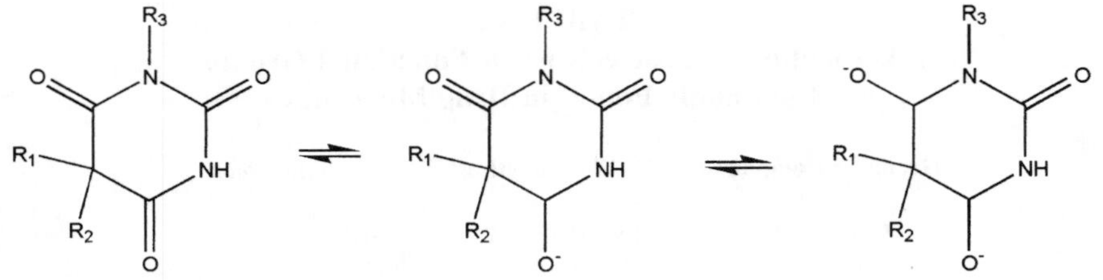

FIGURE 4.4   Species of barbiturates formed in alkaline solution.

4.2.2   COLORIMETRIC MEASUREMENTS

It has been stated that direct measurement of the UV-visible absorption spectrum is of little help as a qualitative test for drugs in biological fluids because these spectra are very similar for compounds having the same chromophores. However, there is a considerable body of literature (especially in forensic toxicology) on spot tests for the identification of drugs in body fluids[327] (Table 4.3). Some of these tests involve a chemical reaction producing very specific colors, and can be adapted for qualitative analysis of the drugs by measuring the change in absorption in the visible region. This process is generally termed colorimetry, although the qualitative step is exactly the same as for the ultraviolet spectroscopic methods already described. Some typical colorimetric assays are described below and a comprehensive list is shown in Table 4.4. Because many colorimetric spot tests are very dependent on such factors as pH, temperature, and time, these parameters are very critical in the development and application of quantitative colorimetric assays.

The advantages of preparing derivatives for absorption measurements are that the sensitivity of detection is improved and the wavelength chosen for measurement is more likely to be free of interference from endogenous components. An alternative to chemical conjugation as described above is to transform the molecular structure of the drug to improve its spectral characteristics. This is a permanent transformation as opposed to the reversible changes effected by changes in pH. In one such procedure, carbamazepine is converted by heating at 150°C for 10 min with hydrochloric acid to an acridine derivative, with a strong absorption maximum at 258 nm (Figure 4.5).[346]

4.2.2.1   Phenylbutazone in Serum

Jahnchen and Levy[230] have described an assay for phenylbutazone which avoids the usual interference by barbiturates in the simpler procedure by Burns et al.[347] In this procedure, permanganate oxidation is used to form azobenzene, absorbing at 314 nm (Figure 4.6).

4.2.2.2   Acetaminophen in Serum

This colorimetric method for acetaminophen was first proposed by Chavetz et al.[337] and modified by Glyn and Kendal[338] and Walberg.[339] Despite more sophisticated assays, this simple method is still applicable, particularly in pediatric practice. In the procedure, the aromatic ring is nitrated with the mild nitrating agent, nitrous acid; the product, in alkaline solution, displays a strong chromophore and the absorption is measured at 430 nm (Figure 4.7).

TABLE 4.3
Qualitative Colorimetric Tests for Drug Substances

Reagent	Drugs	Color
Cold H_2SO_4	Tetracyclines	Purple-blue
	Diphenhydramine, cyclizine, diphenylpyraline, griseofulvin, orphenadrine, pipradol, prednisone	Yellow
	Amitriptyline, chlorprothixene, methacycline	Orange
H_2SO_4-HNO_2	Phenols	Violet-blue
(Liebermann's reagent)	Diphenylamine, mefanamic acid, yohimbine	Blue
	Naphthols, colchicine, hydrastine	Green
	Penicillins, cocaines	Yellow
	Acetanilide, amphetamine, atropine, caramiphen, ephedrine, glutethimide, barbiturates, phenytoin, sulfinpyrazone, warfarin	Orange
	Acepromazine, brucine, oxytetracycline	Red
	Tetracyclines, cotarnine, cresols, ethinamate, methylpentynol, noscapine	Brown
	Morphines, emetine, narceine, papaverine	Black
HNO_3	Desipramine, imipramine, thioridazine	Blue
(Vitali's reagent)	Phenols	Yellow
	Antazoline, bialamicol, brucine	Red
HNO_3-$Hg(NO_3)_2$	Phenols	Red
(Millon's reagent)		
HCl	Chlordiazepoxide, triacetyloleandomycin	Yellow
HCl-furfuraldehyde	Allantoin	Violet
	Carbamates	Black
HCl-$NaNO_2$	Amidopyrine	Violet
	Phenazone	Green
	Sulfaphenazole, sulfasomizole	Orange
HCl-$KClO_4$	Xanthines	Purple
	Chloramphenicol, nitrofurantoin, nitrofurazone	Orange
	Phenolphthalein	Red
$FeCl_3$	Salicylates, saccharin	Violet
	Gentisic acid, morphine, parachlorophenol, acetaminophen	Blue
	Hydroquinone, adrenalin, ethylnoradrenaline, isoprenaline	Green
	Phenazone	Red
p-Dimethylaminobenzaldehyde	Ergot alkaloids	Purple
	Primary aromatic amines, carbamates	Yellow
	Cannabis, phenazone	Red
HNO_2-β-naphthol (diazotisation)	Primary aromatic amines	Red

The method is very specific for the drug although salicylates will interfere. However, in severely uremic patients the original procedure gives falsely high values, possibly due to ring nitration of endogenous phenolic acids. Bailey has proposed a prior ether extraction of serum saturated with ammonium sulfate before processing the ether extract for use in severely uremic patients.[340] Bailey's procedure also has the advantage that strong acids such as salicylates are not extracted.

TABLE 4.4
Quantitative Colorimetric Assays for Drugs in Biological Fluids

Reagent	Drug	Wavelength	Reference
Ninhydrin	Cefalothin	440	328
4-(4-nitrobenzyl)pyridine	Cytostatic agents	560	329
$FeCl_3$/2,4,6-tri(2-pyridyl)-1,3,5-triazine	Acetaminophen	593	330
$FeCl_2$	Bromazepam	580	331
$Na_3Co(NO_2)_6$	Biphenyl-2,2'-diol	410	332
Diazotization/β-naphthol	Acetaminophen	492	333
Methyl orange	Ketamine	420	334
Chloramine-T	3,3'-Dichlorobenzidine	457	335
Ellman's reagent	Thioethers	412	336
Nitration	Acetaminophen	430	337–340
Folin-Ciocalteau	Acetaminophen	660	341
9-Chloroacridine	Isoniazid	500	342
Bratton-Marshall	Procainamide	550	343, 344
Vanillin	Isoniazid	365	345
HCl	Carbamazepine	258	346
$KMnO_4$	Phenylbutazone	314	230

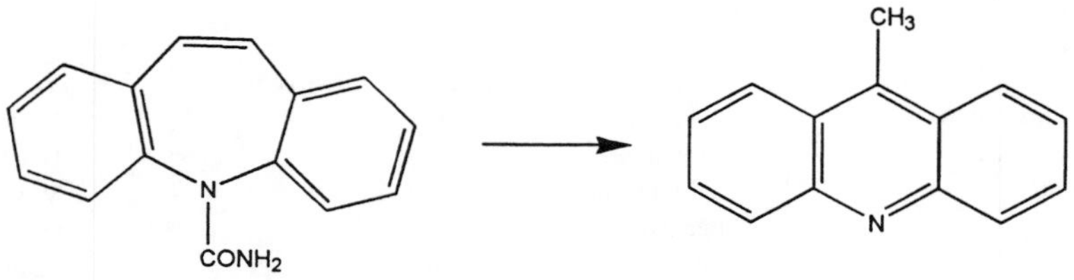

FIGURE 4.5   Conversion of carbamazepine to a fluorescent acridine.

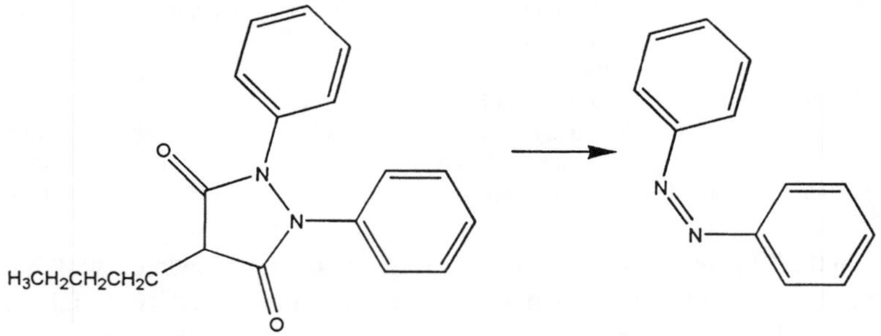

FIGURE 4.6   Permanganate oxidation of phenylbutazone to azobenzene.

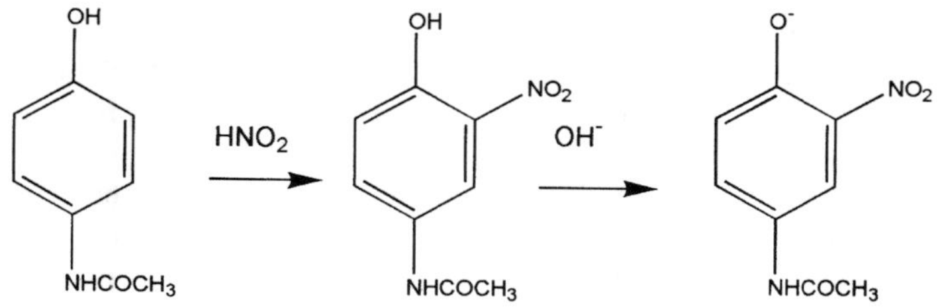

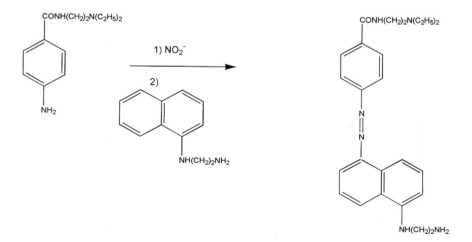

FIGURE 4.7 Nitration of acetaminophen and formation of a strongly absorbing species in alkaline solution.

4.2.2.3 Procainamide in plasma

Bratton and Marshall[343] introduced the use of N-(1-naphthyl)-ethylenediamine as a coupling agent following diazotization for the characterization of aromatic acids as long ago as 1939. The reagent has been used extensively and adapted for the analysis of procainamide in biological fluids. In the version by Sitar et al.,[344] the recommended procedure is briefly as follows. Plasma is made alkaline with a sodium hydroxide-sodium chloride mixture and is extracted with dichloromethane. The organic layer is removed and evaporated to dryness. The residue is dissolved in hydrochloric acid and reacted with sodium nitrite at 0°C, as the diazotization step and then with the Bratton-Marshall reagent to form the complex which absorbs at 550 nm (Figure 4.8). The method is suitable for assay of procainamide in plasma over the range 0.5 to 25 µg ml^{-1}. An important feature of this modification by Sitar et al.,[344] is the carrying out of the diazotization step at 0°C to prevent acid hydrolysis of N-acetylprocainamide, a metabolite of procainamide, and consequent overestimation.

FIGURE 4.8 Analysis of procainamide by use of the Bratton-Marshall reagent.

4.2.2.4   Isoniazid in Urine and Plasma

This procedure depends on the reaction between isoniazid and 9-chloroacridine to form a highly colored orange complex, with an absorption maximum at 500 nm (Figure 4.9).

The efficiency of the reaction depends on the temperature and the time. Stewart and Settle[342] examined these parameters and showed that although the reaction occurred at room temperature, the extent of color production was dependent on the concentration of isoniazid; that is, a linear calibration graph was not produced. A temperature of 50°C was found to be a minimum for maximum sensitivity and acceptable linearity.

For analysis of isoniazid in urine it is necessary to hydrolyze the pyruvic acid and α-ketoglutaric acid hydrazones to free isoniazid before assay as total isoniazid. The method is not suitable for the assay of free isoniazid in urine, because in the presence of these hydrazones the variable degree of their hydrolysis during the color development step will lead to spurious results.

4.2.2.5   9-Chloroacridine

The reagent 9-chloroacridine also forms a colored derivative with aromatic amines and aromatic hydroxylamines. However, these derivatives have their absorption maxima at about 435 nm and Stewart and Settle[342] suggest that this is sufficiently different from isoniazid to be estimated in the presence of other drugs containing primary amines.

4.2.2.6   Folin-Ciocalteau Reagent

This is a popular reagent for forming colored complexes with certain drugs and is used as a spray reagent in thin-layer chromatography (see Chapter 5). Swanson and Walters[341] described the use of this reagent for the quantification of acetaminophen. Serum is mixed with an equal volume of phosphate buffer (pH 7.0) and extracted with ethyl acetate. The ethyl acetate layer is mixed with 0.25 M Na_2CO_3 while Folin-Ciocalteau reagent is added. Acetaminophen forms a blue complex which remains in the aqueous phase on centrifugation. Absorbance is measured at 600 nm. An important feature of this method is that it avoids interference by the presence of salicylates.

4.2.2.7   Reaction with Aromatic Aldehydes

A variety of aromatic aldehydes has been used as spray reagents for paper and thin-layer chromatography, including anisaldehyde, and vanillin. Boxenbaum and Riegelman[345] adapted the vanillin reaction for an assay of isoniazid (Figure 4.10). The resulting hydrazone absorbs at 365 to 370 nm.

4.2.3   OTHER ABSORPTIOMETRIC TECHNIQUES

4.2.3.1   Difference Spectra

Although UV and visible absorption spectroscopy can provide much information on the presence of groups of drugs in biological samples, for quantitative use they are very limiting in sensitivity. The methods so far described are for drugs that appear

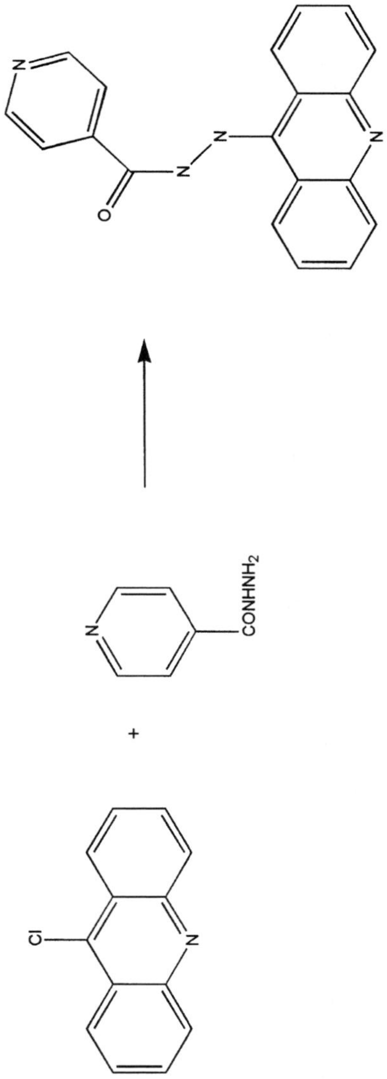

FIGURE 4.9 Reaction of isoniazid with 9-chloroacridine to form a highly colored orange complex.

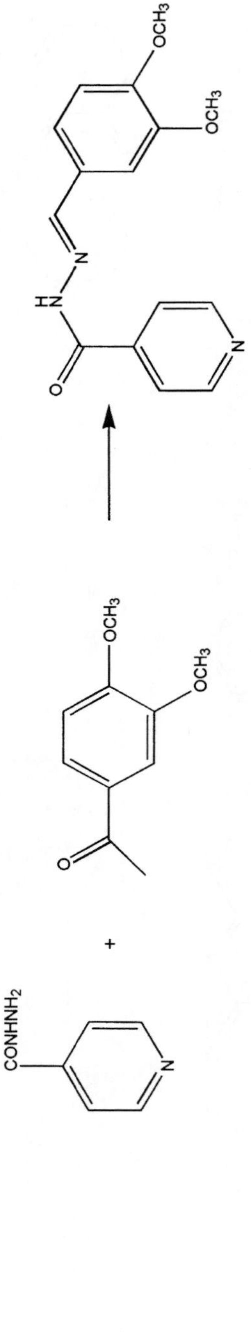

FIGURE 4.10 Reaction of isoniazid with vanillin.

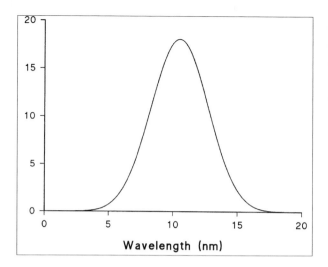

FIGURE 4.11 Symmetrical absorption band.

in the μg ml^{-1} range following therapeutic doses, or are only applicable in screening overdose cases. The use of chromatographic separations prior to detection with absorption spectrophotometers provides a very powerful technique which increases specificity (separation from closely related compounds) and sensitivity (reduction of background absorption). However, these chromatographic systems are either relatively expensive (high-pressure liquid chromatography) or require some measure of skill (thin-layer chromatography). Accordingly, special techniques have been developed to maximize the information available from the absorption spectra. One such technique is the use of difference spectrometry.[348,349] In this technique a double-beam spectrometer is used and the untreated sample is used as the reference for a sample which has been subjected to a chemical change or derivatization. The difference spectrum produced will, of course, show positive and negative peaks in the trace. The advantage of this method is that nonspecific absorption bands are eliminated and even turbid solutions can be assayed. Overlapping bands from other absorbing species will be eliminated, providing that this absorption is not affected by the chemical reaction used for derivatization.

4.2.3.2 Derivative Spectroscopy

The technique used to obtain additional information from UV-visible spectra to make the most impact in recent years is the technique of derivative spectroscopy. The theoretical aspects were expounded as long ago as 1953 by Hammond and Price[350] and further developed by Giese and French[351] for the analysis of overlapping spectral bands. The theory may be briefly reviewed as follows. A simple asymmetrical absorption band is shown in Figure 4.11. As one scans through the wavelengths the rate of change of absorption also changes and this rate can also be plotted against the wavelength (Figure 4.12) as the first derivative. The principal features of this first

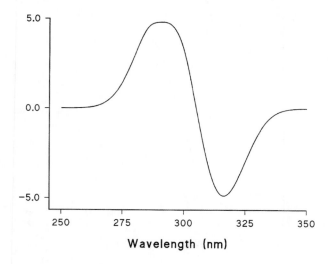

FIGURE 4.12   First derivative of absorption band shown in Figure 4.11.

derivative plot are the minimum and maximum values corresponding to the inflection points of the original spectrum, and the point at which the first derivative plot crosses the zero line, giving an accurate value for the wavelength of maximum absorption.

It is the second derivative, however, which has proved more useful. This plot is shown in Figure 4.13. The original inflection points now appear as maxima and the absorption peak appears as a minimum. For quantitative use, the distance between adjacent maxima and minima is measured, choosing those wavelengths most characteristic of the analyte of interest. Alternatively, adjoining maxima can be joined by

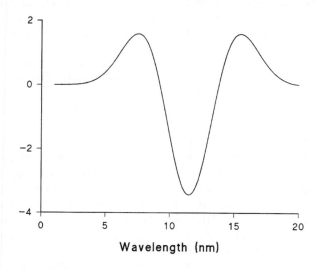

FIGURE 4.13   Second derivative of absorption band shown in Figure 4.11.

TABLE 4.5
Analysis of Drugs in Biological Fluids Using
Derivative Spectroscopy

Drug	Fluid	Sensitivity	Reference
Barbiturates	Serum	0.1 μg ml⁻¹	355
Benzodiazepines	Blood		352
Clonazepam	Urine	150 ng ml⁻¹	356
Diazepam	Plasma	1 ng ml⁻¹	354
Methaqualone	Plasma	2.5 ng ml⁻¹	354
Oxphenbutazone	Plasma	3 ng ml⁻¹	357
Paraquat	Plasma		353
Phenylbutazone	Plasma	1 ng ml⁻¹	357

a straight line and the vertical distance from this line to the minimum can be measured. Either of these two measurements should be proportional to the concentration of analyte, with the second method introducing a correction factor for background absorption.

Apart from the features already mentioned, derivative spectroscopy offers a better resolution of spectra, as shoulders in the normal spectra are converted into sharp peaks. Thus, Martinez and Paz Gimenez[352] described the use of derivative spectra for the assay of benzodiazepines in blood, and Jarvie et al.[353] described a rapid assay for emergency analysis of paraquat in plasma. The method seems to find particular application in determining levels of drugs in biological fluids of coadministered drugs, such as the determination of diazepam and methaqualone,[354] and where there is need for very simple and rapid determination of relatively large concentrations in overdose cases as for paraquat poisoning mentioned above (Table 4.5). As far as instrumentation is concerned for derivative spectroscopy, a normal spectrometer fitted with a means of collecting data points that can be fed into a personal computer is all that is required, although it is more usual to use commercial instruments fitted with their own computing facilities.

4.3 FLUORIMETRY

4.3.1 PRINCIPLES

In the foregoing discussion on absorption spectrometry, the drug being analyzed is quantified on the basis of the degree of dimunition of the incident light. However, when the drug is present in very low concentrations the analyst is trying to detect the difference between two large values, always an unsatisfactory situation in analytical work. In contrast, in fluorimetry, the final measurement is not of transmitted light but of emitted light at right angles to the incident light, and hence the analyst is looking for the difference between zero and small values of emitted light. Thus, the sensitivity of fluorescence methods is better than absorbance methods due to the lower noise limitation.

The fluorescence process is one consequence of the absorption process; some of the energy absorbed by organic molecules is released by the emission process and is always at a longer wavelength than that absorbed. Thus, in the standard fluorescence

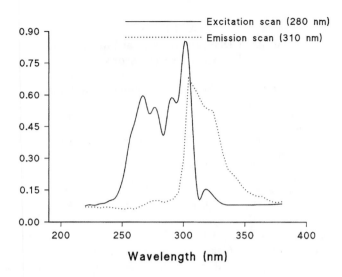

FIGURE 4.14   Excitation and emission spectra of a fluorescent drug.

assay the sample is excited at one selected wavelength and the intensity is measured at another appropriate wavelength. Because of this ability to select two specific wavelengths, the method has a high potential for specificity.

The fluorescence of a molecule is characterized by two spectra; the excitation spectrum and the emission spectrum. It is therefore a good idea to establish both spectra before deciding the best conditions for assay. The two spectra for triamterene are shown in Figure 4.14. In theory, the excitation spectrum is identical to the absorption spectrum, but slight differences will be seen due to differences in instrumentation. For production of the emission spectrum it is usual to set the excitation wavelength at the excitation (or absorption) maximum and scan the emission wavelengths; this reduces the effect of scattered light and gives a cleaner spectrum than excitation with full-spectrum white light.

In principle, all compounds which absorb light will also be fluorescent. However, not all light absorbed is emitted as fluorescence, and it is usual to speak of quantum yield in assessing fluorescence due to compounds or to chemical groupings. The quantum yield is defined as the ratio of the quanta emitted to the quanta absorbed and the range of such yields is shown in Table 4.6.

There are many factors that may reduce the quantum yield, including the solvent used and the concentration of the compound. Teale and Weber[358] determined the quantum yield of a number of organic structures, the most efficiently fluorescing species being 9-aminoacridine, Rhodamine B, and fluorescein (Table 4.6). The fact that not all organic molecules exhibit usable fluorescence, even though they may have strong absorbance, is due to a very low fluorescence efficiency; in such molecules the absorbed energy is dissipated by processes referred to as internal conversion. For the analyst interested in increasing the sensitivity of the detection of drugs, however, the mechanism of such fluorescence quenching is of secondary importance to the end effect and the author shall confine the discussion to some

TABLE 4.6
Quantum Yields of Fluorescence in Solution

Compound	Solvent	Quantum yield
9-Aminoacridine	Water	0.98
Rhodamine B	Ethanol	0.97
Fluorescein	0.1 N NaOH	0.92
1-Dimethylaminonaphthalene-7-sulfonate	Water	0.75
2-Methoxy-6-chloro-9-*N*-glycylacridine	Water	0.74
Acriflavin	Water	0.54
1-Dimethylaminonaphthalene-5-sulfonate	Water	0.53
Fluorene	Ethanol	0.53
1-Dimethylaminonaphthalene-4-sulfonate	Water	0.48
Indole	Water	0.45
Skatole	Water	0.42
Anthracene	Benzene	0.29
Sodium salicylate	Water	0.28
Riboflavin	Water	0.26
Chlorophyll a	Ethanol	0.23
Phenol	Water	0.22
Eosin	0.1 N NaOH	0.19
Pheophytin	Benzene	0.18
1-Aminonaphthalene-3,6,8-sulfonate	Water	0.15
Naphthalene	Ethanol	0.12
Chlorophyll b	Ethanol	0.10
Phenanthrene	Ethanol	0.10
Sodium sulfanilate	Water	0.07
Sodium *p*-toluenesulfonate	Water	0.05
Uranyl acetate	Water	0.04
1-Dimethylaminonaphthalene-8-sulfonate	Water	0.03

structures that have usable fluorescence. A great deal of work has gone into attempts to correlate structure with fluorescence of organic molecules, and it is evident that fluorescence is markedly influenced by structure, as evidenced in the papers by Williams[359] and Williams and Bridges.[360]

The first consideration is that the compound must have strong absorption characteristics to provide the energy for eventual emission. The next consideration is the quantum yield of fluorescence. For consideration of structures with high quantum yields the various review papers should be consulted. A few general rules on structures that may be expected to yield high fluorescence may be drawn up, but it would be wise to investigate any new drug for unexpected phenomena. Aromaticity appears to be a prerequisite for fluorescence, with most simple aromatic compounds having quantum yields of approximately 0.20. This fluorescence is modified by ring substitution (Table 4.7). The efficiency of fluorescence, however, by no means follows the degree of absorbance, as evidenced by the low fluorescence of nitrobenzene, the nitro group apparently enhancing the degree of internal conversion. The pH of the solution has a dramatic influence on fluorescence, as fluorescent compounds are often used as pH indicators.

The key characteristic of a substituent to increase the fluorescence of an aromatic ring is its ability to donate electrons. Hence phenols and aromatic amines are highly

TABLE 4.7
Effect of Ring Substitution on
Fluorescence of Benzene

Substituents giving useful fluorescence	Substituents not giving useful fluorescence
–OH	–NO$_2$
–O–CH$_3$	–COOH
–NH$_2$	–CH$_2$COOH
–NH–CH$_3$	–(NH$_3$)$_3^+$
–F	–Cl, –Br, –I
	–O$^-$
	–NH$_3^+$

From Williams, R. T., *J. Royal Instit. Chem.*, 83, 611, 1959.

fluorescent, but this fluorescence is lost on acetylation, the acyl group having an electron-withdrawing effect. The fluorescence is not lost if the amine or hydroxyl function is alkylated.

External factors other than pH, can also affect observed fluorescence. These include the presence of halide ions which will quench fluorescence. The antimalarial quinine has its maximal fluorescence at pH 1 as would be expected from its structure (Figure 4.15). However, if the measurement is performed in hydrochloric acid, rather than sulfuric acid, then the fluorescence is virtually abolished. Similarly, solutions containing dissolved oxygen will also show reduced fluorescence, and solvents used in fluorimetry should be degassed prior to use.

4.3.2 DIRECT MEASUREMENTS

For the analyst assigned to developing a method of assay of a particular drug in biological fluids, it is very heartening to discover that a compound is highly fluorescent, whether this fluorescence is native in organic solvents or whether it has to be generated in strong acid or alkaline media. A few examples of earlier assay procedures where straightforward fluorescence of the compound was used will indicate the simplicity of the method in the right conditions.

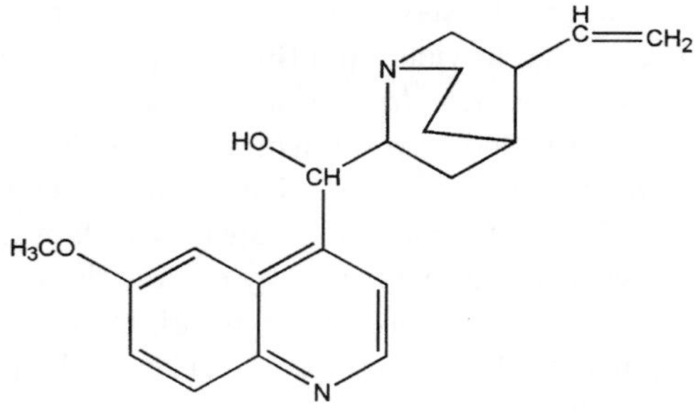

FIGURE 4.15 Structure of quinine.

4.3.2.1 Triamterene[211]

Alkaline plasma or urine is extracted with ethylene dichloride, the organic extract diluted with an equal volume of *n*-hexane and re-extracted with phosphate buffer (pH 6.0), the resultant aqueous extract being assayed by its fluorescence at 440 nm following excitation at 366 nm.

4.3.2.2 Furosemide[361]

Acidified plasma is extracted with diethyl ether, and the ether phase back-extracted with phosphate buffer (pH 7.5). The aqueous layer is reacidified with 1 M HCl to pH 1.5 to 2.5, and the fluorescence measured at 407 nm following excitation at 337 nm.

4.3.2.3 Quinine[362]

Serum or urine (10 ml) is shaken with an equal volume of 0.1 M NaOH and 30 ml benzene. Isoamyl alcohol (1 ml) is added to the benzene layer which is then extracted with 0.1 M H_2SO_4. The fluorescence of the acid phase is determined at 450 nm following excitation at 350 nm. With this method, as little as 1 ng ml^{-1} quinine can be estimated in biological fluids and metabolites have been shown not to interfere with the determination.

4.3.2.4 Imipramine[363]

Plasma is buffered to pH 11 with borate buffer and extracted with heptane. The heptane is further extracted with 0.1 M HCl. The acid phase is then made alkaline with two equivalents of 0.1 M NaOH and the fluorescence is measured at 400 nm following excitation at 290 nm.

4.3.2.5 Propranolol[364]

Total (free plus conjugated) propranolol in plasma can be determined by direct dilution of plasma with a dimethylsulfoxide-water mixture (1:2) and measurement of the fluorescence of this mixture at 340 nm following excitation at 317 nm. The method measures as little as 10 to 20 ng ml^{-1} propranolol, depending on the sample. The main metabolite of propranolol does not interfere with the process. For measuring free propranolol, it is necessary to extract the alkalinized sample with heptane containing 1.5% isobutanol, and back-extract with 0.1 M HCl. The fluorescence of the acid phase is measured at 340 nm following excitation at 295 nm.

4.3.2.6 Butaperazine[365]

Alkaline plasma is extracted with heptane containing 10% isopropyl alcohol and the fluorescence of the organic layer is measured directly at 505 nm following excitation at 275 nm. The sensitivity of the method is 8 ng ml^{-1} and the calibration graph is rectilinear up to 300 ng ml^{-1}. The sulfoxide and sulfone metabolites do not interfere with the determination of unchanged parent drug.

4.3.2.7 Flecainide[366]

Alkaline plasma is extracted with heptane. The organic phase is back-extracted into 0.25 M NaH_2PO_4 and the fluorescence of this phase is measured at 370 nm

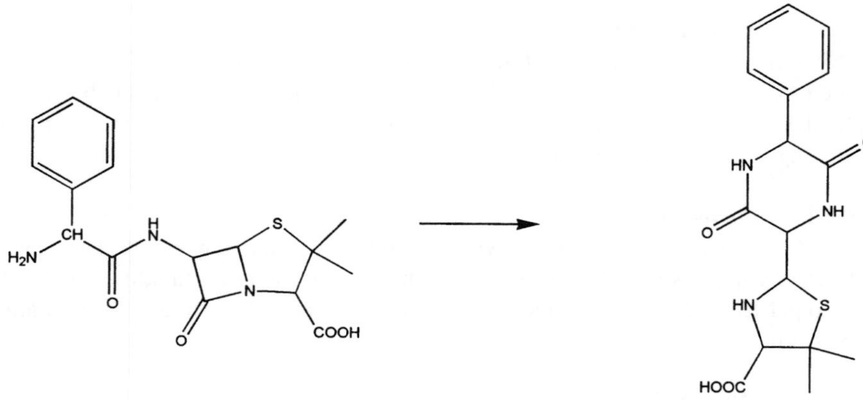

FIGURE 4.16   Conversion of ampicillin to a fluorescent diketopiperazine.

following excitation at 300 nm. The detection limit is 85 ng ml^{-1} and the fluorescence intensity was rectilinear up to 10 μg ml^{-1}.

4.3.2.8   Lysergic Acid Diethylamide (LSD)[367]

Serum or urine is alkalinized with 1 M NaOH, saturated with NaCl and extracted with *n*-heptane containing 2% amyl alcohol. The organic layer is extracted with dilute HCl and the fluorescence of the acid solution determined at 445 nm following excitation at 325 nm. The sensitivity of this method is about 3 ng ml^{-1} and metabolites do not interfere with the process.

4.3.2.9   Temafloxacin[368]

Serum is mixed with 0.05M H$_2$SO$_4$ in a sample cuvette and measured directly at 460 nm following excitation at 276 nm. Limit of detection was approximately 20 ng ml^{-1}.

4.3.3   INDUCED FLUORESCENCE
4.3.3.1   Chemically Induced Fluorescence

The preceding illustrates clearly the use of native fluorescence of several drugs. Unfortunately, few drugs are so accommodating in their spectral characteristics. It may be possible, however, to transform the drugs to fluorescent species either by direct chemical action or by the formation of fluorescent derivatives with appropriate coupling agents. Such transformations need not be confined to increasing fluorescent intensity but may be utilized to shift the excitation wavelength to a more usable mercury line or to enlarge the separation of excitation and emission wavelengths to reduce interference by scattered light. Jusko[369]has described a simple assay for ampicillin which is based on converting it by heating in acid solution to a highly fluorescent diketopiperazine (Figure 4.16). Phenothiazines may be oxidized with hydrogen peroxide to fluorescent products.[370] The actual excitation and emission wavelengths used for measurement depend on the particular phenothiazine being studied. Oxidation of diphenylhydantoin with alkaline potassium permanganate gives

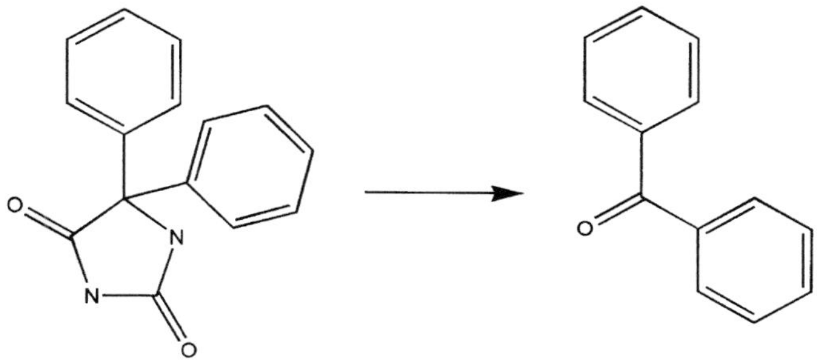

FIGURE 4.17 Oxidation of diphenylhydantoin to a fluorescent benzophenone.

the fluorescent benzophenone (Figure 4.17).[371] Methyldopa may be oxidized with potassium ferricyanide and subsequently rearranged with alkali to give a fluorescent indole (Figure 4.18).[372] Chlordiazepoxide may be oxidized to a lactam which can be assayed by its fluorescence (Figure 4.19).[373]

4.3.3.2   Fluorescence Induced by Irradiation

Chemical transformation may also be achieved by using strong short-wavelength UV light as described by Brodie et al.[374] The original method was used to analyze chloroquine and related compounds by converting the drug to unidentified fluorescent compounds with an absorption maximum closer to the 365 nm mercury line. In the original apparatus the mercury arc lamp was set in the center of a circular rack containing the sample tubes. Because of the dimensions of the apparatus and the intensity of the UV light, an irradiation time of 3 h was required and the whole apparatus needed to be air cooled to keep the temperature below 35°C to minimize thermally induced side reactions.

Hajdu and Damm[375] developed the irradiation device further by using a more intense source of short-wavelength UV light. A general method was proposed for determining drugs in biological fluids, the method being optimized for individual drugs by varying the irradiation time or distance and the time allowed to develop the fluorescence following the initial irradiation. These principles led to the development of a method for analyzing fendosal in plasma.[376] In this method, plasma is mixed with

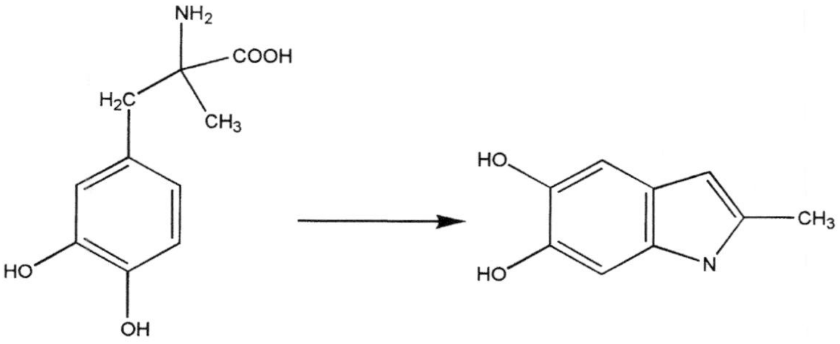

FIGURE 4.18 Conversion of methyldopa to a fluorescent indole.

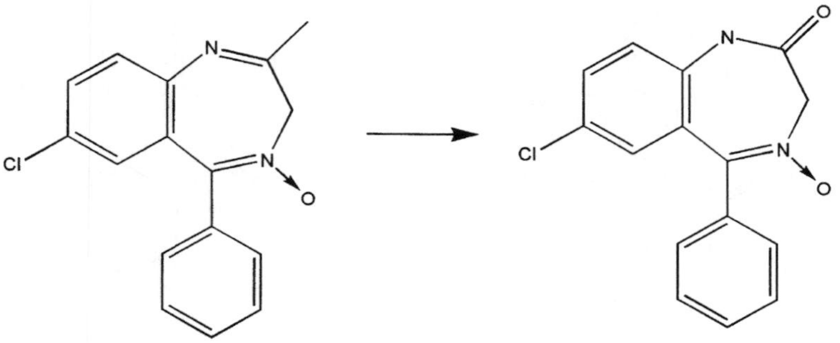

FIGURE 4.19   Conversion of chlordiazepoxide to a fluorescent lactam (demoxepam).

acetate buffer to pH 4.6 and extracted with diethyl ether. The ether layer is transferred to a centrifuge tube and evaporated to dryness. The residue is dissolved in ethanol and the solution transferred to a 1-cm cuvette. The cuvette is irradiated for exactly 4 min with UV light using a low-pressure mercury arc with an intense spectral line at 254 nm. The distance from the source to the center of the cuvette is 3.5 cm, which is conveniently standardized by mounting the source and the cuvettes in a specially constructed aluminum or plastic block (Figure 4.20). The cuvettes are left in daylight for 1 h to develop optimal fluorescence. The fluorescence of the samples is measured both before and after irradiation, with emission being at 404 nm following excitation at 348 nm. The change in fluorescence between the two measurements is proportional to the fendosal concentration. When plasma from normal, undosed volunteers was processed according to this method no induced fluorescence was observed. The presence of up to 200 µg ml^{-1} salicylic acid in plasma did not contribute to either pre- or postirradiation fluorescence. Moreover, when the method was applied to plasma obtained from subjects dosed with aspirin, the strongly fluorescing aspirin metabolites did not change their fluorescence on irradiation. The known metabolites of fendosal do not interfere significantly, being left behind in the aqueous phase during the extraction step. The method was successfully used to study fendosal pharmacokinetics, even being able to detect evidence of enterohepatic circulation in humans, as seen from the double peak in Figure 4.21. The limit of sensitivity of the method was 0.1 µg ml$^{-1.}$

Induced fluorescence (either by chemical means or by irradiation), can be made to be extremely specific by careful choice of pre-extraction conditions and the reaction conditions. However, the nature of the fluorescing species is often uncertain, particularly with irradiation-induced fluorescence. Another approach to increase the fluorescence of a drug in the biological sample is to form a fluorescing complex or to link the drug covalently with a second molecule to produce a fluorescent conjugate. Chelation of drugs with metal ions will often increase fluorescence, possibly by increasing the rigidity of the molecule and reducing internal conversion. For example, tetracycline has been assayed by its fluorescence when chelated with calcium and barbituric acid (Figure 4.22),[377] or with magnesium,[378] and oxytetracycline similarly by chelation with magnesium and ethylene diamine tetra-acetate.[379]

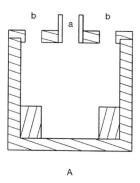

A

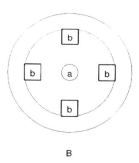

B

FIGURE 4.20 Apparatus for the reproducible irradiation of samples by short-wave UV-light. (A) side view; (B) top view; (a) space for UV-probe; (b) spaces for sample cuvettes.

4.3.4 COUPLING TO FLUORESCENT REAGENTS
4.3.4.1 Fluorescamine

The more popular method of producing a fluorescent species has been the chemical coupling or condensation method (Table 4.8). This has been a productive approach in the assay of amines especially using fluorescamine (Figure 4.23). A commercial version of the reagent is available (Fluram) as described for the analysis of the biogenic amines tryptophan, noradrenaline, dopamine, and the drug, daunorubicin.[380]

The author adapted this reagent for the difficult task of assaying a volatile amine, trifluoromethylaniline, a suspected metabolite of a candidate drug. The advantage of forming a derivative was that the sample could be concentrated without losing any compound through evaporation. In the developed procedure, the alkalinized plasma was extracted with an equal volume of diethyl ether. The ether was rapidly mixed with Fluram to form the fluorescent derivative. The excess reagent was decomposed by the water present in the ether. In our studies we wished to measure trifluoromethylaniline at the ng ml⁻¹ level and for this sensitivity it was necessary to use HPLC for the final quantitative analysis. The advantage of this reagent is that it is not itself fluorescent and excess fluorescamine in the reaction

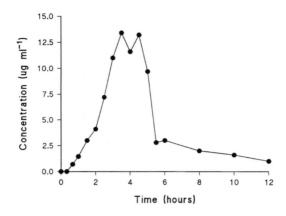

FIGURE 4.21   Plasma concentration of fendosal following a single dose of 100 mg to normal volunteers (mean values from six subjects).

can also be hydrolyzed to nonfluorescing species, thus providing low blanks in the method.

4.3.4.2   Dansyl Reagents

Dansyl chloride (dimethylaminonaphthalene sulfonyl chloride), is a popular reagent for the formation of a fluorescent conjugate with primary and secondary amines or hydroxyl groups. The reagent was first developed as a fluorescent label for protein studies. The quantum yield of fluorescence is the same for all amino groups (60%) but is only about 10% for phenols. The dansyl chloride reagent was used to

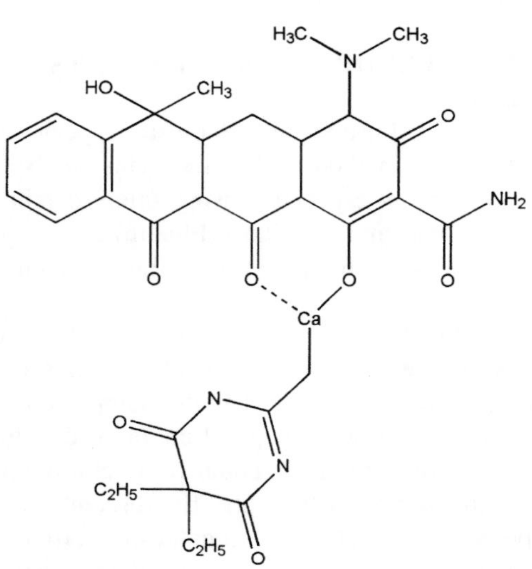

FIGURE 4.22   Fluorescent complex formed by tetracycline, barbituric acid, and calcium.

TABLE 4.8
Reagents Used for Derivatization to Form Fluorescent Conjugates Used for Analysis of Drugs in Biological Fluids

Drug	Reagent	Sensitivity	References
Amiloride	9,10-Phenanthraquinone	20 ng ml⁻¹	384
Amphetamines	4-Chloro-7-nitrobenzo(c)-1,2,5-oxadiazole (NBD chloride)		385
Daunorubicin	Fluorescamine		380
Guanethidine	9,10-Phenanthraquinone	20 ng ml⁻¹	384
Guanethidine	Benzoin	20 ng ml⁻¹	386
Guanfacine	Benzoin	20 ng ml⁻¹	386
Guanoclor	Benzoin	20 ng ml⁻¹	386
Guanoxan	9,10-Phenanthraquinone	20 ng ml⁻¹	384
Guanoxan	Benzoin	20 ng ml⁻¹	386
Praziquantel	Dansyl chloride		381

determine praziquantel in plasma and urine. In the method described by Pütter[381] the drug is extracted from alkaline plasma with a benzene-hexane mixture (1:4). The organic layer is reduced in volume and then successively washed with dilute alkali and dilute acid. The mixture is then hydrolyzed with strong alkali to expose the amine functions which can then react with the dansyl reagent (Figure 4.24). Although in theory two dansyl residues could be coupled to the praziquantel hydrolysis product, it is thought that only one residue is coupled in this case. Other dansyl reagents used in drug analysis include dansyl hydrazine for ketones[382] and dansylaziridine for sulfhydryl compounds.[383]

The fluorescent derivatives prepared with any particular reagent will have similar spectral characteristics unless the coupling of the reagent extends conjugated systems in the drug molecule. Thus, the strategy of forming such derivatives is more usefully employed for increasing sensitivity rather than specificity. Consequently such techniques have found greatest use in combination with specific separation procedures such as paper, thin-layer, or column chromatography.

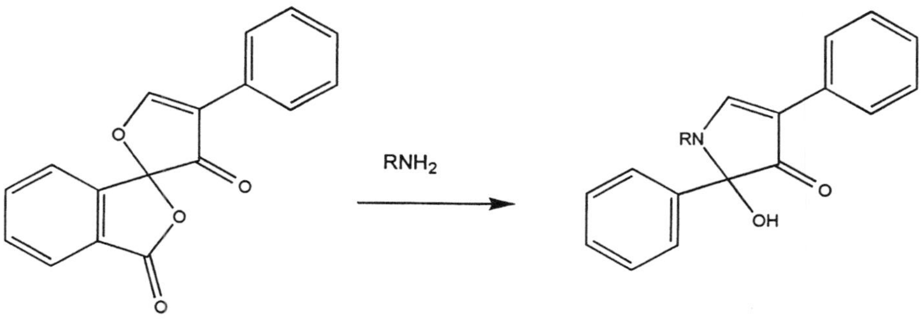

FIGURE 4.23 Reaction of a primary amine with fluorescamine to form a fluorescent pyrrolinone.

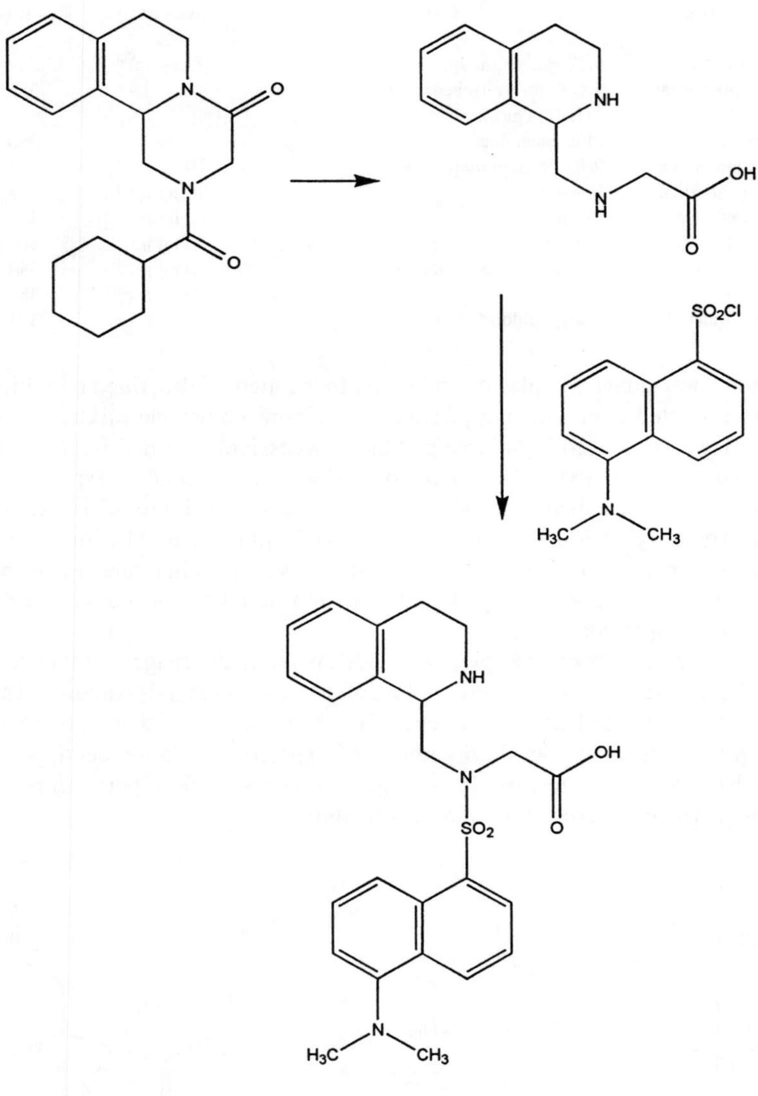

FIGURE 4.24 Hydrolysis of praziquantel and subsequent coupling with dansyl chloride.

5 Planar Chromatography

CONTENTS

5.1 GENERAL INTRODUCTION

Chromatography is the common laboratory method for both quantitative and qualitative analysis of drugs, particularly new investigational drugs, in biological fluids. This is because chromatographic methods can combine a powerful separation step with a sensitive and specific detection system. This chapter will review the basic principles of chromatography and its original development to what is now called planar chromatography. Subsequent chapters will describe, in more detail, the various aspects of the different types of chromatography that the analyst has at his disposal, explaining the advantages of each type in the modern analytical laboratory.

The basic principle underlying all types of chromatography is that the system consists of two components, one stationary and one mobile. This can be exemplified by the original system by Tswett, who described the chromatography of colored pigments extracted from plants. The stationary phase consisted of calcium carbonate packed into a glass column. The extract was placed at the top of the column and washed through with an organic solvent. The components to be separated would be absorbed on the calcium carbonate and dissolved in the organic solvent. The rate at which the individual components (the different pigments) move down the column depends on the relative affinity of the component for the two phases (i.e., stationary and mobile), and it is this relative property which effects the required separation. Tswett's original papers were in Russian and were published in 1906; an excellent review of his early work in chromatography has been written by Ettre.[387]

The preceding discussion is the basis of all types of chromatography. The mobile phase is always a fluid but may be a liquid or a gas; the stationary phase may be a liquid or a solid. If the stationary phase is a solid, the affinity for this phase may be governed by several different factors, including absorption (as in Tswett's original columns) and the ability of the pores in the solid matrix to accommodate the molecules to be separated; this accommodation may depend on the analyte's shape as well as its size. It is important to distinguish between a support, which is always a solid, and the stationary phase, which may be the support itself, or may be a coating on the support, or an adsorbed liquid. If this distinction is not appreciated, then the mechanisms of separation may be obscured, with the result that the rational design of a separation will not be so readily realized. Thus, even in Tswett's system, which at first glance is a straightforward absorption system, if an immiscible liquid is trapped in the calcium carbonate then it may act as a liquid stationary phase. The result is that the system would be, at least partially, a liquid-liquid partition system rather than a liquid-solid absorption system. The affinity of solute for the stationary phase would depend on solubility, which can be controlled by factors such as pH, ionic strength, and polarity.

5.2   PAPER CHROMATOGRAPHY

Although Tswett's original column method proved very useful for preparative separations in natural product and organic chemistry, it did not find extensive favor as an analytical method. Paper chromatography was probably the earliest type of chromatography that was enthusiastically developed for analytical purposes. Its advantage over the classical column technique was that the components could be readily detected by viewing the paper after development of the chromatogram. The general method of paper chromatography consists of spotting a solution of the sample to be analyzed directly on the paper. The solvent is allowed to evaporate and the paper is then subjected to a migrating mobile phase using either an ascending or descending technique (Figure 5.1). For descending chromatography, the end of the paper is held in a trough of mobile phase, usually by the weight of a glass rod, and the mobile phase flows down the paper by a combination of capillary attraction and gravity. This process makes for relatively rapid chromatography. Due to the wetting of the paper as the chromatography proceeds, the solvent front will be clearly visible in this type of chromatography. It is usual to allow the solvent front to flow to within a few centimeters of the end of the paper before terminating the chromatography, so as to characterize the components of the sample by relating their distance from the origin to that of the solvent front — the Rf value. In some applications, the solvent may be allowed to run off the end of the paper, but this raises questions as to when to stop the chromatography before analytes are lost by elution, and how to characterize unknown components.

Using the ascending technique, the paper can be formed into a cylinder and placed with its lower 1 to 2 cm in the mobile phase a few centimeters away from the spotted

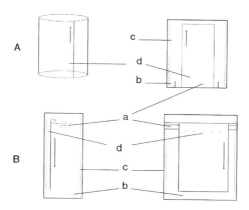

FIGURE 5.1 Paper chromatography. (A) Ascending technique; (B) descending technique. (a) Mobile phase; (b) stationary phase equilibrated against mobile phase; (c) filter paper lining to aid saturation of tank with vapors; (d) samples spotted.

sample. This method takes longer than the equivalent descending system, particularly if a heavy solvent such as dichloromethane is used in the mobile phase. Although this simple type of paper chromatography would appear to be absorption chromatography, there is usually sufficient water absorbed into the cellulose fibers to make the separation depend partly on an organic-aqueous partition. To make the chromatography more reproducible, and to make the system completely liquid-liquid partition, the paper is usually pre-equilibrated with the vapor of the stationary phase, which should be saturated with the mobile phase. The usual practice is to enclose the whole system in an airtight jar containing a mixture of the phases, and after a suitable period of equilibration, perform the elution with the mobile phase saturated with the stationary phase. For partition chromatography, temperature can be critical, and even in modern, temperature-controlled laboratories, local air currents can distort the flow of the mobile phase, making the chromatography irreproducible. Thus, most paper chromatography requires temperature control, usually by placing the solvent tank in an incubator. This not only allows reproducible chromatography, but the mass transfer between phases is also more rapid and chromatographic efficiency is improved.

After developing the chromatography, the paper is removed from the tank and dried of all solvents. The advantages of paper chromatography include its simplicity and the sturdiness of the completed chromatogram. The chromatogram can be rolled, or folded for storage, and even written on for documentation. In many cases the separated components can be viewed directly under ultraviolet light. Further detection or identification can be performed using various spray reagents to detect specific compounds or functional groups. If a nondestructive method of detection is used, then the "spot" can be cut out, eluted with a suitable solvent, and further investigated as a partially purified component. Because, on the analytical scale, such a procedure may introduce impurities form the paper itself, the paper should be washed with mobile phase before use in chromatography.

One of the simplest paper chromatography systems described for drug screening is that for barbiturates. The paper is pre-equilibrated with aqueous ammonia and

TABLE 5.1
Paper Chromatography Systems Used in Drug Analysis

Preparation of paper	Mobile phase	Application	Reference
Dipped in 5% sodium dihydrogen citrate, dried	Citric acid:n-butanol, ascending	Basic drugs	390
Dipped in 20% formamide	5N NH$_4$OH:benzene chloroform, descending	Barbiturates	
Equilibrated against aqueous methanol	Light petroleum, descending	Barbiturates	388
Untreated	Strong NH$_4$OH:pentanol, descending	Neutral drugs, carbamates	391
Dipped in 4% oleic acid in acetone, dried	Acetic acid:water, ascending	Neutral drugs, carbamates	391
Untreated	3M NH$_4$OH:n-butanol, descending	Neutral drugs	392
Untreated	Acetic acid:benzene:water, descending	Neutral drugs	392
Untreated	NH$_4$OH:n-butanol, descending	Salicylates	393
Dipped in 40% formamide in methanol	Formamide:chloroform, formamide:benzene	Corticosteroids	394
Untreated	NH$_4$:amyl alcohol, ascending	Thiazide diuretics	395
Dipped in 10% tributyrin in acetone, dried	Acetate buffer pH 6.4, phosphate buffer	General screening	396–399

developed with light petroleum to provide a straightforward organic-aqueous partition system.[388] The basicity of the aqueous phase is instrumental in effecting the separation of the barbiturates by their different partitioning between the organic phase and the alkaline phase.

The stationary phase can also be an organic solvent with the mobile phase being aqueous. This is usually done by impregnating the paper with a high-boiling organic solvent such as tributyrin or liquid paraffin, usually with the aid of a more volatile solvent which is evaporated from the paper to leave a uniform impregnation of the stationary phase. The paper is then developed with a suitable aqueous phase.[389] Table 5.1 shows a number of paper chromatography systems that found favor in drug analysis, mostly as screening techniques in toxicological and forensic applications.

Paper chromatography was essentially a qualitative technique, although there were notable attempts to make it quantitative, such as in automatic amino acid analyzers and the steroid hormone analysis system Chromatogram Automatic Soaking and Digital Recording Apparatus (CASSANDRA).[400] Man's ingenuity would, no doubt, have produced successful automated quantitative analytical systems using paper chromatography if the technique had not been so dramatically superseded by thin-layer chromatography and gas chromatography.

Because of its simplicity, paper chromatography may still have a role to play in screening systems for drugs, and the modern analyst should at least be aware of its existence, and the part it has played in the development of analytical science; however keeping an eye on the literature over the last 15 years has revealed only one reference to the use of the technique in analyzing drugs in biological fluids.[401]

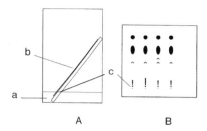

A   B

FIGURE 5.2   Thin-layer chromatography. (A) Plate developing in mobile phase by the ascending technique. (a) Solvent; (b) layer of absorbent; (c) samples spotted; (B) developed plate.

5.3   THIN-LAYER CHROMATOGRAPHY

5.3.1   INTRODUCTION

Although the first description of thin-layer chromatography (TLC) was made as long ago as 1938,[402] it was not until the late 1950s and early 1960s that the technique came into its own. In this type of chromatography the stationary phase, or support, is spread as a thin layer on a flat glass plate. The sample is spotted, as in paper chromatography, near one edge of the plate which is then placed upright in a small amount of the mobile phase in the bottom of a chromatography tank as shown in Figure 5.2. The mobile phase creeps up the plate by capillary attraction, and develops the chromatogram by the ascending technique. Although other methods were described to allow descending chromatography, this usually involved unsatisfactory wick devices and had little advantage over the ascending method, unless the mobile phase was untypically dense and development by capillary attraction was unacceptably slow.

Most of the first TLC systems used an adsorption type of chromatography, with the stationary layer being either silica gel or alumina. In the case of alumina, it was usual to activate the plates (i.e. drive off excess moisture) in an oven and then store them in a desiccator prior to use. If this was not done, the moisture in the plate would either deactivate the adsorption properties of the layer, or there would be an element of partition chromatography, leading to unpredictable and irreproducible chromatography. As in paper chromatography, the separated components in the thin-layer plates can often be detected by viewing under UV light or by using suitable spray reagents. In this respect, conventional TLC is better than paper chromatography because very severe spray reagents can be used for detection. An extreme example is the use of ethanolic sulfuric acid followed by heating in an oven to char organic material, making a truly universal detection system.

TLC is also well suited to two-dimensional chromatography. A single sample is spotted near the corner of a square plate which is developed in the usual way, dried, turned through 90°C, and developed with a different solvent. If the two solvent systems are complementary to each other, this can be a powerful method of identification, because the detected position will be very characteristic of the individual components. Alternatively, the first channel of development can be reacted with a derivatizing agent, for example by acetylation of hydroxyl groups, and the same

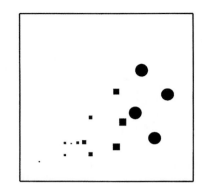

FIGURE 5.3 Characterization of organic alcohols by acetylation and thin-layer chromatography. Unreacted compounds (nonalcohols) have the same retention in the second development; the percentage change in Rf values for organic alcohols is proportional to the number of hydroxy groups that are acetylated.

solvent system used for the second development. Figure 5.3 shows the characterization of organic alcohols using such an approach.

However, if very large numbers of samples are to be screened using two-dimensional TLC, the cost per sample, using conventional plates may be too expensive and the two-dimensional method may be reserved for special research problems or metabolism studies. Recently, the availability of cheap, highly efficient microplates may make the use of single plates for single assays more appealing. In the meantime, most workers would prefer to divide samples into two portions for separate analysis in complementary solvent systems, with 10 to 20 samples being analyzed on a single plate.

Derivatization is relatively rare for subsequent chromatography by TLC, presumably because the usual reasons for derivatization do not apply to TLC; that is, suitable systems can usually be devised to give good chromatography of the native materials, and detection systems are not usually particularly sensitive even for compounds with favorable structures. Roveri et al.[403] prepared fluorescent derivatives of aliphatic thiols, but the method has not been applied to biological samples.

The 20-year delay in the development of TLC has been mentioned and the subsequent historical development is interesting because it shows that a particular technique does not evolve in isolation. The breakthrough occurred in 1958 when Stahl described (and patented) a mechanical device for producing uniform thin layers of alumina or silica gel on glass plates.[404] Stahl also described the use of a binder (calcium sulfate) to provide better mechanical strength to the thin layer. Once Stahl's plates had been introduced, then the applications of the technique became apparent and many impecunious researchers devised several ingenious methods of producing their own thin layers without purchasing the relatively expensive patented equipment. Eventually, the technique became so popular that commercially produced plates were introduced, first on glass supports, and then on plastic and aluminum foil

TABLE 5.2
Optimisation Parameters for TLC Separations

Mobile phase
    Composition (amount of polar, nonpolar organic solvents)
    pH
    Ion-pair reagents
    Other additives
Stationary phase
    Adsorptive properties
    Capacity for aqueous phase
    Capacity for organic phase
    Irreversibly bound organic phases
    Ion exchange
    Size exclusion
Environment
    Temperature
    Humidity
    Ascending, descending, or horizontal development

backings. The convenience and reproducibility of the commercial plates purchased from the same source has now completely superseded the use of homemade plates. Apart from the obvious advantages of plastic and foil-backed plates, such plates could also be easily cut with scissors into any size (or shape!), just like the old paper chromatography equivalents. Plates are even produced with cellulose layers that are even closer to the original paper chromatography systems.

5.3.2   NEW DEVELOPMENTS

Despite the overwhelming success of other types of chromatography, various research efforts continue to be made to improve the technical aspects of the technique, either to improve the resolution available, or to increase the speed of analysis. Some of these advances are summarized below.

5.3.2.1   Increasing Resolution

Many of the aspects of increasing resolution are to be found in examining the practical aspects of the technique itself. Thus, for particular separation problems the researcher can optimize separations by variation of the nature of the mobile phase and the stationary phase or absorbent. Table 5.2 lists the variables that can be considered in effecting particular operations in TLC. It can be seen that the analyst has a wide choice of conditions which may be useful in particular separations.

The parameters shown in Table 5.2 are factors which are within the analyst's control, and which can be applied to optimize theoretical separations. In practical terms, the analyst may be limited by the equipment available and even by his own skills. Thus, a separation may be quite feasible in principle, but may require inordinately long plates or long development times. Another factor is the size of the initial spot that contains the sample applied to the plate; the smaller the spot the better chance of complete separation of the components in the subsequent development.

Commercial developments in improving separations have been made in the provision of high-performance thin-layer chromatography plates, which are referred to by the ugly, and confusing acronym, HPTLC.[405] Typically, such plates have layers with particle sizes close to 5 µm rather than 20 µm, 30 to 40 samples per plate rather than 10 to 12, and sensitivities of less than 100 pg where more than 1 ng would have been necessary on conventional plates.[406] HPTLC plates are best used with automated systems for sample application.

5.3.2.2   Chiral TLC

The problem of separating compounds as closely related as enantiomers has been mainly attacked by using high-performance liquid chromatography (see Chapter 7). Nevertheless, there have been several attempts to use TLC for the same purpose and these need to be mentioned here. The R and S enantiomers of β-blocking drugs could be separated and quantified on chiral TLC plates (Lichrosorb Diol F_{254} HPTLC plates). These included the enantiomers of metoprolol, propranolol, and alprenolol, with 5 mM *N*-benzyloxycarbonylglycyl-L-proline as the chiral additive.[407] Racemic mixtures of hyoscamine and colchicine could be separated with L-aspartic acid-impregnated silica gel layers.[408]

5.3.2.3   Overpressured Layer Chromatography

Almost all of the TLC applications in the analysis of drugs in biological fluids have relied on capillary attraction for the motive force in moving the solvent through the stationary phase. A number of ingenious techniques have been proposed for moving the solvent by other means, leading to a category of forced-flow planar chromatography.

In overpressured layer chromatography,[409-411] the vapor phase is eliminated entirely, by enclosing the stationary phase in a membrane to which pressure can be applied as the chromatogram is developed. Chromatography is more reproducible in this closed system, the analysis is rapid and only a few milliliters of solvent is required. Commercial equipment, both for linear traditional TLC and for circular TLC, is also available which makes the whole technique more reliable and reproducible.[412] Kovacs-Hadady[413] used overpressured layer chromatography as a rapid means to study the retention behavior of various benzodiazepine derivatives on layers impregnated with tricaprylmethylammonium chloride. Overpressured layer chromatography has also been used for the determination of potassium canrenoate by reversed-phase ion-pair chromatography[414] and for determination of some narcotic and toxic alkaloidal compounds.[415]

5.3.2.4   Rotation Planar Chromatography

Using circular chromatography plates, the mobile phase can be forced through the stationary phase using centrifugal force generated when the plate is rotated. Centrifugal force as an aid in analytical TLC was suggested in the first edition of the present work[416] and has been developed so that up to 72 samples can be analyzed on a single circular plate, the centrifugal force driving solvent from the center of the plate to the periphery.[417,418]

5.3.3 APPLICATIONS

TLC has found application in the analysis of drugs in biological fluids in three main areas: qualitative screening, metabolism studies, and as a quantitative analytical method. These are discussed briefly in the following sections.

5.3.3.1 Qualitative Screening

TLC is suitable for screening of samples for the presence of drugs, such as for toxicological screening,[419] because the whole sample can be visualized on the same system, unlike other chromatographic systems depending on elution of components from columns before detection. In forensic toxicology, samples are usually extracted under different conditions of pH to provide acidic, basic, and neutral fractions for further investigation. The relative movement of components in specific solvent systems and the characteristic colors formed with spray reagents are then used to characterize components of interest.

Table 5.3 lists some of the systems used for TLC screening of different classes of drugs. Used as a screening method, a large number of specific and nonspecific spray reagents have been utilized, many adapted from colorimetric tests mentioned in Chapter 4; a comprehensive listing of spray reagents is provided in Table 5.4.

5.3.3.2 Metabolism Studies

The use of TLC analysis in drug metabolism studies is an extremely cheap and reliable method for elucidating drug metabolic pathways. Using conventional absorption plates (usually silica gel), it is general practice to chose a solvent system giving an R_f value of about 0.8 for the parent drug; the more polar metabolites should then be spread over the rest of the plate for optimal separation. If a radiolabeled version of the drug is available, then the metabolite pattern can be produced in a number of ways. The cheapest way is to scrape the silica gel zones into counting vials; small segments (1 mm) can be taken over the complete chromatographic channel and on counting the samples for radioactivity, the complete quantitative radiochromatogram can be reconstructed. This method has the advantage that the counting step is very sensitive, particularly as counting times can be extended, so that even only a few counts above background can be detected. The process of scraping zones into counting vials has also been automated.

The second way to obtain radiochromatograms is to scan the plate with a commercial radioscanner and a third way is to place the plate in contact with X-ray film so that the radioactivity is revealed on the developed film. This autoradiographic method is only semiquantitative but can be extremely sensitive since exposure times of several weeks are feasible.

If a radiolabeled drug is not available, then considerable ingenuity may be required, but TLC is still a powerful ally. If drug and metabolites are present in high concentrations (i.e., 10 to 100 μg ml^{-1}), then nonspecific methods of detection such as the use of iodine vapor or sulfuric acid charring may suffice to reveal the metabolic patterns. Usually, however, a nondestructive method such as ultraviolet absorption or fluorescence is used. If these methods prove successful, then the appropriate areas of the chromatogram can be isolated and the components characterized by further

TABLE 5.3
TLC for Screening of Different Classes of Drugs

Class	Plate	Mobile phase	References
Antihistamines		Acetic acid:butanol:butyl ether (10:80:40)	420
		Ammonium hydroxide:benzene:dioxan (5:60:35)	420
		Acetic acid:ethanol:water (30:50:20)	420
		n-Butanol:methanol (40:60)	420
Barbiturates		Ammonia:benzene:dioxan (5:75:20)	420
		Acetone:chloroform (1:9)	420, 421
		Acetic acid:benzene (1:9)	420
β-Antagonists			422
Cannabinoids		Acetone:chloroform:triethylamine (10:20:1)	423
Cardiac glycosides		Benzene:ethanol (7:3)	424
		Chloroform:acetone:methanol (16:7:2)	425
Corticosteroids		Chloroform:methanol (1:1)	426
		Methylene chloride:dioxan:water (100:50:50)	426
Estrogens		Dichloromethane:methanol:propanol (97:1:2)	427
Ergot alkaloids		Chloroform:ethanol (9:1)	428
Narcotic analgesics		Ammonia:benzene:dioxan:ethanol (5:50:40:5)	429
		Acetic acid:ethanol:water (30:60:10)	429
Phenothiazines		Ammonium acetate:methanol (20:100)	430
Opiates		Chloroform:ethyl acetate:methanol:propylamine (7:9:1:1)	431
		Methanol:ammonia (200:3)	432
		Ethyl acetate:methanol:ammonia (17:2:1)	432
Quaternary amines	Cellulose	Ammonium formate:formic acid:water:tetrahydrofuran	327
Quaternary amines	Silica gel	Methanol: 0.2M HCl	433
Salicylates		Acetic acid:benzene:ether:methanol	434
		Toluene:acetic acid:ether:methanol (120:18:60:1)	435
Sulfonamides	Silica gel	Hexanol	436
Sulfonamides	Alumina	Acetone:ammonium solution	436
		Chloroform:ethanol:heptane:water (33:33:33:1.5)	437
Pesticides	Silica gel	Cyclohexane:acetone:chloroform	438
Lysergic acid diethylamide			439

investigation using spectrophotometry, other chromatographic methods, and the usual techniques of physical chemistry for full elucidation of structure.

5.3.3.3   Quantitative Analysis

When large numbers of samples need to be analyzed, TLC is just as attractive for quantitative use as it is in screening procedures. However, just as for screening procedures described above, the amount of drug in the sample limits the usefulness of the technique and quantitative analysis using TLC is really only possible using well-calibrated, reproducible procedures. Almost all the successful methods for quantifying drugs by TLC have used either fluorescence or UV absorption detection techniques as shown in Table 5.5.

TABLE 5.4
Spray Reagents Used for Visualising Drugs on TLC Plates

Reagent	Examples	Colors	Reference
Ninhydrin	Primary amines	Violet, pink	
	Secondary amines	Yellow	
	Cephalosporins		440
FPN reagent	Phenothiazines	Red	
	Dibenzazepines	Blue	
Dragendorff			
Acidified iodoplatinate	Quaternary ammonium compounds	Violet	433
Mandelin's reagent			
Marquis reagent	Morphine	Black	
Acidified potassium permanganate	Barbiturates	Yellow on violet background	
	Sulfonamides	Yellow-brown	
Van Urk reagent	Sulfonamides, meprobamate	Yellow	
	Ergot alkaloids	Blue	
	Phenazone	Pink, violet	
Ferric chloride	Phenols	Blue, violet	
Mercurous nitrate	Barbiturates	Dark, slowly fading	
Furfuraldehyde			
Chromic acid	Nonsteroidal anti-inflammatory agents	Variety of colours	
Ludy Tenger reagent	Nonsteroidal anti-inflammatory agents	Orange	
Mercuric chloride-diphenyl-carbazone	Barbiturates	White on violet background	
	Sulfonamides	Blue	
Zwicker's reagent	Barbiturates	Pink, green	
Fluorescein	Barbiturates	Pink	
Fast Blue	Cannabidiol	Orange	
	Cannabinol	Violet	
	9-Tetrahydrocannabinol	Red	
Duquenois reagent	Cannabinoids	Blue, violet	
Perchloric acid	Most compounds	Charring agent	
p-Anisaldehyde	Cardiac glycosides	Blue	
Naphthylethylenediamine	Diuretics		
Naphthaquinone sulphonate	Ergot alkaloids	Red-violet	
Nitroso-naphthol	Ergot alkaloids	Blue-black	
Cobalt thiocyanate	Quaternary ammonium compounds		433
Sulfuric acid-ethanol	Steroids	General charring reagent, various colors	
	Cephalosporins		440
p-Toluenesulfonic acid	Steroids	Various colors	
Copper sulfate			

TABLE 5.5
Quantitative Thin-Layer Chromatographic Analysis of
Drugs in Biological Fluids

Drug	Detection and quantification	Detection limit	References
Acebutolol	Paraffin fluorescence	100 ng ml^{-1}	441
Alprazolam	Densitometry at 256 nm	0.05 µg	442
Amiodorine	Reflectance, 240 nm	0.1 µg ml^{-1}	443
Benzodiazepines			444
Carbamazepine	Induced fluorescence	0.1 µg ml^{-1}	445, 446
Carbamazepine	UV at 285 nm	0.5 µg ml^{-1}	447
Cinromide	UV at 270 nm	0.05 µg ml^{-1}	448
Diazepam	Induced fluorescence, 450 nm	0.05 µg ml^{-1}	449
Dihydroergotamine	Fluorescence	10 nM	450
Disopyramide	UV at 254 nm	0.5 µg ml^{-1}	451
Frusemide	Fluorescence	0.02 µg ml^{-1}	452
Ifosfamide	Photography-densitometry	1 µg ml^{-1}	453
Melphalan	UV at >400 nm	20 nM	454
Metronidazole	UV at 320 nm	0.5 µg ml^{-1}	455
Morphine	Fluorescent plates, 287 nm		456
Morphine	Iodoplatinate, sulfuric acid	c20 ng	457
Nadolol	Fluorescence, 313 nm	5 ng ml^{-1}	458
Nalidixic acid	Induced fluorescence, 430 nm	160 ng ml^{-1}	459
Netilmicin	Fluorescence	0.2 µg ml^{-1}	460
Perazine	UV at 250 nm	20 ng ml^{-1}	461
Phenazone	Densitometry at 254 nm		462
Phenobarbitone	UV at 215 nm	2 µg ml^{-1}	447
Phenytoin	UV at 215 nm	2 µg ml^{-1}	447
Primidone	UV at 215 nm	2 µg ml^{-1}	447
Quinidine	Fluorescence 445 nm	1 µg ml^{-1}	463
Salbutamol			464, 465
Salicylic acid	Fluorescence, 402 nm	10 µg ml^{-1}	449
Sulfadiazine	UV at 268 nm	2.5 µg ml^{-1}	466
Sulfamethoxazole	UV at 268 nm	2.5 µg ml^{-1}	466
Theophylline	UV at 270 nm	5 µg ml^{-1}	467
Theophylline	UV at 273 nm	0.25 µg ml^{-1}	468
Tienillic acid	UV at 300 nm	0.3 µg ml^{-1}	469

In the simplest cases the drug will have sufficient ultraviolet, or even visible absorption for the analyst to be able to use straightforward densitometric scanning for quantitation. Some of these are listed in Table 5.5, along with their limits of detection. For TLC, as for other forms of chromatography, the limit of detection is determined by efficiency of chromatography and it is the efficiency of the HPTLC plates which makes them suitable for quantitative work. An important factor is the size of the spot that can be applied. The analyst using manual methods of application can control this factor by very careful manipulation, using a stream of warm air to help evaporate the solvent before it can spread. Other methods for reducing the spread of the initial sample on the plate include predevelopment of the sample for a few millimeters with a solvent which moves all the sample to a new, narrow starting line, or the use of specially made plates with a thicker layer of adsorbent at the origin.

Fluorimetry is also used for quantitation. However, as explained in Chapter 4, the fluorescence of a drug is critically affected by its environment, and it is often useful to treat the developed plate with acid, alkali, or even organic solvents prior to examination with a scanning fluorodensitometer. For example, in the analysis of nalidixic acid,[459] the separation is by use of an ammoniacal mobile phase, whereas fluorescence is only developed after exposure of the plate to gaseous hydrochloric acid and irradiation with short-wave ultraviolet light. In the analysis of frusemide, the plates are developed in an acetic acid mobile phase containing chloroform; the chloroform must be entirely removed from the plate before being recoated in aqueous acetic acid for fluorescence evaluation.[452] In the method for nadolol, the developed plate is impregnated with liquid paraffin to improve the fluorescence.[458]

In principle, drugs can be quantified on thin-layer plates following the use of specific spray reagents. The successful use of this approach as a manual technique is very dependent on the skill of the analyst. An elegant procedure has been described for the determination of sulfadiazine and sulfamethoxazole in urine and plasma.[466] The method requires only 200 μl of biological fluid and after extraction and chromatography, the plate is sprayed with Bratton-Marshall reagent (see also Chapter 4) and densitometry carried out at 575 nm. The relationship between peak height and amount of substance was rectilinear over the required range, and the limit of detection ranged from 2 to 10 μg ml^{-1}. The method was also able to separate the known metabolites of the two drugs. From the point of view of convenience, a less satisfactory procedure is to elute the drug from the chromatographic plate and to quantify it separately by other quantitative techniques such as spectrophotometry, fluorescence, or even other types of chromatography.

6 Gas Chromatography

CONTENTS

6.1 INTRODUCTION

The two methods of chromatography described earlier employ relatively simple and cheap equipment, and also have in common the fact that the final chromatogram is a complete picture of the separated components. Gas chromatography (Figure 6.1), although depending on the same basic separation principles, is characterized by continuous detection of the components eluting from the column, and the chromatogram is the recorded trace rather than the paper or the thin-layer plate (Figure 6.2). In gas chromatography the compounds to be separated are partitioned between a stationary phase, usually a high-boiling liquid coated onto an inert support (such as celite, diatomaceous earth, firebrick, etc.) and the gaseous carrier gas. The variable parameters affecting separation in gas chromatography are: (1) the temperature of the column, usually maintained in a thermostated oven; (2) the flow rate of the carrier gas; and (3) the nature of the stationary phase. Although gas-solid chromatography has been described (where the absorption of solute onto a solid surface is the separation determinant), this technique has not found general application in the analysis of drugs.

The advantage of gas chromatography is that gases emerging from the column can be monitored with great sensitivity using a variety of universal and specific detectors. This makes it a very suitable technique for quantitative analysis. However, the technique is limited to those compounds that can be vaporized at reasonable temperatures without decomposition. Thus, low molecular weight compounds

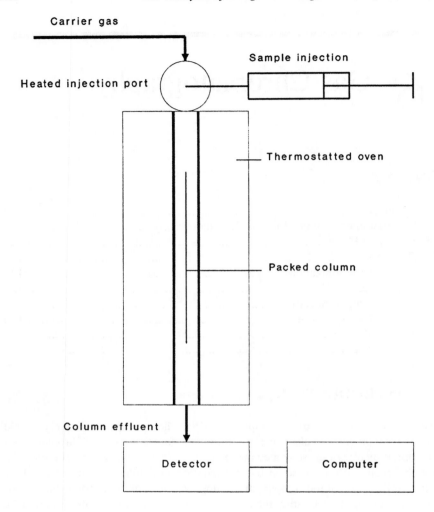

FIGURE 6.1   Schematic diagram of gas chromatography.

in their nonionic forms with no intermolecular bonding should be readily volatilized and hence amenable to this type of chromatography. Volatility in gas chromatography sense is different from the normal concept of volatility in organic chemistry; compounds which are commonly solids even at elevated ambient temperatures may have sufficiently high vapor pressures at 200 to 300°C to be gas chromatographed successfully. Among the drugs with molecular weights in the neighborhood of 300°C are steroids, benzodiazepines, tricyclic compounds, and barbiturates. With higher molecular weights, higher temperatures are required and this can lead to decomposition. In some cases, decomposition to a single component which itself chromatographs with high efficiency may form an acceptable basis of an analytical method. For example, a simple method for the qualitative identification and quantitative

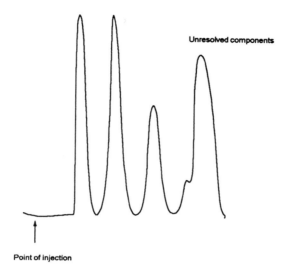

Unresolved components

Point of injection

FIGURE 6.2 Typical tracing of separated components using gas chromatography.

determination of macrolide antibiotics has been proposed using pyrolysis and flame-ionization detection.[470] However, this practice appears less relevant because now there are so many alternative techniques that do not rely on such decomposition, particularly when one realizes that many such decomposition products could be metabolites, such as the decomposition of spironolactone to canrenone on the injection port of the chromatograph.[471] Many such decompositions are catalyzed by the hot metal or other active surfaces in the equipment, and whereas some laboratories are able to overcome such volatilization without decomposition, this problem remains a potential difficulty in transferring methods to other laboratories.

In the development of a gas chromatography method, it is often instructive to demonstrate the structural integrity of the eluted components to ensure that mistakes in interpretation are not made. A convenient way to do this for gas chromatography is to link the gas chromatograph with a mass spectrometer and compare the mass spectrum of the eluting material with the mass spectrum of the authentic compound injected directly into the mass spectrometer. This procedure should be checked for the parent compound, as well as for identification of metabolites.

6.2   PREPARATION OF SAMPLES FOR GAS CHROMATOGRAPHY

In cases where poor volatility is due to intermolecular hydrogen bonding, the volatility can be increased by masking such bonding groups by forming appropriate derivatives, such as acylation of amines or hydroxy groups and esterification of organic acids. For esterification the classical methylation procedure using diazomethane has largely been replaced by the less hazardous boron trifluoride-methanol reagent. For esters other than methyl esters, other alcohols can be used

TABLE 6.1
Trimethylsilylating Reagents in Increasing Order of Activity

Reagent	Structure	Application	Reference
Hexamethyldisilazane	$(CH_3)_3Si$ NH $Si(CH_3)_3$	Penicillin	476
		Spectinomycin	477
Trimethylchlorosilane	$(CH_3)_3Si$ Cl	Cannabinoids	478
N-Trimethylsilyldiethylamine	$(CH_3)_3SiN(C_2H_5)_2$	Neomycin	479
N,O-Bis(trimethylsilyl)-acetamide	$(CH_3)_3Si$ N=C (CH_3) O Si $(CH_3)_3$	Apomorphine	480
		Chloramphenicol	481
N,O-Bis(trimethylsilyl)-trifluoroacetamide	$(CH_3)_3Si$ N=C (CF_3) O Si $(CH_3)_3$	Morphine	482
N-Trimethylsilylimidazole	$(CH_3)_3$ $(C_3H_3N_2)$	Ampicillin	483

which is particularly useful if the methyl ester may be present naturally in the sample. Barbiturates can be N-methylated using diazomethane,[472] dimethylsulfate,[473] or trimethylphenylammonium hydroxide.[474]

A convenenient method of methylation is the flash or on-column method where the reagent and analyte are coinjected and the temperature of the injection port completes the reaction. For example, a mixture of anticonvulsants (e.g., carbamazepine, phenytoin, phenobarbitone, and primidone) was dissolved in a solution of trimethylammonium hydroxide and the sample injected via an injection port at 200°C, the drugs then chromatographing as their corresponding N-methyl derivatives.[475] Total time to prepare the sample was about 20 min and the drugs could be quantified over the range 1 to 5 µg ml^{-1}. Similar methods have been used for phenobarbitone and phenytoin using methanolic tetramethylammonium hydroxide as the coinjected reagent.

The most common class of reagent for increasing the volatility of organic compounds, including drugs, is the group of silylating agents typified by trimethylchlorosilane. This reagent reacts with an active hydrogen:

$$(CH_3)_3SiCl + H\text{–}R \rightarrow (CH_3)_3Si\text{–}R + HCl \qquad (1)$$

Polar and active functional groups are converted to nonpolar and inert compounds, which is an ideal situation for successful gas chromatography. Some of the popular silylating agents are listed in Table 6.1. All these reagents are used to form the trimethylsilyl derivatives of the drug, the different reagents having different degrees of activity. Typically, the drug (or an extract containing the drug) is mixed with the reagent and a suitable solvent in a screw-capped vial fitted with a Teflon-faced septum. The reaction is usually instantaneous and the mixture can be sampled directly and injected into the gas chromatograph. As the derivatives are often unstable in the presence of water, the ploy of containing the sample in a sealed tube throughout ensures the derivatization remains complete. Trimethylsilylation is suitable for primary and secondary amines, alcohols and carboxylic acids. Fuller discussions on silylation reactions have been presented by Pierce[484] and by Knapp.[485]

Amines as well as alcohols may give poor peak shapes due to absorption onto the column support. This can be particularly bad for primary and secondary amines

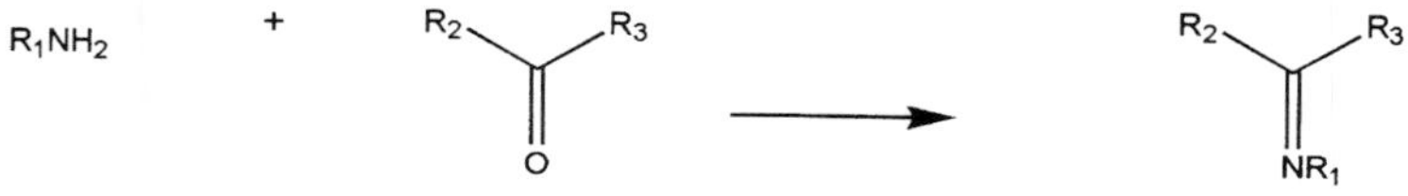

$$R_1NH_2 \quad + \quad R_2 \overset{O}{\underset{}{\diagdown}} R_3 \quad \longrightarrow \quad R_2 \overset{NR_1}{\underset{}{\diagdown}} R_3$$

FIGURE 6.3   Reaction of primary amines and ketones to form Schiff's bases.

where absorption may be so bad that if small quantities are injected then absorption will be total and no peaks at all will be seen. Even worse, subsequent injections of solvent or of samples containing no drug may dislodge the adsorbed material and show spurious peaks in subsequent analyses. Some workers recommend silylation of all surfaces, but elimination of adsorption by derivatization of the amine is probably safer and more reproducible. Delargy and Temple[486] extended the method of de Silva et al.[487] for flurazepam and its desmethyl metabolite down to 1 ng ml^{-1} by daily treatment of the column with hexamethyldisilazane and by overloading the column with sample before the analytical runs. Similarly, a gas chromatographic method for clobazam and desmethylclobazam was only suitable for the metabolite by strict attention to silylation of all parts of the system.[488]

Primary amines can be reacted with ketones to form Schiff's bases (Figure 6.3). This reaction can be performed prior to gas chromatographic analysis, but more simply by injecting the primary amine in acetone solution, the reaction taking place at the elevated temperature of the injection port. The reaction can also be used as a characterization of primary amines by injecting the sample in a nonketonic solvent and observing the characteristic difference in relative retention times of the eluting peaks. In addition to increasing the volatility of the drug, or covering absorbent groups, derivatives are also prepared to increase the response of the detector to the drug, or to improve separation of drug from interfering components of the sample. These are more fully discussed in later sections.

6.3   STATIONARY PHASES

The need for high temperatures to volatilize drugs for gas chromatographic analysis has necessitated the development of special stationary phases which themselves are neither volatile nor unstable at the operating temperatures. The first such series of stationary phases were waxes, gums, or silicone oils at room temperature but liquids at high temperatures. Problems can arise as the temperature is raised if the stationary phase bleeds off the column. This bleeding will result in contamination of the detector and depletion of the stationary phase with consequent deterioration of the performance of both. The stationary phase also covers absorption sites on the column. If the column bleed is extensive, this stripping of the column will result in increased absorption (poor peak shape) and decreased retention compared with the original column, and the exposed active sites may also catalyse decomposition of the analyte. Table 6.2 shows some of the popular thermally stable stationary phases available for analyzing drugs.

Obviously, high loadings of stationary phase will help solve the problem of absorption on the support, but such high loadings may not be desirable for the required chromatography. It is therefore more usual to inactivate the support itself,

TABLE 6.2
Stationary Phases Used in the Gas Chromatographic Analysis of
Drugs in Biological Fluids

Phase	Description	Temperature range (°C)	Example	References
OV-1	Methylsilicones regarded	100–350	Benzodiazepines	489
OV-101	as nonpolar stationary	20–350	Clioquinol	490
SE-30	phases	50–300	Hydroxyzine	491
SP-2100		0–350	Barbiturates	492
OV-7	Increasing substitution of	20–350	Loxapine	493
OV-17	methyl groups of methyl-	20–350	Carbamazepine	494
SP-2250	silicones with phenyl	25–275	Phencyclidine	495
OV-250	to increase polarity	20–300	Amitriptyline	496
OV-210	Very polar phases with	20–300	Acetaminophen	497
QF-1	up to 50% substitution	20–250	Isosorbide dinitrate	498
SP-2401	with trifluoropropyl groups	0–275	Primaquine	499
OC-225	Very polar phases with	20–275	Cannabidiol	500
XE-60	25% cyanopropyl and 25% phenyl substitution	20–275	Amygdalin	501
OV-275	Most polar stationary phase of the series	250	Cyclophosphamide	502
Dexsil 300	Thermally stable phase	40–450	Alphadolone	503
Carbowax 20M	High molecular weight polyethylene glycol	60–225	Propranolol	504
Neopropylglycol succinate	Fatty acid glycol ester	50–240	Guanabenz	505

prior to coating with stationary phase, by silylating with hexamethyldisilazane or similar reagent. The glass column itself should be similarly treated.

The particle size of the packing material is important for good chromatography. The particles have to be small to provide a large surface area per unit length of column and hence high efficiencies, but not too small to create too high a back-pressure for the carrier gas. The preferred particle size is not less than 60 μm and not more than 120 μm.

Although many laboratories in the early days of gas chromatography prepared and packed their own supports, this practice is now virtually extinct, with the commercial availability of highly efficient, highly reproducible, and highly reliable packing materials and packed columns.

The type of stationary phase has a bearing on the separations obtained for various classes of compounds. The stationary phases can be divided into two classes: selective and nonselective. The nonselective phases separate the analytes more by virtue of molecular size and shape rather than by chemical properties. Thus, very closely related compounds do not separate well on nonselective phases, but such phases are useful for the separation and characterization of homologs. Selective phases will separate according to the the selective retention of certain groups. For example, pentoxifylline and its primary metabolite (Figure 6.4) have virtually identical molecular weights and do not separate on phases such as OV-1, but separate on phases

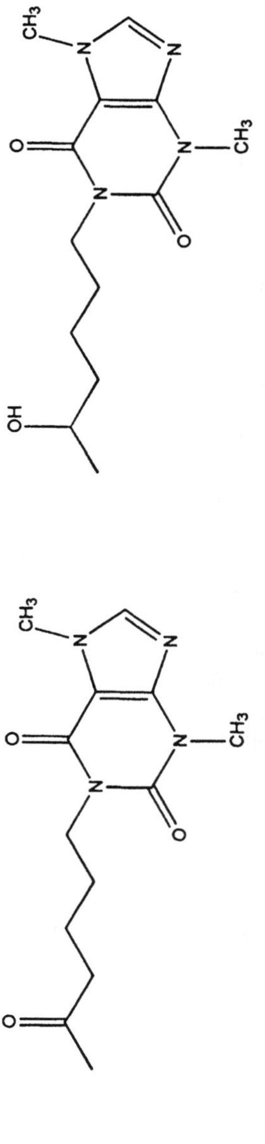

FIGURE 6.4 Structures of (a) pentoxifylline and (b) its primary metabolite.

such as OV-101, which selectively retains oxo compounds compared with the corresponding hydroxy compound. Specific mention should also be made of the group of column packings designed for low molecular weight analogs (such as alcohol). These columns (of which the Porapaks and Chromosorbs are the most widely used) separate the analytes by molecular size exclusion rather than by a partitioning effect.

6.4   DETECTORS

The development of gas chromatography was facilitated by the availability of mass detectors for detecting changes in the composition of a flowing gas. However, the original detectors were too insensitive for successful use because of the high sensitivity usually required for drug analysis in biological fluids. The following discussion is limited to those few gas chromatographic detectors that are in general use for drug analysis.

6.4.1   FLAME IONIZATION

The flame-ionization detector (FID) was introduced by McWilliam and Dewar[506] and although there have been improvements in design, the principle remains the same. Hydrogen and air are mixed with the effluent gas and burned in a small flame. A potential difference is applied across the flame and when an organic compound is burnt in the flame, the resulting increase in ions causes an increase in current. This change in current is measured and converted to the typical gas-chromatographic trace. The detector is insensitive to water, which makes it useful for analyzing aqueous samples, and it gives a relative response, roughly equal to the number of carbon atoms in the molecule, although the presence of oxygen or nitrogen atoms will tend to lower the relative response. Thus, the detector must be calibrated for all individual compounds in an analysis using this detector. Using packed columns, the measurement of drugs in biological fluids is typically in the $\mu g \ ml^{-1}$ range. If the preparation of the sample is extensive, especially to remove lipids from plasma and serum, then measurements in the submicrogram range are feasible. As with all chromatography, the limit of detection depends also on the efficiency of the column and for highly efficient columns such as capillary columns, the limit of detection can stretch into the nanogram range. Table 6.3 lists a number of drugs that have been assayed by gas chromatography using a flame ionization detector.

6.4.2   ELECTRON CAPTURE

The original electron-capture detector was described by Lovelock in 1960.[525] In the original design, the principle was that a current was set-up using the β-particles of a radioactive source as the primary ion current. When a strongly electron-absorbing compound was present in the effluent gas from the column, this primary source was quenched and there was a decrease in the measured current. The electron-capture detector has undergone considerable development since this early design to improve its stability and general working range; the early designs overloaded easily and were thus either unsuitable for quantitative work over wide ranges of analyte concentration, or tended to be put out of action by anything but the most gentle treatment.

TABLE 6.3
Analysis of Drugs in Biological Fluids by Gas Chromatography Using a Flame Ionization Detector

Drug	Approximate plasma during therapy (μg ml^{-1})	Limit of detection (μg ml^{-1})	Reference
Almitrine	5	5	507
Barbiturates	10–20	5	508
Bupivacaine	0.2	0.005	509
Carbamazepine	10	1	510
Disulfiram	0.4	0.01	511
Ethosuximide	150	0.01	512
Fentanyl	0.001	0.0001	513
Ibuprofen	40	—	514
Lignocaine	5	0.03	515
Meprobamate	50	<1	516
Methaqualone	2	0.05	517
Methoxsalen	1	0.1	518
Mexiletine	2	0.05	519
Naltrexone	0.1	0.01	520
Perhexiline	0.1	0.09	521
Phencyclidine	0.5	—	522
Phenobarbitone	50	1	511
Phenytoin	10	1	511
Salicylic acid	70	5	523
Valproic acid	100	0.001	512
Viloxazine	1	0.5	524

The usefulness of the electron-capture detector is its responsiveness to certain types of compounds and its lack of response to even large amounts of other compounds. Thus, the modern electron-capture detector is extremely useful for analyzing drugs in both pharmaceutical preparations and in biological fluids. The most strongly electron capturing compounds are those containing halogens or nitro groups (Table 6.4), or extended conjugated systems, such as for the metabolite of spironolactone, canrenone.[471] A popular general method for benzodiazepines is to convert the benzodiazepine to the corresponding benzophenone and determine the highly conjugated, halogenated product by electron-capture gas chromatography (Figure 6.5).[526] Typically, the electron-capture detector would be expected to measure drugs at the low ng ml^{-1} level in biological fluids.

The use of derivatization to improve chromatographic behavior (greater volatility, less loss by adsorption on the column) has already been mentioned. It is also useful to combine these advantages with an increase in electron-capturing properties. The most obvious derivatization then would be to protect a hydroxyl group with halogenated acid anhydrides and many successful assays have been described using trifluoroacetic anhydride, pentafluoropropionic anhydride, or heptafluorobutyric anhydride (Table 6.5).

When used to analyze drugs in biological fluids, such acylations have two particular problems. Endogenous compounds can also be readily acylated with the derivatizing

TABLE 6.4
**Determination of Underivatized Drugs in Biological Fluids by
Gas Chromatography with Electron-Capture Detection**

Drug	Approximate plasma levels following therapy (ng ml⁻¹)	Limit of detection (ng ml⁻¹)	Reference
Clobazam	500	5	527
Clonazepam	200 mM	3 mM	528
Diazepam	200	4	529
Diazepam	200	125	530
Diclofenac	5000	1	531
Flurazepam	5	1	532
Flunitrazepam	50	0.5	533
		15 nM	534
Isosorbide dinitrate	10	0.5	535
		0.1	536
Midazoloam	100	4	537
Nifedipine	70	0.5	538
Nitroglycerine	1	0.2 nM	539
		0.4	540
Nitromethaqualone	—	1	541
Oxiconazole	250	1	542
Pyrimethamine	400	50	543
Temazepam	500	1–5	544
Triazolam	5	0.05	545
		0.02	546
		5 nM	534

agents and may interfer with the chromatography, perhaps by being even more sensitive to the detector than the targeted analyte; a suitable pre-extraction scheme would therefore need to be introduced to eliminate such compounds. When the drug is present in unknown amounts, it is necessary to ensure that the reagent is present in excess, and failure to remove excess reagent which is itself electron capturing may also cause a large unwanted signal or even overloading of the detector. Thus, volatile

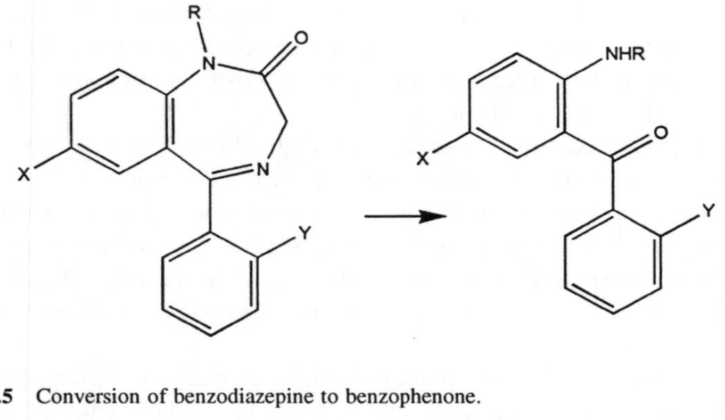

FIGURE 6.5 Conversion of benzodiazepine to benzophenone.

TABLE 6.5
Formation of Halogenated Derivatives of
Drugs to Improve Limits of Detection by
Electron-Capture Gas Chromatography

Derivatizing reagent	Example	Reference
2,4-Dichlorobenzene boronic acid	Alprenolol	547
Chlorodifluoroacetic acid	Moroxydine	548
Heptafluorobutyric anhydride	Tocainide	549
Heptafluorobutyric anhydride	Fluoxetine	550
N-Heptafluorobutyrylimidazole	Phentolamine	551
Hexafluoroacetone	3-Hydroxyguanfacine	552
Pentafluorobenzoyl chloride	Amphetamine	553
Pentafluoropropionic acid	Fluoxetine	554
Pentafluoropropionylimidazole	Pipothiazine	555
Trichloroacetyl chloride	Amantadine	556
Trifluoroacetic anhydride	Pseudoephedrine	557
Trifluoroacetylimidazole	Pindolol	558

reagents are very acceptable, which explains the popularity of perfluorinated acid anhydrides and trimethylsilylating reagents where one or more of the hydrogen atoms of the methyl groups is replaced with a halogen. The use of electron-capturing esters of carboxylic acids seems to be rarely employed, probably because organic acids as drugs tend to be given in relatively large amounts and thus sensitivity of detection is rarely a problem, and the more robust detectors will suffice. For drugs with oxo functions, electron-capturing derivatives can be formed with fluorinated hydrazines, forming Schiff's bases; conversely, amine drugs can be reacted with suitable oxo reagents such as hexafluoroacetone.[552]

The sensitivity of the electron-capture detector toward eluted compounds also depends on factors such as flow rate and temperature, and for any particular drug, such parameters need to be optimized to obtain the best results from the detector. In addition, particular precautions need to be taken to protect the detector by careful sample preparation and even by ensuring the correct environment. The author knows of one laboratory where the electron capture detectors became inoperable on the days when an adjoining canteen had onions on the menu!

6.4.3   NITROGEN SPECIFIC

There is a story of a young research chemist who took his newly created compound to his older and more experienced supervisor for approval. The supervisor looked at the structure and proclaimed that it was likely to be an excellent drug. The young man asked him why, expecting a penetrating insight into structure-activity relationships. "Bags of nitrogen, lad! Bags of nitrogen!" was the reply. Considering one of the most successful drugs of all time is cimetidine which is 33% nitrogen, the old-timer may well have had a point.

Bags of nitrogen in a compound was what the analyst also liked to see when the nitrogen-specific detector first made its mark. The detector is a development of the alkali-flame detector introduced by Karmen and Giuffrida[559] in 1964 as a phosphorous-halogen detector for pesticide monitoring. The detector depends on the thermionic emission of an alkali source rather than the flow of ions in the conventional flame ionization detector. The presence of certain molecules in the flame reduces the alkali metal salt concentration, increasing the temperature of the flame and causing a consequent increase in the measured thermionic emission. The energy supplied to the alkali source can be by a conventional flame or by heating a bead with an electrical current. The operating conditions of alkali-flame or thermionic emission detectors are extremely critical. In the early days of development of this detector this was seen as a drawback because operating conditions were so unpredictable. However, the modern detector can be so well controlled that it is now considered a detector that can be made specific for a number of elements, the most common being nitrogen, phosphorous, and the halogens. In drug analysis, the most common mode is a nitrogen-specific detector rather than a thermionic emission detector.

As its name suggests, the usefulness of the nitrogen-specific detector lies in its selectivity for nitrogen compared with carbon. In a biological extract such as plasma, there are large amounts of lipids but not much nitrogen-containing compounds. Many drugs, on the other hand, contain one or more nitrogen atoms which enables the drug to be determined with considerable sensitivity even in the presence of large amounts of, cholesterol, for example. In suitable cases, underivatized drugs can be analyzed directly in relatively crude extracts at levels where the conventional FID would be swamped by residual amounts of cholesterol.

Just as large amounts of nitrogen are no guarantee of pharmacological activity, not all nitrogen-containing compounds are readily detected with a nitrogen-specific detector. No clear rules exist for predicting which structures will give useful responses, although it has been suggested that there must be a potential for the formation of cyanide radicals in the detector.[560] Of course, it is useful to prepare derivatives to improve chromatographic performance, but it has also been pointed out that the alkylation of amines will also increase the detector response, presumably by increasing the potential to form cyanide radicals. Although the use of nitrogen-containing reagents to form derivatives that are more amenable to detection with a nitrogen-specific detector is an attractive proposition, in practice there have been few such applications. This is probably because the detector has such wide applicability without going to extraordinary lengths, and the preparation entails the usual problems associated with the use of electron-capturing derivatives.

The selectivity of the detector toward nitrogen-containing compounds enables the detector to be used for the analysis of very small samples such as 50 μl pediatric blood specimens; the determination of theophylline in such cases is a widespread application. In more conventional drug analyses, limits of detection of less than 1 ng ml^{-1} can be routinely achieved (Table 6.6).

6.4.4   MASS SPECTROMETRY

A detector is often thought of as a relatively cheap part of the overall gas chromatographic instrument. It is rather surprising then to find that the mass spectrometer (MS) has now become one of the most popular detectors for analyzing drugs in biological fluids (Table 6.7).

TABLE 6.6
Analysis of Drugs in Plasma and Serum by Gas
Chromatography Using a Nitrogen-Specific Detector

Drug	Peak therapeutic concentration (ng ml^{-1})	Limit of detection (ng ml^{-1})	Reference
Almitrene	500	1	561
Apovincaminic acid	1000	2	562
Barbiturates	1–50 µg ml^{-1}	0.5	563
Bupivacaine	1000	0.01	564
Butanilicaine	500	5	565
Buptriptyline	100	2	566
Caramiphen	20	2.5	567
Cinnarizine	100	0.5	568
Cotinine	100	0.4	569
Desipramine	20	5	570
Fluphenazine	5	0.5	571
Flunarizine	50	0.5	568
Lidocaine	5000	2.5	572
Methadone	70	5	573
Mexiletine	2000	5	574
Mianserin	30	1	575
Nefopam	30	5	576
Neostigmine	100	5	577
Nicotine	50	0.1	569
Oxycodone	20	2	578
Pethidine	1000	5	579
Phencyclidine	500	5	580
Phenoperidine	10	2	581
Prajmalium	100	5	582
Procyclidine	60	1	583
Pyrodostigmine	17	5	577
Trimipramine	50	3	584
Tranylcypromine	60	0.5	585
Trifluoperazine	2	0.1	586

When used as a detector, the mass spectrometer may be set to detect a single fragment ion specific for the compound being analyzed. The sensitivity can be optimized by making this ion one of the most abundant in the spectrum, which can be optimized by using an appropriate chemical ionization technique.

In many of the published descriptions of gas chromatography using mass spectrometric detection the specificity of the detector is being utilized rather than its potential for detecting very small amounts of analyte. The literature shows a number of applications for the analysis of drugs in biological fluids and other samples where the limits of detection using GC-MS are no better than and sometimes inferior to those that could be achieved using electron-capture or nitrogen-specific detectors (compare Table 6.7 with Tables 6.4 and 6.6). However, where absolute specificity is a principal requirement, then the mass spectrometer as a detector comes into its own.

TABLE 6.7
The Use of a Mass Spectrometer as a Detector in Gas Chromatography for the Quantitation of Drugs in Biological Fluids

Drug	Sensitivity	Reference
Anabolic steroids	1 ng ml^{-1}	587
Amphetamine	12 ng ml^{-1}	588
Benzoylecgonine	40 ng ml^{-1}	589
Bifemelane	—	590
Bromovalerylurea	10 ng ml^{-1}	591
Buflomedil	—	592
Cannabinoids	—	593
Cocaine	40 ng ml^{-1}	589
Codeine	30 nM	594
Codeine	10 ng ml^{-1}	595
Diazepam	50 ng ml^{-1}	596
Diazepam	7 pg ml^{-1}	597
Diuretics	100 ng ml^{-1}	598
Enalapril	200 pg ml^{-1}	599
Eperisone	0.2 ng ml^{-1}	600
Ethylmorphine	30 nM	594
5-Fluorouracil	—	601
Heroin and its metabolites	—	602
Levoprotiline, N-desmethyl metabolite	0,2 pmol	603
Methamphetamine	12 ng ml^{-1}	588
Methandienone	—	604
Metoprolol	—	605
Morphine	30 nM	594
Morphine	10 ng ml^{-1}	595
Oxazepam	50 ng ml^{-1}	596
Oxazepam	7 pg ml^{-1}	597
Propranolol	—	605
Quinapril	10 ng ml^{-1}	606
Theophylline	10 ng ml^{-1}	607
Valproic acid	3.5 µg ml^{-1}	608
Valproic acid, and 14 metabolites		609

The technique is now almost mandatory for validating other methods[610] which may be used routinely, and also for confirming the identity of separated components in less specific but more rapid screening methods, both in forensic toxicology and in testing for drugs of abuse in suspects' samples.[611] In such cases, the single-ion mode will be replaced by methods which collect the complete spectrum of the separated components, a procedure which has become possible only by extensive use of computers to collect and store thousands of signals from the instrument, so that any spectrum can be retrospectively examined and compared with spectra of authentic compounds. In this mode, the GC-MS system is also widely used in drug metabolism studies as a convenient way of obtaining mass spectrometric data for characterization

TABLE 6.8
Stable Isotopes Used as Carriers for the Determination of Drugs in Biological Fluids by Gas Chromatography-Mass Spectrometry

Stable isotope	Example	Limit of detection	Reference
^2H	Trifluoperazine	0.5 ng μl^{-1}	620
^{13}C	2,3,7,8-Tetrachlorodibenzo-*p*-dioxin	10 pg	621
^{15}N	Fluorouracil	1 ng ml^{-1}	622
$^{15}N^{13}C$	Guanfacine	5 pg ml^{-1}	623
	Theophylline		607
$^{15}N^2H$	Glyceryl trinitrate	0.1 ng μl^{-1}	624

of metabolites not only in development studies but also in clinical samples, as for terfenadine,[612] quinapril,[613] imidapril,[614] and aphidicolin.[615]

For laboratories well equipped with GC-MS systems the use of the mass spectrometer as a specific detector may not be particularly expensive when the capital cost of the instrument is not considered, and such a use of the instrument even for simple assays may be justified. However, the mass spectrometric detector is better indicated for its sensitivity as a necessary method. For this purpose, it is desirable that the fragmentation pattern results in high abundance ions characteristic of the drug under investigation. This is one of the reasons why the use of derivatives giving simple fragmentation patterns is popular in this field.

To fully use the highly sensitive properties of the mass spectrometer, it is essential also to ensure that the small amounts of drug are not absorbed onto the column. The high quality of the commercially available column packings makes this less of a problem, but nevertheless it is worth noting that mass spectrometric detectors are able to solve this problem in a way not available to other methods of detection. This method is by adding large amounts of carrier to the sample at the beginning of the analysis — the carrier being isotopically labeled forms of the drug itself; separation of carrier and analyte is not necessary because the mass spectrometer can be focused on ions which are unique to the analyte; additionally, the carrier can be used as an internal standard by focusing on the ions unique to the carrier. Alternatively, a suitable analog can be used as the internal standard so the single-ion mode can still be used to produce a more conventional chromatogram, without switching from one ion to another.[616]

The most usual isotopes used as carriers are deuterium and ^{13}C, although ^{15}N has also been used (Table 6.8). There is a limit to how much carrier can be added because of the amount of naturally occurring isotope, but this can be minimized by labeling more than one atom in the molecule as in an assay for ascorbate where six carbon atoms were replaced by ^{13}C.[617] Mention should also be made of a related use of isotopically labeled drug where the intravenous and oral pharmacokinetics of a drug can be investigated simultaneously in the same individual by administering one isotope intravenously and the other orally and then analyzing the same chromatograms obtained from blood and urine samples.[618] A similar rationale lies behind the determination of the pharmacokinetics of enantiomers of a drug where one enantiomer

TABLE 6.9
Drugs Determined in Biological Fluids Using
Chemical Ionization-Mass Spectrometry as a
Detection Method for Gas Chromatography

Drug	Reference
Amphetamine	625
Desmethyldeprenyl	625
Enalapril	599
Lovastatin	626
Methamphetamine	625
Pravastatin	626
Quinapril	606
Simvastatin	626

can be isotopically labeled, while the other is not, and both forms are administered simultaneously as in the enantiospecific quantification of hexobarbitone and its metabolites in biological fluids by gas-chromatography/electron-capture negative-ion chemical/ionization mass spectrometry[619]; in this method, the R-(–)-hexobarbitone was labeled with deuterium and the S-(–)-hexobarbitone was unlabeled.

For the operation of the mass spectrometer, the sample needs to be at low pressure in the source. Consequently, using conventional packed columns the carrier gas has to be separated from the analyte. Several ingenious separators have been designed to enable the separation to be as efficient as possible to ensure that most of the sample reaches the detector to maintain sensitivity. To obtain the best sensitivity, the use of chemical ionization has become widespread (Table 6.9)

After a rather uncertain start, the use of capillary columns is now a reality in all types of gas chromatography. In this technique, the stationary phase is coated onto the inside of a glass column. Because the column is not packed, very long columns without unacceptable back-pressures are feasible and thus it is possible to prepare long columns with very high powers of resolution. Such columns have been used extensively in GC-MS combining two powerful routes to specificity of analysis.

The preceeding discussion has concentrated on the use of a mass spectrometer as a specific detector for specific quantitative analysis at very low concencentrations. The combination of gas chromatography and mass spectrometry, however, has found its widest use in drug metabolism studies and it is appropriate to refer to this aspect here. Mass spectrometry has been a useful tool in determining structures of metabolites isolated from biological samples usually by a mixture of extraction and chromatographic techniques. Such a method as applied in the past has been tedious, has been limited to those samples that had been partially purified, and has usually required relatively large amounts of sample. With data capture methods using computers, one could feasibly collect and store complete spectra for hundreds of fractions of eluent from chromatographic columns. In this way, it has been possible to elucidate complex patterns of metabolism by studying and interpreting data obtained from a handful of chromatographic analyses.[627,628]

6.4.5   OTHER DETECTORS

The success of the mass spectrometer as a detection and characterization system for gas chromatography has led to consideration of other powerful physical chemistry methods to be used in a similar fashion. Thus, both infra-red[629-631] and NMR have been investigated as on-line techniques after gas-chromatographic separation of drugs from their metabolites or from complex matrices. Kalasinsky et al.[629] investigated the feasibility of using GC-FTIR for drug analysis in forensic toxicology and concluded that although the technique was less sensitive than GC-MS it had the advantage of a much greater capacity for the identification of separated compounds.

6.5   CHIRAL GC

Mention must be made of the aproaches to chiral separation using GC. As for other chiral analytical methods, the overall problem is approached in one of two ways: (1) prepare derivatives which can be separated according to their physical chemical properties, or (2) use a separation method, which itself has chiral characteristics. For the former, the compounds to be separated are derivatized with optically active reagents, such as in a study on the pharmacokinetics and inversion of the enantiomers of suprofen by reaction with S-(–)-1-naphthylethylamine followed by GC-MS,[632,633] the determination of the enantiomers of fluoxetine and norfluoxetine by derivatization with (S)-(–)-(trifluoroacetyl)propyl chloride,[634] the determination by gas chromatography with electron-capture detection of enantiomers of ketoprofen and ibuprofen,[635] and the specific determination of the labetalol metabolite, 3-amino-1-phenylbutane derivatized with (–)-α-methoxy-α-trifluoromethylphenylacetyl chloride[636] and assayed by GC-MS. All of these examples rely on conventional gas chromatography to separate the diastereoisomers formed in the reactions.

More relevant to the specific topic of gas chromatography are methods utilizing chirally active stationary phases. Although there has been less success in this area than in the field of liquid chromatography, there have been interesting developments which may have their particular application. The use of chiral columns using capillary columns predates the development of high-pressure liquid chromatography being well established in the mid-1960s. There are two main types of chiral stationary phases used in gas chromatography: (1) diamide phases, particularly those modified by polysiloxanes such as Chirasil-val[637] and XE-60-L-valine-phenylethylamide[638]; and (2) cyclodextrin derivatives introduced as chiral stationary phases for gas chromatography by Koskielski et al.[639]

The diamide phases belong to the class of separations dependent on the three-point attachment hypothesis covered more fully in Chapter 7 on liquid chromatography. Cyclodextrins are cyclic glucose oligomers with 6 (α-cyclodextrin), 7 (β-cyclodextrin) or 8 (γ-cyclodextrin) linked glucose units, and separate compounds by size inclusion of molecules in the cavity formed by the cyclic molecule and the chiral property arises from the unique topology formed by the chirality of the glucose units (Figure 6.6). The stability and selectivity of cyclodextrins has been improved by

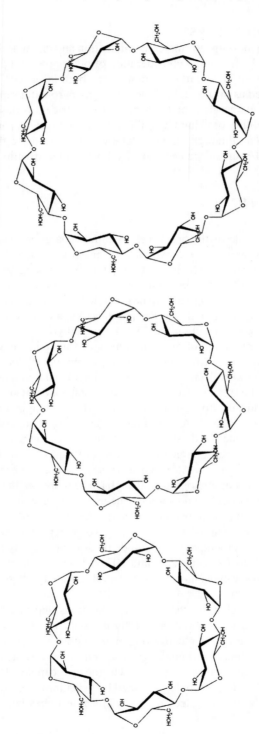

FIGURE 6.6   Structures of cyclodextrins.

TABLE 6.10
Chiral Gas Chromatography Phases Used in the Analysis of Drugs in Biological Fluids

Drug	Phase	Reference
Arabinitol	Hexakis(2,3,6-tri-*O*-pentyl)-α-cyclodextrin	643
Flecainide	Hexakis(2,3,6-tri-*O*-pentyl)-α-cyclodextrin	637
Ifosfamide	Chirasil-val	645
Vigabatrin	Chirasil-val	646

formation of suitable alkyl and acyl derivatives[640,641,644] and by combination with siloxanes.[642]

Table 6.10 lists applications where a chiral gas chromatography phase has been used to analyze drugs in biological fluids.

7 High-Performance Liquid Chromatography

CONTENTS

7.1 DEVELOPMENT OF HPLC

Just as paper chromatography was displaced by thin-layer chromatography, gas chromatography has been displaced to a significant extent by HPLC. This is such a familiar term in the analytical laboratory that there is often uncertainty, and indeed disagreement, on the meaning of the letters. They may stand for either "high-pressure liquid chromatography" or for "high-performance liquid chromatography." Personally, the author prefers the former which seems to be a true description of the technique rather than the latter which implies a degree of subjectivity. Recently, the major chromatographic journals have seemed to settle for the less satisfactory (in my view) "high-performance," so to avoid a semantic discussion; this volume will settle for the simple HPLC.

In essence, HPLC is no more than a development, although rather a late one, of Tswett's classical experiment. Chemists and analysts belatedly recognized that their chromatography in columns of adsorbent such as alumina could be speeded up by having a large head of solvent in the eluting reservoir. This could be taken to extremes by having the solvent reservoir in the same building several stories above the chromatographic column! One point realized early was that, for efficient absorption chromatography a large surface area was required and this meant using very fine particles. However, the consequence of using fine particles was that the adsorbent presented a very high resistance to the solvent flow and the pressure, or head of solvent needed to be increased again. It was only at the end of the 1960s when reproducible, small particles were combined with constant flow (or constant pressure) pumps that the dramatic development of HPLC took place.

7.2   COLUMN PACKINGS

The early columns, being essentially similar to the experiments of Tswett, were based on absorption chromatography. Much effort was put into obtaining reproducible packing materials, and separations tended to follow those obtained on standard TLC plates. Indeed, one of HPLC's stated advantages was that the well-established systems developed for TLC could be directly adapted for HPLC. In addition to adsorbent phases, ion-exchange phases were also developed to separate ionizable drugs using their charge.

Reversed-phase chromatography was also tried on HPLC columns, usually by impregnating an adsorbent support with a nonvolatile organic phase such as paraffin oil, but this had the disadvantage that the stationary phase was gradually stripped from the column and the characteristics of the column gradually changed from partition behavior to the familiar adsorbent behavior. The real breakthrough in HPLC came with the introduction of packing materials in which a stationary phase was covalently linked to the silica support. This enabled reproducible and durable reverse-phase columns to be manufactured. The usual method is to react the silica with organochlorosilane or similar reagent to make a stationary phase of any desired polarity.

$$-SiOH + Cl-SiR_3 \rightarrow -SiO-SiR_3 \qquad (1)$$

The most popular supports have been where R is octadecyl or octyl. The former is suitable for all applications where it is important to not have any residual interaction with the silicon atoms, which are sterically shielded. Octyl phases are less lipophilic and retain drug molecules less strongly; separations can then be faster for equivalent

TABLE 7.1
Reverse-Phase HPLC

Drug	Support or stationary phase	Sensitivity	References
Bupivacaine	Supelco C_8	25 ng ml^{-1}	647
Clomipramine	Nucleosil RP-phenyl		648
Codeine	Zorbax CN	5 ng ml^{-1}	649
Cyclosporin A	Nova-Pak C_{18}		650
Dapoxetine	Zorbax RX-C_8	20 ng ml^{-1}	651
Diltiazem	Supelcosil LC-8-DB	3 ng ml^{-1}	652
Dothiepin	Cyano	50 ng ml^{-1}	653
Flumequine	Cp-Spher-C_8 (Chrompack)	50 ng ml^{-1}	654
5-Fluorouracil	Supelcosil LC-18-5, ODS		655
Fluoxetine	Ultrapore C_8 RPMC	2 ng ml^{-1}	656
Fluvoxamine	Hypersil MOS C_8	10 ng ml^{-1}	657
Haloperidol	Develosil ODS-5	5 ng ml^{-1}	658
Indomethacin	Inertsil ODS-2	0.5 μg ml^{-1}	659
Lorazepam	Apex octadecyl	0.5 ng ml^{-1}	660
Melphalan	Bensil 5C_{18} ODS	10 ng ml^{-1}	661
Methotrexate	Inertsil ODS-2		662
Nimodipine	Hypersil ODS	10 ng ml^{-1}	663
Oxantrazole	Zorbax RX-C_8	50 ng ml^{-1}	664
Oxytetracycline	Lichrosorb RP$_8$	5 ng ml^{-1}	665
Pipamperone	Nucleosil C_{18}	2 ng ml^{-1}	666
Pindolol	Rinin Dynamax Microsorb C_{18}	2 ng ml^{-1}	667
Piroxicam	Techsil C_{10} CN	50 ng ml^{-1}	668
Praziquantel	Ultrasphere ODS C_{18}	30 ng ml^{-1}	669
Rapamycin	Spherisorb C_8	1 ng ml^{-1}	670
Rufloxacin	Viosfer LC-RP-18	30 ng ml^{-1}	671
Stavudine	μBondapak Phenyl	10 ng ml^{-1}	672
Tacrolimus	Nucleosil C_{18}		673
Yohimbine	Lichrosphere 100 RP$_{18}$	100 ng ml^{-1}	674
Zidovudine	Novapak C_{18}		675

lengths of column. In either case, the phase should be "capped" with a silylating agent such as trimethylchlorosilane after the main derivatization to ensure the chromatography is not partly of the absorption type. Such packings are commercially available under a variety of names and some of these are listed in Table 7.1. Although packings from different manufacturers appear to have the same description on paper, the different methods of manufacture and treatment, as well as purity of reagents can lead to different performance characteristics. Thus, the transfer of methods from one laboratory to another may not be as straightforward as would appear on the surface and the analyst should be aware of the potential differences.

Apart from the robustness of the bonded stationary phases, the new columns were eminently suited to drug analysis in biological fluids. First, the drug is likely to be one of the most lipophilic components of the sample and hence is retained on the column longer than most endogenous compounds. This includes metabolites, which generally are more polar than the parent drug. Second, the reverse-phase mode means that the mobile phase is aqueous which is a more appropriate medium for analyzing

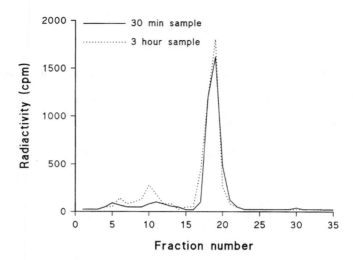

FIGURE 7.1   Metabolic pattern produced by collection and counting of radioactivity in fractions obtained following HPLC analysis of plasma extracts. The plasma was obtained from a dog 30 min and 3 h after an oral dose of ^{14}C-labeled drug.

an aqueous sample. The use of a partition system rather than an absorption system gives sharper peaks with less tailing.[676] Add to these advantages the fact that the technique can be used for involatile drugs, then it can be seen why HPLC is now so well accepted. Before HPLC, it was usual to release drugs and their metabolites from their conjugates prior to analysis (see Chapter 2), and the measurement of conjugates, where this was considered important, depended on differential measurements of parent drug or metabolite. Using HPLC it became feasible to assay conjugates directly,[179] and this is now often done with the consequent realization of the importance of some of these conjugates in their own right, rather than merely means of disposal of foreign compounds. HPLC is widely used in drug metabolism studies, particularly where a labeled drug is used; fractions of eluent may be collected and counted in scintillation vials, or on-line counters can be used to produce an informative metabolic pattern in the early stages of metabolism work (Figure 7.1). The eluted fractions containing metabolites can be easily collected for further characterization.

An interesting addition to the range of stationary phases is the use of restricted access media. Brewster et al.[677] evaluated three such phases for the determination of sulfonamides in bovine serum:

GFF-II: internal surface reversed-phase column
Hisep: shielded hydrophobic phase
SPS-18:   semipermeable surface column

Improvements in the performance of reversed-phase systems have been claimed by the use of radial compression chromatographic columns; these are intended to eliminate the channeling that can occur at the periphery of the column, by compressing the column inward, and assays using such columns have been described for azathioprine,[678] lomefloxacin,[679] and galocitabine.[680]

7.3   MOBILE PHASES

In gas chromatography, as we have seen, the separation of components is achieved using suitable stationary phases and perhaps derivatization. In HPLC, it is now more common to settle on a single reverse-phase column and achieve the required separations by controlling the partition between the organic stationary phase and the mobile aqueous phase.

In its simplest form, the mobile phase is water. Such a simple system would be extremely limited; however, the addition of a water-miscible organic solvent, such as methanol, ethanol, or acetonitrile, may be used to modify the polarity of the organic phase. Over the short range of concentrations usually used for the organic modifier, the retention of the analyte will be proportional to the logarithm of the percentage amount of modifier in the mobile phase; over wider ranges the relationship is more likely to be quadratic.

For drugs which are weak acids or weak bases the partition can be adjusted by altering the pH of the aqueous phase with suitable buffers. Just as the extraction of drugs which are organic acids or bases is controlled by the pH of the aqueous phase, so can the same principle be applied to changing the partition between the stationary and mobile phases in HPLC. In the simplest applications, small amounts of acid (usually acetic or phosphoric but sometimes nitric or sulfuric) are added to the aqueous mobile phase. This will cause increased retention of organic acids as the ionization is suppressed and decreased retention of bases as the ionic form is preferred. Available reverse-phase columns do not take kindly to alkali, but as the overall retention can be controlled using organic modifiers, most separations can be achieved without resorting to high pH values. A more precise control of pH can be achieved with buffered mobile phases using acetate or phosphate buffers in combination with methanol or acetonitrile.

Another classical method for effecting extraction of charged compounds from aqueous phases is the technique of ion-pairing, which is also used in altering partition coefficients in HPLC. In the classical method, a counter-ion is added to the solution containing the ion to be extracted and association with a reduced charge is formed; this entity is then extractable with an organic solvent. Thus, for organic acids, a suitable counter-ion would be tetramethylammonium, while for organic bases, such ions as heptane sulfonate have been used. Much of the development of the use of paired ion chromatography is due to the work of Schill[681] and Tomlinson et al.[682] who have written several extensive reviews on the theory and practice of the technique.

Armstrong and Henry[683] first suggested that aqueous micelle solutions could replace organically modified aqueous mobile phases in reversed-phase HPLC. When surfactants in aqueous solution reach a certain critical concentration, the surfactant molecules aggregate into roughly spherical structures, or micelles; the concentration at which this occurs is termed the critical micelle concentration (CMC). The micelles are organized with the lipid tails in the center of the sphere and the hydrophilic heads

TABLE 7.2
Analysis of Drugs in Plasma or Serum by Direct Injection and Use of Mobile Phases with Micelles

Drug	Limit of detection (μg ml^{-1})	Therapeutic range (μg ml^{-1})	Toxic threshold (μg ml^{-1})	Reference
Acetaminophen	0.2	10–20	250	11
Acetylsalicylic acid	2	20–100	300	11
Carbamazepine	2	8–12	12	11
Chloramphenicol	0.7	10–20	25	11
Phenobarbital	2	15–40	40	11
Phenytoin	3	10–20	20	11
Procainamide	1	4–10	10	11
Quinidine	0.3	2–5	5	11
Theophylline	1	10–20	20	11
Theophylline	0.5	10–20	—	688
C_{12} DAPS	—	—	—	689

on the surface. The micelles remain in true solution and cannot be filtered, nor do they scatter incident light. When surfactant is added to mobile phase, there is a dramatic change in analyte behavior at concentrations roughly corresponding to the CMC. Above the CMC, the change in retention time is similar to the changes expected of the conventional organic modifiers.

The change in concentration of micelles in the mobile phase affects only those solutes which interact with the micelles, either by hydrophobic or electrostatic effects, and optimization of the separation of solutes using micelles in the mobile phase by consideration of these factors has been studied extensively.[684,685]

It was recognized long ago that, although separation could be achieved using micelles as organic modifiers, the efficiency of the chromatography tended to be less than for corresponding organic modifiers. This has been attributed to slow rates of mass transfer between the mobile and stationary phases, possible caused by insufficient wetting of the mobile phase by the modifier. In this respect, the use of an organic modifier to aid the wetting process (3% n-propanol) and operating columns at increased temperature (40°C) has proved beneficial in improving peak shape.[686]

One great advantage of micelle chromatography is that the system is capable of solubilizing proteins and hence is capable of accepting direct injections of plasma or serum without precipitating the proteins and causing clogging or deterioration of the column. This is obviously a great advantage in bioanalysis. As well as making a preprecipitation step unnecessary, it is also claimed that the micelles compete with drug for binding sites on the protein, thus releasing the free drug for chromatography.[687] A number of applications of such direct analysis have been described (Table 7.2). The use of micelles has found particular application in capillary electrophoresis (see discussion later).

Although many workers use trial and error in arriving at an appropriate stationary phase for liquid chromatography, there have been many valiant attempts to bring some rational basis to the choice of components and their relative amounts. For example Nicholls et al.[690] described the optimization of the separation of anthrocyclines and their metabolites using reversed-phase liquid chromatography, and the use of a retention index library has been initiated for screening by reversed-phase HPLC.[691]

7.4 DETECTORS

7.4.1 INTRODUCTION

In the early development of HPLC the two most popular detectors were the refractive index detector and the ultraviolet absorption detector. The former was looked on as a universal detector because it would be sensitive to any change in the composition of eluant. However, the refractive index detector has a high limit of detection and is rarely used in the demanding applications discussed in this volume.

7.4.2 ULTRAVIOLET ABSORPTION

The most popular detector by far for drug analysis and other applications has been the UV absorption detector. Although for very sensitive applications, the fluorescence detector is also very popular, UV detectors are the workhorses of HPLC analysis for applications which are not very demanding. In its simplest form, the UV detector consists of a single wavelength source (the 254 nm mercury line). The detector monitors the effluent from the column and will produce a response for all compounds which have some absorption at this wavelength. It is not necessary that the compound should exhibit its maximum absorption at this wavelength; UV spectra display wide absorption bands, and hence in practice the use of fixed wavelength detectors is not as limited as might first appear. The limit of sensitivity of detection with this detector depends on the extinction coefficient of the analyte at 254 nm and also on the efficiency of the column. It is essential that the concentration of the drug in the flow cell be as high as possible. If the chromatographic peak is broad, the measured drug is considerably diluted.

The limits of detection for particular components can be extended by using a variable wavelength detector. This can be done by setting the wavelength at the absorption maximum of the drug; however, it is often just as useful to chose a wavelength that minimizes interference, rather than maximizes response, and this wavelength may well be set away from the maximum absorption wavelength of the analyte. The two factors have to be balanced to optimize the use of variable wavelength detectors.

In screening methods, where drugs may have absorption maxima at widely different wavelengths, some of the sensitivity of the detector may be lost, and detectors which can simultaneously monitor column effluent at several different wavelengths have proved particularly useful. The advent of microcomputers to process large amounts of information very rapidly and the design of multiple diode array detectors made it possible to scan complete spectra of the chromatographic effluent every few seconds. The data can be manipulated as desired to obtain the maximum information on all the components so detected. Because of its relative cheapness, the multiple diode array detector has now passed from its specialized uses in drug metabolism studies and in screening assays to being a routine tool in the drug assay laboratory (Table 7.3).

TABLE 7.3
Analysis of Drugs in Biological Fluids Using a
Multiple Diode Array Detector

Drug	Application	Reference
Antidepressants, neuroleptics	Toxicological screening	692
Butoprozine	Metabolites in bile	693, 694
Ferulic acid	Pharmacokinetics	695
Phenylalkylamines	Assay	696
Valproic acid	Therapeutic drug monitoring	697
Nalbuphine	Assay	698
Benzodiazepines	Assay	699
Morphine metabolites	Assay	700
Anticonvulsants	Screening	701
Neuroleptics	Screening	702
Zopiclone, zolpidem, suriclone, alpidem	Screening	703
Pyrimethamine	Monitoring in overdose	704

In a number of commercially available multiple-diode array detectors the detector incorporates as many as 256 separate diodes set at wavelengths over the range of ultraviolet and visible wavelengths. Information provided by the computer can include checking the known relative absorption of authentic compounds at different wavelengths for confirmation of identity and the resolution of overlapping components.

An alternative to the multiple diode array detector is the use of a rotating filter disk. The development of a cheap and very effective system for identifying and assaying barbiturates and other drugs in formulations (although not in biological fluids) has been described.[705]

Naturally the drug molecule needs to have UV-absorbing groups to be used with a UV detector. Unsaturated ketones, such as in the steroid sex hormones and cortisone provide excellent absorbing species, but even saturated ketones can be detected with reasonable sensitivity. If a drug is apparently lacking in a suitable absorption band, then a variable wavelength detector can still be used at very low wavelengths such as 200 nm where even isolated double bonds will have some absorption. At such wavelengths the purity of the solvent used as mobile phase, and the design of the detector become crucially important, but even if the compound can be detected, the specificity is likely to be poor.

In principle, derivatives can be made of non-UV absorbing compounds to increase their detectability with UV detectors, or to shift the absorption to a more favorable wavelength. An example is the conversion of tobramycin to a trinitrophenyl derivative by reaction with 2,4,6-trinitrophenylsulfonic acid and HPLC with detection at 340 nm.[706] However, the use of derivatization to take advantage of the specific property of UV detectors has been much less exploited in HPLC than corresponding specific detectors in gas chromatography. A few of the applications using derivatives for UV detection are listed in Table 7.4.

The derivatives may be prepared as part of the sample preparation prior to chromatography, but the concept has been more widely developed by forming the derivative between the end of the column and the detector, such as in the classical

TABLE 7.4
Rare Use of Derivatization Followed
by UV Detection

Drug	Derivative	Sensitivity	Reference
Amikacin	Phthalaldehyde	0.5 µg ml^{-1}	707
Amphetamine	Polymer reagent	15 ng ml^{-1}	708
Benzylpenicillin	1,2,4-triazole-HgCl$_2$	5 ng ml^{-1}	709
Tobramycin	Trinitrophenyl	—	706
	Phthalaldehyde	—	710

amino acid analyzers developed by Stein and Moore. An example in the field of drug bioanalysis is the determination of clodronate in aqueous solutions by HPLC using postcolumn derivatization[711] The method was applied in urine without interference from phosphate; derivatization was with a Th-EDTA-xylenol orange mixed-ligand complex enabling UV detection at 550 nm, with a detection limit of 200 µg ml^{-1}.

This process has the advantage that small structural differences between closely related compounds are not eliminated by derivatization with bulky groups, thus depriving the analyst of the resolution afforded by the chromatography. In addition, precolumn derivatization requires efficient conversion to a single species, whereas postcolumn reactions may result in mixtures, the only requirement being a good color yield in the reaction.

A particular case of derivatization is the use of an ion-pairing reagent to impart more favorable detection characteristics to the drug, as well as better chromatographic properties. Thus an anthracene sulfonate was used as the counter-ion in the determination of secoverine giving a limit of sensitivity of 20 pg for this compound.[712] In addition, such derivatives allow the collection of the effluent for recovery of intact, original analyte without chemical intervention.

7.4.3 FLUORESCENCE

Fluorimetry has also found great favor as a detection system in HPLC. Many drugs have native fluorescence and as described in Chapter 4, fluorimetry provides an assay with a low limit of detection. However, if the parent drug is fluorescent, then its metabolites are also probably fluorescent. A particular example is triamterene. The principal metabolites of this diuretic are hydroxytriamterene and the sulfate conjugate. The early assay methods for this drug in biological fluids were based on straightforward fluorescence methods.[713] It was only when HPLC was used that it was seen that most of the fluorescence was due to these metabolites and the kinetics of the drug had to be reinterpreted[714] (Figure 2.21). The limits of detection afforded by the fluorescence detector depend on the nature of the drug and also on the solvent, as previously emphasized in Chapter 4. In favorable cases the limits of detection are less than 5 ng ml^{-1}, for example, for triamterene and others are listed in Table 7.5. Using lasers as the excitation source a sensitivity of 36 fmol was claimed for an assay of methionine enkephalin by reversed-phase liquid chromatography.[715] It is likely that new instruments based on laser-induced fluorescence will further improve the detection limits of drugs using fluorescence, as in an analysis for ivermectin.[716]

TABLE 7.5
Analysis of Drugs in Biological Fluids by HPLC Utilising
Their Native Fluorescence

Drug	Wavelengths (nm) Ex.	Em.	Approximate plasma levels following therapy (ng ml⁻¹)	Limit of detection (ng ml⁻¹)	Reference
Acyclovir	375	260	500	10	717
Amiloride	360	413	1000	7.5 pg	718
Bromolasalocid	215	370	1000	100	719
Celipropolol	265	418	20	10	720
Diclofenac	—	—	1000	—	721
Debrisoquine	210	290	20 μg ml⁻¹	100 ng ml⁻¹	722
Ellipticine	360	445	100	5	723
Etoposide	230	328	—	50	724
Extramustine	195	250	40	40	725
Flecainamide	300	370	500	50	726
Furosemide	233	389	1000	20	727
Furosemide	360	413	1000	100 pg	718
Glibenclamide	308	360	200	10	728
Labetalol	335	370	100	1	729
N-l-Leucyldoxorubicin	480	580	1 μm	1.3 n*M*	730
Melphalan	260	350	100	5	731
Meptazinol	282	300	30	3	732
Moxisylyte	195	—	80	25	733, 734
Nebivolol	288	310	0.5	0.1 ng ml⁻¹	735
Pyrimethamine	290	345	200	10	736
Prenalterol	220	320	5	1	737
Rufloxacin	350	510	15 μg ml⁻¹	—	738
Salbutamol	200	—	100	—	739
Salmeterol	230	305	10	1	740
Terbutaline	200	—	5	—	739
Triamterene	366	440	10	1	714
Verapamil	203	320	100	1	741
Cinnarizine	245	310	60	0.5	742

The formation of fluorescent derivatives of drugs for subsequent detection using a fluorescent detector has received more attention than the use of derivatives to increase absorption in the UV or visible wavelengths. This is perhaps because the rewards are greater when a good fluorescent derivative is obtained. Cyclosporin does not have a readily derivatizable functional group and is lacking in UV absorption. It is first converted to isocyclosporins and then dansylated with dansyl chloride. The fluorescent derivative is used on HPLC, giving a sensitivity of c10 ng ml⁻¹ (CV 19% at this concentration) suggesting it is not sensitive enough for monitoring, but holding promise for more sensitive fluorescent reagents.[743]

Favored reagents for forming such fluorescent derivatives include dansylation reagents, phthalaldehyde, and fluorescamine. Recently, many novel reagents have been proposed and some of these are listed in Table 7.6.

There have also been a number of attempts to include the derivatization step in reaction columns prior to passing sample onto the analytical column; however the

TABLE 7.6
Formation of Fluorescent Derivatives for Subsequent HPLC Analysis
and Detection With a Fluorescence Detector

Drug	Derivative	Sensitivity	References
N-Acetylglucosaminides	9-Anthroyl cyanide	100 fmol	744
Aminohydroxypropylidene biphosphonate	Fluorescein		745
Amphetamines	4-(*N,N*-dimethylaminosulfonyl)-7-fluoro-2,1,3-benzoxadiazole	~100 fmol	746
Anabolic steroids	Dansyl		747
Dideoxyadenosine	2-(5-chlorocarbonyl-2-oxazoyl)-5,6-methylenedioxybenzofuran		748
Metformin	Desyl chloride	50 nm	749
	Benzoin/mercaptoethanol	50 nm	749
Fluoxetine	Dansyl chloride	3 ng ml^{-1}	750
Gentamicin	Fluorescamine		751
Medroxyprogesterone acetate	4-(N,N-dimethylaminosulfonyl)-7-hydrazino-2,1,3-benzoxadiazole		752
Mercaptopurine	4-Ethylmorpholine/monobromobimane	3 nM	753
Mexiletine	Dansyl chloride	20 ng ml^{-1}	754
Phenylpropanolamine	Phthalaldehyde		755
Perhexiline	Dansyl chloride	50 ng ml^{-1}	756
Rokitamycin	Dansylhydrazine	20 ng ml^{-1}	757
Tocainide	Fluorescamine		758

formation of derivatives before chromatography always has the effect of reducing differences in the analytes and hence reducing their resolution on the column. A preferred approach is to form the derivatives with the column effluent before passing it through the detector. Some particularly elegant methods involving extraction with fluorescent ion-pair reagents have been described for the analysis of cocaine, benzoylecgonine, ecgonine methyl ester, ethylcocaine, and norcocaine in human urine,[759] and the formation of traditional fluorescent derivatives such as the reaction of phthalaldehyde with amikacin in dog plasma separated by reversed-phase ion-pairing liquid chromatography.[760]

Precolumn reactors are usually designed to form traditional derivatives, such as the phase-transfer-catalyzed dansylation of phenolic compounds, ethynylestradiol (0.8 μM) and paracetamol (0.5 μM), followed by normal-phase liquid chromatography with fluorescence detection,[761] or by the formation of 9-fluoreneacetates using a 9-fluoreneacetyl-tagged solid-phase reagent.[762] Postcolumn reactors, on the other hand, need only increase the fluorescence of the species and therefore chemical methods leading to more than one product or degradation product are quite acceptable.

A method for the determination of thiamine in human plasma for oral bioavailability studies describes postcolumn oxidation with $K_3[Fe(CN)_6]$ to thiochrome. The method allowed complete separation of the various thiamine phosphates in plasma and detection of the total thiamine fluorimetrically by HPLC and postcolumn oxidation to 2 ng ml^{-1} plasma.[763] $K_3[Fe(CN)_6]$ was also used to induce chemiluminescence of dexamethasone in the analysis of this compound by HPLC with a fluorescence detector.[764] Hydrogen peroxide has been used as an oxidizing agent to induce fluorescence in indomethacin prior to detection by a fluorescence detector.[659]

9-Aminocamptothecin in plasma was determined by monitoring the fluorescence induced upon postcolumn acidification,[765] whereas indomethacin was measured by inducing fluorescence with alkali in a postcolumn reaction coil.[766] Oxazepam was also measured in urine by inducing fluorescence in the column effluent by oxidation with acetic acid at 100°C.[767]

The fact that fluorescence of many drugs can be induced by irradiation with short-wave ultraviolet light has formed the basis of a number of reaction detectors. Such a device does not require a mixing or dilution step and furthermore can be literally switched off when the reaction is not required, for example for application to drugs which are already sufficiently fluorescent. Uihlein and Schwab[768] have described such a photochemical reactor and discussed factors influencing its performance, such as contribution to peak broadening and optimization of the photochemical reaction. The maximum radiation at the minimum distance was achieved by knitting a narrow Teflon tube into a sleeve which was placed around the ultraviolet source; effluent from the analytical column was directed through the Teflon tubing and into the fluorescence detector. The described device was successfully applied to the detection following HPLC of clobazam and its metabolites, and to fenbendazole and its metabolites.

An ultraviolet reaction coil was also used in the trace analysis of methotrexate and 7-hydroxymethotrexate in human plasma and urine by HPLC with fluorescence detection,[769] and a similar method was used by Urmos et al.[770] for measurement of cis and trans isomers of clomiphene. Similarly, thermally induced reduction to fluorescent species using knitted reaction coils has been described by Lambert et al.[131] for the determination of phylloquinone in human milk.

7.4.4   ELECTROCHEMICAL

Several detectors have been described which depend on the property of analytes to be oxidized or reduced, the resulting electron flow being the basis of the detection principle. These electrochemical detectors are based on the fact that the current is measured as a function of time with a constant potential applied at a fixed electrode exposed to a moving fluid. In the strict sense, polarographic detectors (see also Chapter 9) are those which use a dropping mercury electrode; these are not widely applied in drug analysis using chromatography.

In the most common form of the electrochemical detector, the drug in the detector cell is converted to its oxidized form, yielding one or more electrons per molecule reacted. The oxidized form is usually unstable, and reacts further to form a stable compound which passes out of the cell. The instantaneous current is proportional to the concentration of drug in the cell; it should be pointed out that not all molecules are oxidized and in the early forms of this detector a conversion of 10% was considered acceptable.[771] Improvements in design and theoretical considerations have made electrochemical detectors more widely used in drug analysis and their sensitivities rival those of fluorescence detectors under favorable conditions, as shown in Table 7.7.

TABLE 7.7
Determination of Drugs by HPLC Using Electrochemical Detectors

Drug	Fluid	Mode of detector	Voltage	Sensitivity	Reference
Acetaminophen	Urine	Carbon paste	+1.1	150 pg	772
Artether	Plasma	Reductive			773
Ascorbic acid					774
Buspirone, metabolites	Plasma			50 pg	775
Benzodiazepines					699
Captopril		Gold-mercury	+0.07	1 pmole	776
Carbidopa		Glassy carbon	+0.70	15 ng ml^{-1}	777
Chlorpromazine	Plasma	Glassy carbon	+0.9	250 ng ml^{-1}	778
Cisplatin	Plasma	Glassy carbon-based wall-jet			779
Desipramine	Plasma				780
	Plasma	Glassy carbon	+1.05	5 ng ml^{-1}	781
Imipramine	Plasma				780
	Plasma	Glassy carbon	+1.05	5 ng ml^{-1}	781
Inulin	Biological fluids	Pulsed amperometric			782
Mefenamic acid	Serum				783
Mianserin	Plasma				784
Mitomycin	Plasma, serum, urine	Mercury drop	+0.70	250 pg	785
6-Monoacetylmorphine	Urine				786
Morphine		Glassy carbon	+0.60	1 ng ml^{-1}	787
Naloxone	Plasma				788
Paracetamol	Blood, plasma	Dual electrode coulo-metric quantification in the redox mode			789
Phenothiazines	Blood, plasma	Glassy carbon	+0.90	100 ng ml^{-1}	790
Promethazine	Serum				791
Pamaquine	Plasma				792
Primaquine, carboxyprimaquine	Plasma				792
Risperidone	Plasma				793
Salbutamol	Plasma	Amperometric			794
	Plasma				795–797
	Plasma	Carbon paste	+0.95	500 pg ml^{-1}	798
Sumatriptan	Plasma, urine				799
Terbutaline	Plasma				796, 800
Trimethoprim		Glassy carbon	+1.2	10 ng ml^{-1}	801
Theophylline	Plasma				802
Pentazocine	Plasma			0.5 ng ml^{-1}	803

7.4.5 MASS SPECTROMETRY

The linking of gas chromatographs to mass spectrometers resulted in the most powerful analytical tool where separation from complex matrices had to be combined with very sensitive detection. The successful development of this tool (as described in another part of this chapter) enabled very rapid analysis of problems in drug metabolism and toxicological screening. Indeed, for the stringent requirements of proper pharmacokinetic analysis for many drugs, the GC-MS became essential. While most development compounds consisted of small molecules, with sufficiently low boiling points, the gas chromatograph was an acceptable means of separation. However, as the need arose for the analysis of less volatile metabolites, particularly of conjugates, and then later the emergence of labile, highly polar drugs such as polypeptides, the advantages of linking HPLC to mass spectrometers became attractive. In HPLC, there is a greater range of parameters to effect any particular separation; there is no need to make derivatives to chromatograph polar compounds and no need for high temperatures which may destroy labile compounds. However, the features of gas chromatography that made the interface relatively simple were absent in HPLC. The mobile phase was a liquid which would produce unreasonably large volumes of vapor — 18 ml of water would produce 22.4 l of vapor at standard temperature and pressure!

Early attempts to provide an HPLC-mass spectrometer interface included enrichment of the effluent using a membrane separator,[804] vaporization of the entire sample in an ionization source,[805] laser vaporization of the solvent,[806] splitting the eluant and allowing the minimum amount of sample to be ionized (and using the carrier solvent as the reagent in the chemical ionization mode),[807] and transport of the effluent through vacuum locks on a wire, belt, or ribbon.[808,809] Some of these early systems were commercially exploited with only moderate success although analytical methods were reported for thiazides,[810] ranitidine and its metabolites,[811] sulfa drugs,[812] and nonsteroidal anti-inflammatories.[813]

Development of interfaces has been so successful recently that one laboratory even described HPLC-MS as the method of choice for nonvolatile or thermally labile drugs in biological fluids.[814] The most successful interface has been the thermospray device or other procedures involving atmospheric nebulization of sample and a few of the applications are listed in Table 7.8. Despite optimism in certain quarters, this remains a difficult and expensive technique, and it is unlikely to be routine in the near future.

7.5   HPLC AND NMR

Nuclear magnetic resonance (NMR) has long been used as a powerful tool in structural analysis in the chemistry laboratory. Use of NMR in identifying or characterizing drugs in biological fluids, like mass spectrometry, was little used until a decade ago, due to the relative insensitivity of the method; on those occasions where definitive characterization of a metabolite was required (and chemical practices demand an NMR spectrum as final proof), it would be necessary to process large amounts of sample through several chromatographic processes to obtain a suitably large and suitably pure, sample. However, as the technique was refined to obtain greater and greater resolution of the complex NMR signals, the amount of information that could be obtained from such spectra, and the resolution of signals from drug and endogenous material became possible, then NMR became feasible as a screening method for particular compounds in urine using the unique signals from the compounds of interest.[830]

TABLE 7.8
Analysis of Drugs in Biological Fluids Using
Thermospray and Nebulising Methods to
Interface HPLC With Mass Spectrometry

Drug	Fluid	Reference
Albanoquil	Blood	815
Benzodiazepines	Biological materials	699
	Whole-blood	816
Budesonide	Plasma	817
Busulphan	Serum, cerebrospinal fluid	818
Cefaclor	Serum	819
Ceftibuten	Sputum	814
Cyclosporins	Blood	820
2′,3′-Dideoxycytidine	Plasma	821
L 654066	Plasma	822
Oltipraz	Serum	823
Rogletimide	Plasma	824
Taxol	Yew, bark and needles	825
Tenidap	Serum	826
Timolol	Plasma	827
Trenbolone	Bile, feces	828
Trimethoprim	Serum	829

This method of analysis is now well established, for example, in the investigation of the excretion of paracetamol metabolites in rat and human urine and in rat bile.[831,832] However, the signals from parent drug and metabolites would often overlap and some method of prior separation was necessary and the coupling of NMR spectrometers, like the coupling of mass spectrometers, appeared a very attractive proposition. Some of the technical problems with this coupling seem to have been overcome[833] and the use of HPLC-NMR has been proved to be a rapid method for the detection and identification of drug metabolites in biological fluids at a considerable saving in time and costs. At present, the technique only seems to have been applied to drugs given in relatively large amounts, such as paracetamol,[834] antipyrine,[835] and ibuprofen,[833] and only for these well-established drugs as demonstrations of its feasability, but its supporters see wide applications particularly when combined with supercritical fluid chromatography, or use with other NMR-sensitive nuclei.

7.6 CAPILLARY ELECTROPHORESIS

The usual form of liquid chromatography is to provide the moving force either by gravity or by applied pressure; in capillary electrophoresis the molecules are moved by the application of a potential difference, the molecules moving through the stationary phase at rates depending on their charge and their molecular weight. Almost all organic drugs can exist as acids or bases and therefore the pH has a major effect on the selectivity of the separation. For specific compounds, it is useful to set the pH near the pK_a value because any changes around this value have the greatest

TABLE 7.9
Applications of Capillary Electrophoresis to
Analysis of Drugs in Biological Fluids

Drug	Fluid	Sensitivity	Reference
Acyclovir	Serum, plasma		841
Amoxycillin			842
Antiepileptics	Plasma		843
Bendroflumethazide	Urine	0.1 µg ml^{-1}	844
Carbamazepine	Plasma		843
Chlorthalidone	Urine	0.1 µg ml^{-1}	844
Cimetidine	Urine		845
Dextromethorphan	Urine		846
Ephedrine			847
Ethosuximide	Plasma		843
Fosfomycin	Serum		848
Pentobarbital	Serum		849
Phenobarbital	Plasma		843
Phenytoin	Plasma		843
Pilocarpine			850, 851
Primidone	Plasma		843
Proteins			852
Suramin			853
β-Blockers			838
β-Blockers	Urine	25 µg ml^{-1}	854
Theophylline	Plasma		855
Theophylline	Serum		856
Valproic acid	Plasma		843
Phenobarbital	Serum		857
Ethosuximide	Serum		857
Primidone	Serum		857

effect on mobility and hence the separation can be fine-tuned most effectively by changes in pH.[836]

Further changes in selectivity can be achieved in capillary electrophoresis by complexation, by the use of molecular sieves, by the addition of organic solvents,[837] by changing temperature,[838] and most significantly by the use of micelles. In recent reviews of capillary electrophoresis, the strategy for best application of the technique to the analysis of drugs has been outlined.[839,840] Table 7.9 shows some of the applications of capillary electrophoresis to analysis of drugs in biological fluids.

The wide range of application of capillary electrophoresis is illustrated by the development of an assay for protein drug substances by fluorescamine derivatization and capillary electrophoresis.[852] The method has also been used for debrisoquin metabolic phenotyping[846] by simultaneous assay of dextromethorphan and dextrorphan in urine. Just as for conventional HPLC, the technique coupled to mass-spectrometry has been investigated.[858]

7.7   SUPERCRITICAL FLUID CHROMATOGRAPHY

Everyday experience with compounds that can exist as gases, liquids, or solids suggests that heating a solid (for example, ice) converts it to a liquid (water) and then

TABLE 7.10
Analysis of Drugs by Supercritical Fluid Chromatography

Drug	Stationary phase	Mobile phase	Detector	Reference
Anti-inflammatories	Cyanopropyl	CO_2	FID	863
Benzodiazepine	Cyanopropyl	CO_2	FID	864, 865
Estrogens	Cyanopropyl	CO_2	FID	866
Miprostol	Cyanopropyl	CO_2/methanol	UV	867
Omeprazole	Aminopropylsilica	CO_2/methanol/triethylamine	UV	868
β-Blockers	Chiralcel OD	CO_2/methanol/propylamine	UV	869
β-Blockers	ChyRoSine	Methanol/CO_2	UV	870

to a gas (water-vapor). However, the state of the compound depends not only on the temperature, but also on the pressure; carbon dioxide as dry ice will convert straight to the gaseous form without becoming a liquid because the ambient pressures are below the so-called critical pressure. If the pressure is increased, then carbon dioxide can also exist as a liquid; this is the supercritical fluid state of the compound. The properties of the supercritical fluid, including its ability to dissolve other compounds, can be modified by varying the temperature and pressure. Supercritical fluids have wide application in industry as extractants, one of the great advantages of using supercritical carbon dioxide is that the solvent can be readily removed at room temperature and pressure. This property has also been utilized to prepare blood samples for the trace analysis of flavone,[859] with subsequent HPLC analysis. The use of supercritical fluids as chromatographic mobile phases was first described by Klesper et al.,[860] but was little developed until 1981, when Novotny et al.[861] described the use of supercritical fluids in capillary chromatography. The apparent use of liquid chromatography principles using a piece of equipment that is a mainstay of the more advanced gas chromatography neatly sums up the position of supercritical fluid chromatography between the two types of fluid chromatography. Recent developments in the technique have run in parallel with developments in HPLC and gas chromatography and have made use of improvements in solvent modifiers, equipment design, and linking the technique with diverse detection methods. This ability of supercritical fluid chromatography to borrow from the older types of chromatography was demonstrated by Roberts and Wilson.[862] Table 7.10 summarizes some recent publications that demonstrate the same point.

7.8   CHIRAL HPLC

7.8.1   INTRODUCTION

The most dramatic breakthroughs in HPLC have been made in the last 10 years in the development of methods for the resolution of enantiomers. The resolution may depend on the chiral nature of the stationary phase, the chiral nature of the mobile phase, or may be effected by preparing diastereoisomers with appropriate chiral reagents with subsequent chromatography on achiral systems. Within any of these approaches, the analyst has the usual array of choices to optimize the separation of the targeted compounds.

FIGURE 7.2   The three-point theory to explain stereoselective selection in chromatography. One isomer can interact closely with all three binding centers; the second isomer can only interact closely with two binding centers at any one time.

7.8.2   CHIRAL STATIONARY PHASES

Wainer proposed dividing the chiral stationary phases into five classes in his detailed discourse on the mechanism and use of these phases. Some of Wainer's classes overlap and some phases will consist of a mixture of the phase types. As a result, the author will confine his discussion to three broad types of phase without attempting a rigid classification.

7.8.3   SYNTHETIC CUSTOM-DESIGNED COLUMNS

It has been more than 40 years since Dalgleish[872] postulated a mechanism of chiral recognition, whereby at least three simultaneous interactions take place between the chiral molecule and the recognition site. In chromatography, these entities are represented by the analyte and stationary phase, respectively. Because of the unique distribution of the three (or more) points of interaction of the stationary phase, the simultaneous interaction will be more likely for one enantiomer than the other (Figure 7.2). Thus the enantiomer that is most strongly bound will be retained on the column more readily than the less strongly bound enantiomer. The interactions do not have to be attractive forces but could equally well be repulsive forces; it is the total effect that is the important factor.

Pirkle devised a series of chiral stationary phases based on this concept — the three-point interaction — exemplified by the production of *N*-(3,5-dinitrobenzoyl) amino acids. According to his theory such phases are appropriate for enantiomers where there is an aromatic group adjacent to the chiral center. A large number of these so-called Pirkle stationary phases have been prepared and have been successfully applied in the separation of enantiomers in biological fluids (Table 7.11).

7.8.4   INCLUSION COMPLEXES

Chiral environments can exist in stationary phases in the form of irregularly shaped cavities, which may accommodate different molecules more effectively than others according to their shape. Thus one enantiomer may fit more comfortably into the cavity than the other and its progress through the stationary phase will be retarded, thereby effecting resolution of the enantiomers.

The best known of these types of stationary phases are the cyclodextrins (see Figure 6.6). These are cyclic oligosaccharides composed of α-D-glucose units linked

TABLE 7.11
Some Chiral Separations on Pirkle Columns

Compounds	Column	Reference
Glutethimide and 4-hydroxymetabolites	35DNBleucine	873
Debrisoquine and the R/S metabolites	35DNBphenylglycine	874
Sulfoxide metabolite of albendazole	Tyrosine derivative	875
Mefloquine	Naphthylethylurea	876
Benzodiazepines	(S)-*N*-(3,5-dinitrobenzoyl)phenylalanine	877
Naproxen	α-Acid glycoprotein	878
	{3-[1(3,5-dinitrobenzamido)-1,2,3,4-tetrahydrophenanthren-2-yl]-propyldimethylsily}silica	879
Benzodiazepines	HSA	880, 881

through the 1,4 position, the α-, β- and γ-cyclodextrins containing six, seven and eight glucose units, respectively. To prepare suitable stationary phases the toroidal structure is attached to silica, with the hydrophobic cavity allowing a variety of water-soluble and insoluble solutes to enter. Because of the irregular shape of the cavity, the phase can be used to separate geometric isomers and structural isomers[882] and sugar anomers[883] as well as to separate enantiomers. Some separations using β-cyclodextrins as column material are shown in Table 7.12.

Cellulose is a highly crystalline polymer composed of chains of linked D-β-glucose units. Cavities exist within the structure of cellulose as the spaces between the chains and because of the chiral nature of the glucose units, this cavity is able to discriminate between presenting enantiomers. Cellulose was one of the first components of analytical chromatography in the form of paper chromatography. Its properties make it ideal as a chiral stationary phase. However, the native forms of cellulose are too mechanically fragile to be used with the usual pressures of HPLC. To overcome this deficiency, triacetyl derivatives of cellulose were prepared with some success.[890] The advantages of such triacetylated cellulose phases are that they

TABLE 7.12
Separation of Enantiomers Using β-Cyclodextrin Columns

Enantiomeric pair	Phase	Reference
Atenolol	β-Cyclodextrin derivative	884
Levomepromazine	β-Cyclodextrin	885
Ibuprofen	β-Cyclodextrin-diol phase silica	886
Lorazepam glucuronides	β-Cyclodextrin	887
Nordiazepam glucuronides	β-Cyclodextrin	887
Oxprenolol	β-Cyclodextrin-, γ-cyclodextrin-bonded phases (Cyclobond I, Cyclobond II)	888
Temazepam glucuronides	β-Cyclodextrin	887
Warfarin	β-cyclodextrin	889

TABLE 7.13
Separation of Enantiomers Using Cellulose Columns

Enantiomeric pair	Phase	Reference
Carazolol	Chiralcel OD	892
Celiprolol	Chiralcel OD	892
Cyclohexylaminoglutethimide	Cellulose-based chiral	893
Dihydropyridines	Cellulose tris(3,5-dimethylphenylcarbamate)	894
Felodipine	Chiralcel OJ	895
Flurbiprofen	Chiralcel OJ	896
Hydroxyoxcarbazepine	Chiralcel OD	897
Ketamine	Daicel CA-1	898
Manidipine	Chiralcel OJ	899
Metoprolol	Chiralcel OD (Daicel)	900
Mianserin	Microcrystalline triacetate cellulose	901
Nimodipine	Daicel OJ	902
Penbutolol	Chiralcel OD	892
Propranolol	Chiralcel OD	903
Tramadol	Chiralpak AD, Chiralcel OD	904
Warfarin	Chiralcel OC	905

are easy to prepare, the source is plentiful, they have a high loading capacity, they are very discriminatory for chiral compounds, they have a broad applicability, and they are very cheap.[891] Some applications are detailed in Table 7.13. Much of the literature now describes the use of commercially available modified celluloses and some of the more popular ones are listed in Table 7.14.

7.8.5 PROTEIN-BINDING FOR ENANTIOMERIC SEPARATION

Natural products, and especially proteins, occur in chiral forms and it is not surprising that they have been exploited for their ability to bind enantiomers to differing extents. Hence chromatographic columns have been prepared using drug binding proteins such as α_1-acid glycoprotein. A useful review on the use of protein-

TABLE 7.14
Some Commercially Available
Modified Cellulose Phases

Designation	Modified Form
CTA	Microcrystalline cellulose triacetate
CTB	Microcrystalline cellulose tribenzoate
Chiralcel OA	Cellulose triacetate
Chiralcel OB	Cellulose tribenzoate
Chiralcel OC	Cellulose trisphenylcarbamate
Chiralcel OD	Cellulose tris(3,5-dimethylphenylcarbamate)
Chiralcel OG	Cellulose tris(4-methylphenylcarbamate)
Chiralcel OJ	Cellulose tris(4-methylbenzoate)
Chiralpak AD	Amylose tris(3,5-dimethylphenylcarbamate)
Chiralpak AS	Amylose (S)-α-methylbenzylcarbamate

TABLE 7.15
Separation of Enantiomers on Protein Phases

Drug	Stationary phase	Reference
β-Antagonists	α_1-acid glycoprotein	907
Acenocoumarol	α_1-acid glycoprotein	909
Betaxolol	OD	910
Bupivacaine	α_1-acid glycoprotein	911
Cromakalim		912
Cyclophosphamide	OD	913
Disopyramide		259, 914, 915
Disopyramide	α_1-acid glycoprotein	259, 915
DN 2327		916
Doxazosin	α_1-acid glycoprotein	917
Fenoprofen		918
Flecainide	Ovomucoid	919
Flurbiprofen	Avidin, ovomucoid	920
Folinic acid	HSA	921
Homochlorcyclizine	Ultron ES-OVM	922
Ibuprofen	Ultron ES OVM	923
Ibuprofen	α_1-acid glycoprotein	917
Ibuprofen	Human serum albumin	908
Ifosfamide	OD	913
Lorazepam	Ovomucoid	924
Methadone		925
Midazolam metabolites	Ultron ES-OVM	926
Naproxen		927
Phenprocoumon	α_1-acid glycoprotein	909
Pimobendan	Sumichiral OA-4400	928
Pirarubicin	Lichrocart SupersherRP 8	929
Pirarubicin	Sumipax OA-2500I	930
Propranolol	Ultron ES-OVM	931
Propranolol	Avidin, ovomucoid	920
Propranolol	Cellulose-tris(3,5-dimethylphenylcarbamate)	276, 932
β-Blockers	Ovomucoid, cellulase	933
Tiaprofenic acid	Immobilized HSA	915
Trimipramine metabolites		934
Trofosfamide	OD	913
Verapamil	α_1-acid glycoprotein	935
Warfarin	Avidin, ovomucoid	920
Warfarin	α_1-acid glycoprotein	909

based commercially available chiral stationary phases — α_1-acid glycoprotein, BSA, HSA, ovomucoid, trypsin, chymotrypsin has been published.[906] Examples of separations on protein phases are shown in Table 7.15.

7.8.6 DERIVATIZATION WITH CHIRAL REAGENTS

An alternative to using special phases for the separation of enantiomers is to follow the classical technique used by Pasteur[936] by reacting the enantiomeric mixtures with a chiral reagent to form diastereoisomers which can be separated by

TABLE 7.16
Separation of Enantiomers on Achiral Columns After Formation of Diastereoisomers

Drug	Derivatizing Agent	Reference
Amlodipine	(–)-Menthylchloroformate	937
Albuterol (salbutamol)	2,3,4,6-Tetra-*O*-acetyl-_-D-glucopyranosyl isothiocyanate	938
Atenolol	(–)-Menthylchloroformate	939
Aminotetralin derivative	*R-(+)*-_-Methylbenzylisocyanate	940
Amphetamine	FMOC-L-prolyl solid phase	941
Amphetamine	Tetra-*O*-acetyl-β-D-glucopyranosyl	942
β-Blockers	(–)-Camphanic acid (*S*)-(–)-1-phenylethylisocyanate 2,3,4,6-tetra-*O*-acetyl-β-D-glucopyranosyl isothiocyanate	943
Beclobric acid	Amine of S-flunoxaprofen	944, 945
Betaxolol	(+) or (–)-1-Naphthylethyl isocyanate	946, 947
CS 670 active metabolites		948
Etodolac	(–)-_-Phenylethylamine	949, 950
Furprofen	*S*(–)-1-Phenylethylamine	951
Ibuprofen	*R-(+)*-_-Phenylethylamine	952
Ibuprofen	(–)-2-[4-(1-Aminoethyl)phenyl]-6-methoxybenzoxazole	953
Labetalol	(4*S-cis*)-5-Isothiocyanato-2,2-dimethyl-4-phenyl-1,3-dioxan	954
Metoprolol	*S*(+)-1-(1-Naphthyl)ethyl isocyanate	955
Naproxen	4-Nitro-,4-(*N,N*-dimethylaminosulfonyl)-7-(3-aminopyrrolidin-1-yl)-2,1,3-benzoxadiazole-4-(aminosulfonyl)-7-(3-aminopyrrolidin-1-yl)-2,1,3-benzoxadiazole	956
Methocarbamol	(*S*)-(+)-1-(1-Naphthyl)ethylisocyanate	957
Mexilitine	*N*-Acetylcysteine	958
Propranolol	*S*-2-Octanol	959
Pirprofen	L-Leucinamide	960
Sotalol	(–)-Menthylchloroformate	961
Sotalol	*S*-(–)-_-Methylbenzylisocyanate	962
Tranylcypromine	*N*-Acetylcysteine	963
Warfarin	(–)-1-Menthylchloroformate	964

conventional physicochemical means, including nonchiral chromatography. Some of the enantiomers that have been separated in this way are listed in Table 7.16.

7.8.7   CHIRAL MOBILE PHASES

The chiral properties of the chromatographic system can be invested in the mobile phase rather than the stationary phase, for example by including a chiral counter-ion in ion-pair chromatography, as described in analyses for propafenone enantiomers,[965] or by modification of the mobile phase with exclusion agents such as β-cyclodextrin, which are used for analysis of tipredane,[966] phenylthioproline,[966] barbiturates,[967] and phenytoin.[968] Hsie and Huang used prior purification on normal HPLC and automatic transfer to a ligand-exchange column with subsequent chromatography using a chiral mobile phase octyl-L-prolinamide[969] for the analysis of phenytoin metabolites.

Mitchell and Clark[970] have recently investigated the separation of enantiometric barbiturates using β-cyclodextrin in the mobile phase, with a view to establishing structural relationships and chromatographic behavior.

TABLE 7.17
Chiral Separations in Capillary Electrophoresis

Drug	Chiral agent	Reference
Bupivacaine	Modified cyclodextrin buffers	973
Arylpropionic acids	Maltodextrins	974
Cefalosporins	Maltodextrins	974
Epinephrine	Methyl-β-cyclodextrin	975
Amphetamines	Micelles	976
Leucovorin B	Bovine serum albumin	977
Clenbuterol	Cyclodextrins as buffer additives	978
Picumetrol	Cyclodextrins as buffer additives	978
Epinephrine	Cyclodextrin	979
Isoprenaline	Cyclodextrin	979
Tryptophan	Proteins	980
Benzoin	Proteins	980
Pindolol	Proteins	980
Promethazine	Proteins	980
Warfarin	Proteins	980
Warfarin	Polyimide Methyl-β-cyclodextrin	981
ß-Blockers	Methyl-β-cyclodextrin	982
Chlorpheniramine	Micelles, cyclodextrin	983

Valtcheva et al.[971] have described the use of an enantioselective enzyme, cellobiohydrolase I, as a binding protein added to the mobile phase to separate enantiomers of β-blockers.

7.8.8   CHIRAL DETECTION

Goodall et al.[972] have described an interesting device to determine the ratio of *R* and *S* isomers of ibuprofen in urine. The sample is analyzed and quantified for ibuprofen in the usual way with the effluent from the column being monitored by parallel UV and polarimetric detectors. The signal from the polarimetric detector, in conjunction with the signal from the UV detector can be used to determine the ratio and hence the amount of each isomer.

7.8.9   CAPILLARY ELECTROPHORESIS

Several groups have described successful separation and analysis of enantiomers using capillary electrophoresis and these are summarized in Table 7.17.

8 Saturation Analysis: Radioimmunoassay and Other Ligand Assays

CONTENTS

8.1 INTRODUCTION

The main topic of this chapter is the use of radioimmunoassay (RIA) for the analysis of drugs; RIA being a special case of saturation analysis.

There is an amusing story about how the medieval monks used the principles of saturation analysis to determine the population of Paris. The analyst simply positioned himself on a street corner and counted the number of people passing as well as the number of monks, who would of course be easily recognized by their distinctive robes. The total number of monks in Paris would already be known from the church records, and assuming that on a typical day the monks all roamed free in the city, then the population was simply the number of people counted divided by the number of monks counted and multiplied by the total monk population. For example, the analyst could decide to count off the first 100 people he saw and count the number of these wearing the priestly garb as being say, 3. If he already knew there were 300 monks in Paris, then the population of Paris would be calculated as $100 \times 300 \div 3 = 10,000$.

For the more familiar analytical problem described in this book, saturation analysis may be illustrated by the following example. Imagine a large bucket of water, the volume of which we wish to measure. Add to the bucket a radioactive, water-soluble compound which mixes completely with the water. Take a fixed-volume sample of this water and measure its radioactivity content. The total volume of water in the bucket will then be the volume of the sample multiplied by the amount of radioactivity added, divided by the radioactivity of the sample.

The essential components of any saturation analysis, therefore, are a means of labeling the drug molecule, a saturable compartment and a means of separating the components in the saturated compartment, from those not in the compartment. A very simple type of this assay is that described for the determination of methyl mercury, using a substoichiometric reaction in the presence of a labeled mercury compound and counting the activity of an organic extract.[984]

For RIA, the saturable compartment is a specific antibody for the material to be measured (the antigen), the label is generally a radiolabeled tracer of the same substance, and the method of separation is simply by virtue of the fact that the antigen-antibody complex is usually precipitated, leaving the uncomplexed entity in solution.

RIA was first described by Yalow and Berson,[985] who extended their work on insulin antibodies to devise a technique for quantitative analysis of the pancreatic hormone. Insulin is a relatively large molecule and when injected into an animal (human or laboratory) will cause the natural defense mechanism of the body to raise antibodies against the foreign invader. The animal is then bled to provide serum containing the antibodies which can be used as reagents in subsequent assays. If the serum is rich in insulin antibodies, it can be diluted, usually with a physiological buffer or blank serum to provide a large volume of reagent suitable for many thousands of assays. To perform the analysis, the antiserum is chosen so that there is only enough to react with a fraction of the insulin in the unknown sample. The radioactive tracer is usually insulin which has been labeled with ^{125}I or [131]I. In Yalow and Berson's original method, the insulin-antibody complex was separated from unbound insulin by paper chromatoelectrophoresis. This was a time-consuming procedure and when the technique became popular as an analytical method, the use of a mild absorbent such as dextran-coated charcoal, replaced the paper chromatoelectrophoresis step. The final step was to measure the amount of radioactivity in one or both of the separated phases. In constructing calibration curves using known amounts of insulin added to control sample, the radioactivity was expressed in suitable terms for plotting against the total amount of insulin in the system.

At about the same time that Yalow and Berson were developing RIA for insulin, Ekins was describing a similar technique for the small molecule, thyroxine.[986] In Ekins' method, however, the reagent was the naturally occurring protein, thyroxine-binding globulin; for the technique Ekins used the term saturation analysis. In the early days of the development of saturation analysis, it was the field of endocrinology that consistently led the way and the next few years saw the development of RIA for growth hormone, follicle-stimulating hormone, and luteinizing hormone, as well as a method for determining cortisol in plasma using another naturally occurring, specific binding protein, corticosterone-binding globulin, or transcortin.[987]

The use of corticosteroid-binding globulin enjoyed considerable popularity for the assay of plasma or serum cortisol in the 1960s due to the ready availability of the protein and the relatively simple apparatus needed to perform the assay. The method could also be applied to synthetic steroid hormones such as prednisolone and it is worth pointing out at this stage that the tracer need not be the same as the compound being measured; it is only necessary that the tracer and the analyte compete for the same binding sites on the protein.

The extension of such methods to drugs in general, however, was still hampered by the fact that the competitive protein binding methods were limited to the naturally occurring proteins; small molecules are generally not immunogenic and the reagent for RIA was not so easily prepared for drugs as it was for large molecules such as the polypeptide hormones. The next major development was again in the endocrinological field, where it had been shown that if a small molecule, such as testosterone, was covalently linked to a large molecule, such as bovine serum albumin or polylysine, then antibodies could be raised to this compound that would also act as antibodies to the small molecule alone. Although the first steps in this process were reported for steroids in 1957,[988] it was not until 1969 that steroid RIA methods really began to find widespread acceptance after Abraham prepared a conjugate of estradiol and bovine serum albumin by linking the 17β-hydroxyl group, through a succinic acid, to amino groups in the protein.[989] The next few years saw a large number of assays described for the steroid hormones using these techniques. One problem that was soon recognized was that such antibodies prepared by this technique are relatively lacking in specificity since one of the features of the steroid is effectively masked in the complex and the resulting antibody will tend to react with compounds and metabolites which have variations in their structure close to the group used for linking to the protein. To overcome this problem of nonspecificity, it was often necessary to partially purify biological samples prior to using the sensitivity of detection afforded by the technique.

In order to enable antibodies of enhanced specificity to be prepared, the steroid chemists introduced the idea of linking the steroid molecule through positions distal to the normal functional groups,[990] and also by varying the length of the linking bridge.[991] This enabled the maximum exposure of total steroid molecule without losing the antigenic potential of being part of a large molecule. This philosophy of antigen design to provide highly specific antisera is now routine procedure in the field of drug RIA as exemplified in the development of an RIA for morphine. The selection of the point of linkage will depend on the metabolites of the compound expected to be present in the biological sample. Spector and Parker[992] first used the 3-carboxymethyl derivative, but since this is actually carboxycodeine then it is not surprising that the antibody so formed does not distinguish between morphine, codeine, and heroin. In order to keep the 3-position free to contribute to the antigenic specificity, 6-succinyl morphine was proposed, although there was still considerable cross-reactivity with codeine and heroin, and also with morphine-3-glucuronide.[993]

The various aspects of the radioimmunassay of drugs are discussed in the following sections. In addition, there are sections on saturation analysis techniques which do not depend on an antibody response (receptor assays) and techniques utilizing nonradiolabel techniques (fluorescence labeling, enzyme labeling).

8.2   RADIOIMMUNOASSAY OF DRUGS

8.2.1   SELECTION OF CONJUGATE

The first step in the total development of an RIA for drugs is to enlist the help of a synthetic chemist. The problem is to link the drug molecule through a reactive group to a large protein such as bovine serum albumin. Other proteins, such as thyroglobulin and chicken γ-globulin have been used, but bovine serum albumin, which is obtainable in large amounts, is commonly used as the carrier protein. The availability of free amino and carboxylic acid groups on the protein suggests that a peptide bond is the most suitable type of conjugate and it is therefore desirable to have the corresponding function on the drug or hapten, as it is termed in immunological nomenclature. Thus, conjugates can be directly prepared with drugs having a carboxylic acid function, such as prostaglandins or lysergic acid, and with drugs having an amino function such as amphetamines.[994] Suitable methods for such direct conjugation were described by Erlanger et al.,[988] Halloran and Parker,[995] and Korn et al.[996]

Reactive hydroxyl groups in the drug molecule can be used to form conjugates. This can be by direct reaction with phosgene to form the chlorocarbonate which can then form amides with lysine residues of the albumin, using a Schotten-Baumann reaction. Alternatively, the hydroxyl group can be reacted with succinic anhydride to form the half ester with the free acid available for conjugation with lysine residues. Reactive carbonyl groups can also be utilized to form appropriate derivatives for subsequent conjugation to protein.

Variations of these general procedures are used for individual drug assays. For example, a reactive group can be produced in the terminal glucose residue of digoxin by periodate oxidation and subsequent direct conjugation to albumin, or the whole triose may be removed to reveal a reactive hydroxyl group. The various methods of conjugation can be used to produce different bridge lengths and an optimal bridge length of four carbon atoms (i.e., succinic acid) has been suggested.[994]

All these reactions, however, utilize functional groups already present in the molecule, and masking such groups could well compromise the potential specificity of the assay. If possible, the linkage should be through a position which leaves the maximum number of antigenic determinants exposed. For example, Castro et al.[997] prepared conjugates for phenobarbitone by conjugation through a 4-aminobutyl analog at position 5 and obtained antisera of high specific activity. For the development of RIAs, however, such analogs are unlikely to be available in sufficient quantities. In this area, it will be fruitful to consult with the originating chemist; it is likely that a number of analogs of drugs will have been prepared as part of a series and it possible that a suitable analog can be chosen for producing specific antibodies. This is particularly so for the newer drugs, where the pharmaceutical company is well versed in procedures for producing closely related compounds.

The final chemical parameter that appears important in the quality of the antisera is the number of hapten residues linked to each protein molecule. This can be anything from two or three up to nearly a hundred. There is some evidence to suggest that the smaller the number, the more immunogenic is the conjugate, although James and Jeffcoate[998] suggest that the number of moles of steroid per mole of albumin has little influence on the affinity or specificity of the resulting antisera. Robinson et al.,[994] however, claimed that between 20 and 30 substitutions of drug residues per molecule of bovine serum albumin produce the optimum immunogen. The position of attachment on the protein molecule may also be important for specificity; an antiserum for 9-desglycinamide-8-arginine vasopressin raised using site-specific attachment of the drug to keyhole limpet hemocytanin was claimed to be more specific than the serum raised against conventional drug-thyroglobulin conjugate.[999]

8.2.2 RAISING ANTIBODIES

Once prepared, the hapten-protein conjugate is injected into animals to raise antibodies. The details of suitable procedures for raising antibodies are more properly found in textbooks of immunology than in this type of book. However, it is sufficient to comment that the immune response is by no means predictable, and there is no precise immunization schedule, nor any particular animal species that can be recommended. It is usual to inject animals, first with a primary dose of conjugate, normally about 100 mg, and then with periodic booster injections to prolong exposure of the animals to antigen. An immune response should appear in about 2 months. Animals are bled and the serum is tested for the presence of antibodies to the hapten. If there is a large response, then it will be necessary to dilute the serum; the measure of the strength of the antisera is the degree of dilution necessary before a solution is obtained which binds 50% of the test antigen. Although dilutions of several thousand are common for RIAs of large proteins, as far as drugs are concerned, antisera with dilutions of more than 2000 are rare. For this reason, larger animals, such as sheep or goats, have found favor, because the bleeds will provide large volumes of reagent. This becomes very important, not just for reasons of economy, but because every new bleed has to be reassessed for specificity and titer, due to the irreproducible nature of the immune response.

Although clear guidelines cannot be established for optimum antibody production, normal immunological practice suggests that it is desirable to use an animal of species and strain known to respond well, in good health, and not subject to other antigenic stimuli. So many factors are thought to contribute to the successful raising of antibodies, and yet the testing is so tedious, that few detailed studies have been carried out to elucidate any sort of clear rules.[1000]

Commercial development of RIA, particularly for drugs in widespread use, means that clinical chemistry screening or monitoring laboratories are rarely faced with the problem of raising their own antisera. Nevertheless, the specificity of commercially available material should be checked, especially if the laboratory adapts the material for operations other than those intended in kit forms.[1001]

A successful antibody to a particular drug or hapten raised by conventional methods does not constitute a single chemical entity, but will comprise a range of proteins having different characteristics of binding to the drug. This is the reason why antisera may differ from time to time, even when obtained from bleeds from the same animal. It is possible to increase the specificity of the reagent by fractionation of the antisera to produce the pure protein with a very specific response. Like all proteins, this entity can also be cloned to produce a plentiful supply of antibody with a consistent profile. Such monoclonal antibodies have been used in commercial kits to ensure brand consistency, and to enable highly specific reagents to be obtained.[1002-1004]

8.2.3 THE RADIOLABEL

The early RIA techniques for protein hormones invariably used ^{131}I and later ^{125}I as the radiolabel. This was because a sample of protein could be readily iodinated without loss of immunoreactivity, and the iodine isotopes, being γ-emitters, could readily be counted in γ-counters without recourse to liquid scintillation techniques. However, iodinationation of drugs, which are small molecules, brings a considerable change to the structure and properties of the drug molecule, and such a generally iodinated tracer will be unsuitable. However, the iodine isotope can be used for small molecules if the tracer can be made similar in structure to the conjugate used as the antigen. For example, iodine or an iodine-containing small moiety such as iodophenyl, can be introduced at the same position as the hapten is linked to the protein. Robinson et al.[994] attempted this ploy using tyrosine methyl ester, histamine, and a copolymer of glutamic acid, lysine, alanine, and tyrosine as the iodinatable tags but concluded that the resulting labeled complexes did not have an immunoreactivity comparable with unmodified drugs and could not be used for RIA. Nevertheless, successful RIAs for drugs have been described using the iodine label, mainly by incorporating the label in the link in the carrier protein or in the protein itself, so the drug-protein conjugate is also used as the tracer (Table 8.1).

The radiolabel that would be ideal for drug assays would be ^{14}C, and such RIAs have been described for diazepam,[1026] methadone,[1027] and phenytoin.[1028] Often this tracer is prepared for use in metabolic studies and this isotopic form is, for our purposes, identical to the unlabeled drug. Unfortunately, ^{14}C-labels are limited in their specific activity and when sufficient label is added to the test system for acceptable counting statistics, the mass of added drug is large relative to the amount to be measured. For these reasons, the label which has found the greatest acceptance for drug RIAs is tritium. Tritiated drugs are, in fact, easier to prepare than ^{14}C-labeled drugs because the former can be obtained by tritiation of the original molecule, whereas the latter needs

TABLE 8.1
Radioimmunoassay of Drugs Using Radioisotopes of
Iodine to Label the Tracer

Analyte	Nature of labeled tracer	Reference
4-Acetamidobiphenyl	Tyramine-hemisuccinamido	1005
Barbiturates	3-Iodobutyric acid	1006
Bencyclane	Iodo-4-Hydroxybencyclane	1007
Bromovinyluracil	—	1008
Buprenorphine	Iodinated drug	1009
Bupropion	Iodinated drug	1010
Ceronapril	Iodinated drug	1011
Cyclosporin	Iodinated drug	1012
N',N'-Diacetyldibenzidine	Iodinated drug	1013
Dihydrodigoxin	Iodinated ribonuclease	1014
Epimethyltestosterone	Iodinated histamine-steroid conjugate	1015
Morphine	Iodinated drug	1001
Meobentine	Iodinated drug	1016
Nicergoline	Iodinated drug	1017
Penicillin	Bovine serum albunin	1018
Phencyclidine	Iodinated drug	1019
Phenothiazines	Tyrosine methylester	1020
Plicamycin	Iodinated drug	1021
Sodium chromoglycate	Monotyramide	1022
Spiraprilate	Iodinated drug	1023
Vancomycin	Iodinated drug	1024
Zidovudine	Iodinated derivative	1025

original synthesis to incorporate the C atom. Tritiated molecules prepared in this way are often unsuitable as a tracer in metabolic studies because the tritium may be exchangeable, but this is seldom a problem with RIA methods. Tritium has a much longer half-life than either ^{131}I or ^{125}I, and therefore, once prepared, has a long shelf-life. RIAs using tritium have been developed for almost all types of drug (Table 8.2). The main disadvantage of tritium and ^{14}C is that, since they are β-emitters, liquid scintillation techniques are needed for the measuring step.

8.2.4 OPTIMIZATION OF ASSAY

The heart of the RIA procedure is the incubation of the antiserum reagent, the tracer, and the unknown. It is sensible to assume that the method will be most useful for quantitative analysis if amounts of these three are present such that a small change in the quantity of the unknown will produce a measurable change in the distribution of radioactivity between the free and bound fractions. Figure 8.1 shows the simplest form of plotting the relationship between the bound radiactivity and the absolute amount of hapten present. In this idealized curve, the change in distribution is relatively small for a tenfold increase in hapten concentration in the distribution of radioactivity as the nonbound fraction increases; at the other extreme (i.e., at low concentrations of hapten), the actual measurement of the distribution becomes susceptible to small errors in technique. Although various ways of plotting the binding curves have been proposed

TABLE 8.2
Drugs Analyzed by Radioimmunoassay Using Tritiated Ligands

Drug	Ligand	Sensitivity	Reference
Amphetamines			
Pseudoephedrine	Pseudoephedrine	2.5 ng	1029
Angiotensin-converting enzyme inhibitors			
Spirapril	Spirapril		1023
Zabicipril	Zabicipril	1 ng ml^{-1}	1030
Zabiciprilate	Zabiciprilate	1 ng ml^{-1}	1030
Antibiotics			
Alacinomycin	Alacinomycin	1 pmol	1031
Acyclovir	Succinylacyclovir		1032
Fortimicins	Succinimidyl propionate	0.2 ng	1033
Tetracycline	Tetracycline	20 ng ml^{-1}	1034
Vancomycin	Succinimidyl propionate	4 ng ml^{-1}	1024
Antihistamines			
Terfenadine	Terfenadine	0.25 ng ml^{-1}	1035
Antihypertensives			
Tiamenidine	Tiamenidine	10 pg ml^{-1}	1036
Clonidine	Clonidine	10 pg	1037
Antineoplastics			
Bruceantin	Acetylbruceantin	1 ng ml^{-1}	1038
Cytarabine	Cytarabine	1 ng ml^{-1}	1039
Methotrexate	Methotrexate	55 mM	1040
Vinblastine	Vinblastine		1041
Vincristine	Vincristine		1041
Barbituates and anticonvulsants			
Phenobarbitone	Phenobarbitone	0.2 pmol	1042
Phenytoin	Phenytoin	0.5 ng	1043
β-Antagonists			
Acebutol	Acebutol	10 pg	1044
Diacetolol	Diacetolol	60 nm	1044
Propranolol	Propranolol	0.12 pmol	1045
Benzodiazepines			
Chlordiazepoxide	Diazepam	2 ng ml^{-1}	1046
Diazepam	Diazepam	0.1 ng ml^{-1}	1047
Flunitrazepam	Methylflunitrazepam	0.15 ng ml^{-1}	1048
Oxazepam	Diazepam	10 ng ml^{-1}	1049
Cannabinoids			
d^9-Tetrahydrocannabinol	^{8}-Tetrahydrocannabinol	2 ng ml^{-1}	1050
Ergot alkaloids			
Methylergometrine	Dihydromethylergometrine	0.5 ng ml^{-1}	1051
Dihydroergotoxine	Dihydroergocryptine	0.5 ng ml^{-1}	1051
Cabergoline	Cabergoline	12 pg ml^{-1}	1052
H$_2$-Antagonists			
TG-41	N-Acetyl TG-41	3 ng ml^{-1}	1053
Hyperglycemics			
Gliclazide	Gliclazide	0.1 μg ml^{-1}	1054
Neuroleptics			
Chlorpromazine	Chlorpromazine	0.75 ng	1055
Fluphenazine	Trifluoperazine	0.25 pg ml^{-1}	1056
Haloperidol	Haloperidol	1 ng ml^{-1}	1057
Perphenazine		0.25 ng ml^{-1}	1058

TABLE 8.2 *(continued)*
Drugs Analyzed by Radioimmunoassay Using Tritiated Ligands

Drug	Ligand	Sensitivity	Reference
Opiates			
Alfentanil	Alfentanil	50 pg	1059
Buprenorphine	Buprenorphine	50 pg ml^{-1}	1060
Hydrocodone	Dihydromorphine	10 ng ml^{-1}	1061
Hydromorphone	Dihydromorphine	2.5 ng ml^{-1}	1061
Methadone	Methadone	3 ng ml^{-1}	1062
Phencyclidine	Phencyclidine	0.5 ng ml^{-1}	1063
Sufentanil	Sufentanil	0.5 ng ml^{-1}	1059
Parasympatholytics			
Atropine	Atropine	9 nm	1064
Hyoscamine	Atropine	9 nm	1064
Prostaglandins			
Prostacyclin	6-Oxoprostaglandin F$_{1-}$		1065
Steroids			
Ethinylestradiol	Ethinylestradiol	12.5 pg	1066
Flunisolide	Flunisolide	20 pg ml^{-1}	1067
Norgestrel	Norgestrel		1068
Prednisolone	Prednisolone		1069
Tricyclic antidepressants and related compounds			
Amitryptiline	Nortryptiline	1 ng ml^{-1}	1055
Amitryptiline	Amitryptiline	1 ng ml^{-1}	1070
Bupropion	Bupropion	0.6 ng ml^{-1}	1010
Clomipramine	Clomipramine	0.2 ng ml^{-1}	1071
Desipramine	Desipramine		1072
Doxepin	Imipramine	2.5 ng ml^{-1}	1073
Imipramine	Imipramine	1 ng ml^{-1}	1070
Imipramine	Desipramine	0.1 ng ml^{-1}	1055
Nomifensine	Nomifensine	0.3 ng ml^{-1}	1074
Nortriptyline	Nortriptyline	1 ng ml^{-1}	1055

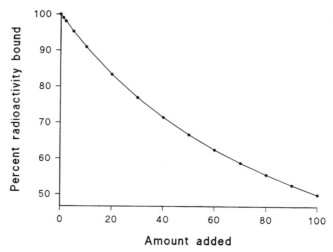

FIGURE 8.1 Simple plot of relationship between percentage binding and amount of added analyte in saturation analysis.

to help optimize the usable range of RIA, Ekins has pointed out that these are more apparent than real and sophisticated computer routines are essential for full optimization of RIA.[1075] In general, however, it is agreed that the RIA is best operated under conditions where approximately half of the specifically bound radiolabel is displaced, and the useable range should be of an order of magnitude that is above and below the concentration of drug at which this happens. Concentrations outside this range should be reassessed using diluted sample or reagent.

8.2.5   THE SEPARATION STEP

For traditional RIA, it is not possible to measure the bound and unbound label directly and the two fractions normally have to be separated prior to using standard counting techniques. Many separation techniques have been used including different migration methods, such as paper chromatoelectrophoresis[1076] and gel filtration,[1077] absorption methods, such as charcoal[1078] and silicates,[1079] fractional precipitation with salts[1080] or organic solvents,[1081] double antibody methods,[1082] and solid phase methods.[1083] However, the most popular techniques for drug RIA are absorption methods, solid phase methods, and the double antibody method. The method chosen should be rapid enough to effect the separation without altering the competitive equilibrium situation set up in the assay.

The use of activated charcoal was introduced by Ekins[987] for separating free thyroxin from the thyroxin bound to thyroid-binding globulin in the competitive protein-binding assay for this hormone. The method was later applied to the RIA of insulin.[1084] The charcoal absorbs small organic molecules, and the charcoal is readily separated from the soluble fraction containing hapten bound to the antibodies by centrifugation; the supernatant can then be counted for radioactive content. The nature of the charcoal used needs to be well controlled and reproducible. Activated charcoal with particle size up to 60 μm is generally satisfactory.[1085] However, untreated activated charcoal is usually too severe for direct use; it may bind the antibody itself or strip the hapten from the antibody. To overcome this, Herbert et al.[1085] suggested that the charcoal be coated with dextran of an appropriate molecular weight to provide pores through which the free hapten will pass and be bound, but will exclude the larger antibody-hapten complex.

In solid phase methods, the antibody is covalently bound to the inside of the assay tube.[1086] After the assay has been carried out up to the separation step, the supernatant is simply poured off, leaving the bound label attached to the antibody on the tube. The tube can then be counted directly for γ-emitters, or after the addition of appropriate scintillation fluid for β-emitters. Alternatively, the antibody may be linked to magnetizable particles which can be subsequently separated with a magnet from the supernatant containing free hapten.[1077]

The notable exception to the need to separate the bound and unbound components for RIA is by using a technique known as scintillation proximity assay.[1087,1088] In this procedure, a reagent consisting of derivatized inorganic fluoromicrospheres based on yttrium silicate, are impregnated with a fluorophore and coated with a second antibody specific for the primary antibodies. Only the radioligand that is bound to the secondary-primary complex is close enough to the fluoromicrospheres to provide

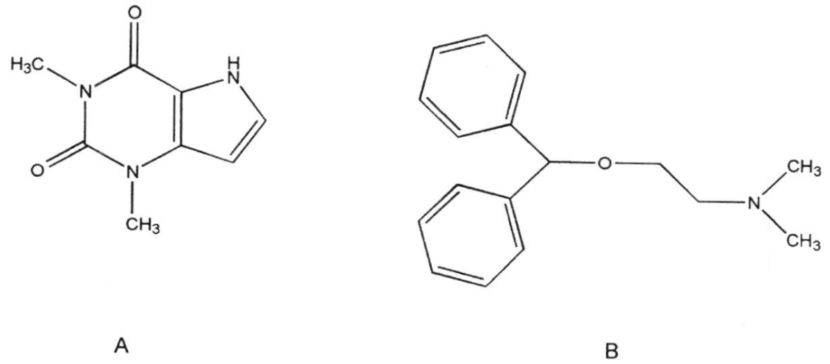

A   B

FIGURE 8.2   Structures of (A) theophylline and (B) diphenhydramine.

energetical particles to activate the fluorescence for counting. A detailed evaluation of such a technique for the assay of low levels of ranitinide has been reported.[1089]

The antibody complex is usually larger than the antibody itself; for drug RIA, however, the complex is still relatively small and remains in solution thus the classical precipitation methods of immunology do not apply. But, if a second antibody which has been raised against the globulins of the animal in which the drug antibody has been raised, is added, then the resulting precipitate can readily be separated by decantation of the free fraction.[1082]

8.2.6   SPECIAL PROBLEMS AND PITFALLS

In contrast to chromatographic, spectrophotometric, and many other methods depending on well-understood physical phenomena, RIA and related saturation techniques always have surprises for the unwary. It is this unpredictability that constitutes a hazard in the development and use of RIA as an analytical tool. For some types of assay, for example receptor assays and those using naturally occurring globulins such as transcortin or thyroid-binding globulin, the unpredictability on the part of the binding protein is perhaps less, because once the method of preparation is determined, the same protein is always obtained. However, when antibodies are raised in individual animals to foreign haptens, there is no guarantee that the resulting antiserum will be consistent in its binding characteristics. For this reason, large animals are desirable for raising antibodies, because there is a large amount of consistent binding reagent available for many thousands of reproducible assays. An alternative approach is to use antisera obtained by pooling samples from a large number of animals. This approach may have the effect of losing specificity, because the resulting pool will contain antibodies of differing specificities, but at least such a serum should provide consistency.

An unpredictability that is met for immunoassay, competitive protein binding assays, and receptor assays is the cross-reactivity of congeners, or sometimes totally unrelated structures. It has been shown, for example that diphenhydramine interferes with one RIA procedure for theophylline even though these two compounds have quite different structures (Figure 8.2),[231] and hordenine has been shown to interfere with some morphine immunoassays.[1089]

Interferences due to similar structures are more normal; it has been shown that steroids in cord blood (including corticosteroids, progesterone, and testosterone) may have digoxin-like immunoreactivity.[1091] Likewise, phenylpropanolamine and ephedrine gave false positives in several immunoassay methods for amphetamines.[1092] Tertiary amines may also cross-react in amphetamine-targeted assays.[1093] Many important drugs have structures which are based on the structures of natural hormones. Antibodies raised to such haptens will often cross-react with their natural counterparts, and this can obviously cause problems in the analysis of real samples. Haak et al.[1094] investigated the characteristics of antisera raised against various synthetic glucocorticoids and their results indicate that those synthetic compounds more closely related to the natural glucocorticoids (prednisolone and 6β-methylprednisolone) resulted in less specific antisera than those raised against the more heavily substituted ones (i.e., betamethasone, dexamethasone, beclamethasone, and triamcinolone). The specificity of assays may also depend on the disease state of the patient; for example, the applicability of digoxin immunoassay specificity for patients with chronic renal failure or hepatic cirrhosis has been questioned.[1095]

Use of cloned material has been proposed to increase the required specificity, for example, in the Boehringer homogeneous assay for digoxin, which relies on the competition between a digoxin-labeled peptide fragment of β-galactosidase and digoxin for digoxin antibodies.[1096] However, even monoclonal antibodies are no guarantee against cross-reactivity in unusual circumstances such as in the assay of cyclosporin in the plasma of cardiac allograft patients where endogenous material peculiar to these patients cross-reacted to the extent of 36%, using a monoclonal antibody raised against cyclosporin.[1097]

Because the RIA procedure is relatively simple (given the appropriate dispensing and counting equipment), the technique is suitable for packaging into commercial kits. This should give results which are reproducible from laboratory to laboratory and also over a relatively long period, as the manufacturers will only market a kit which has a large back-up in terms of reagents. Nevertheless, it is wise to characterize kits received from different manufacturers by carrying out appropriate control assays prior to using them, and before changing suppliers.

In developing any original RIA for drugs, that checking for specificity must be part of the evaluation process. Unfortunately, the reactivity of other compounds cannot be generally predicted; hence, if it desired to perform the assay without any pre-extraction, then the only recourse is to assess the cross-reactivity of individual candidates. Even when this is done there is a potential misinterpretation if the concentration of the various components is ill chosen. Thus it is necessary to test each component over a range of concentrations or dilutions. Often during the development of a drug, the nature and extent of metabolism is unknown and consequently it is impossible to test the cross-reactivity of metabolites in a developed assay. In these cases, it is desirable to have a reference method, of high specificity, which can gauge the extent of cross-reaction of metabolites by carrying out parallel assays for a range of time points following administration of the drug. Thus, a rather tedious, but extremely specific assay for a novel sleep inducer was developed and used to assess the specificity of two antisera prepared against the same compound.

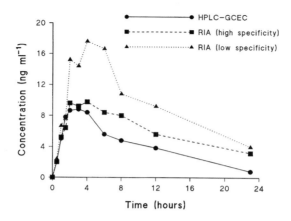

FIGURE 8.3   Determination of loprazolam in plasma of subjects receiving an oral dose (2 mg) of the drug, using a specific HPLC-GCEC method and two RIA methods.

The plasma concentration time profiles obtained using the reference method and using the two antisera in this study are shown in Figure 8.3. Antiserum 42 was raised against a conjugate linked to the drug in a position which maximized exposure of the antigenic determinates of the drug compared with the the conjugate used for antiserum 47 (Figure 8.4).[167]

The method most favored in specificity for reference is one involving chromatography (usually gas chromatography or HPLC) followed by mass-spectrometry using single-ion monitoring, as has been reported for opiates[1098,1099] and ethynylestradiol.[1100]

An interesting but unforeseen problem occurred in one of the author's own laboratories in the use of a radioimmunoassay for a novel diuretic. The same antiserum unaccountably gave differing results using different standard samples of the same drug, even though the samples had passed strict chemical and chromatographic quality control checks. The problem was finally tracked down to the presence of a cross-reacting impurity, too minute to be detected by the quality control assays, yet sufficiently immunoreactive to make the standard drug appear to be present in an amount greater than the weighed value. This problem cannot be overcome by using the same sample for dosing purposes as for calibration of the assay, because the impurity would not necessarily follow the drug through all the bodily processes.

Similarly, a problem was encountered in the radioimmunoassay of a hyperglycemic agent, where the radioimmunassay consistently gave twice the value for drug concentration in the plasma over that given by the corresponding HPLC analysis. The usual explanation is to write off the RIA method as being nonspecific, but the analyst responsible for the development and evaluation of the method had given no indication of such dramatic interference by any metabolites or endogenous material. This is a situation, known to all analysts, that generates much heart-searching, double checking of readings and scouring of equations to locate this factor of 2. Several years later, the author came to the conclusion that the answer was in the use of a racemic drug. The plausible explanation is as follows: the drug was administered as

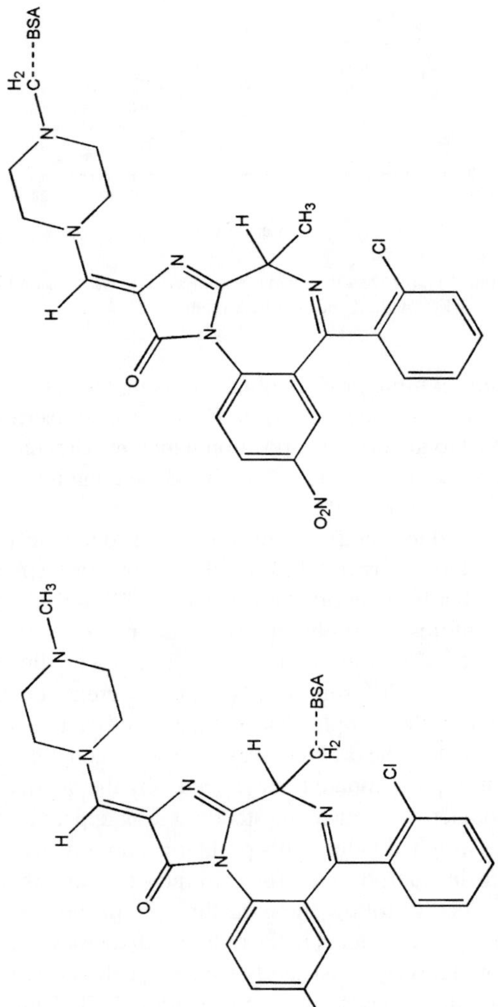

FIGURE 8.4   Structures of conjugates prepared for raising antibodies used for the determination of loprazolam by RIA.

the racemate, but only one entaniomer (the immunoreactive one) appeared in the plasma; the nonchiral HPLC assay gave the correct answer in that it would always give the total of drug, whichever form it appeared in; the RIA method, however, was essentially calibrated against half-strength standard and hence the signal in the calibration was only half of what it should have been, thus resulting in the reported RIA result being high by a factor of two.

The converse of this problem is to prepare chiral immunoassay reagents. However, the potential for using radioimmunoassays for chiral discrimination seems to have been little utilized, even though McKay et al.[1101] demonstrated *trans* vs. *cis* discrimination for doxepin by raising antibodies to a bovine serum albumin conjugate of *trans-N*-(2-carboxyethyl)desmethyldoxepin.

As detailed elsewhere, many drugs are administered as racemates and hence the immunological properties of the enantiomers need to be assessed if RIA is to be considered as an option for the analytical method. When drugs are administered as racemates, an enantiospecific assay such as RIA may be desirable, as indicated by Woedtenborghs et al.[1102] for the antihypertensive drug, nebivolol.

Another potential pitfall in radioimmunoassay is in the purity, both chemical and radiochemical, of the tracer used. If there is a chemical impurity, then the same kind of problem can occur as described above; if the tracer is not radiochemically pure, then the binding characteristics of the impurity may affect the assay.

8.2.7 APPLICATIONS OF RADIOIMMUNOASSAYS

Immunoassays find considerable application in forensic analysis. Consequently, methods are developed for small samples, often collected under unfavorable conditions. Thus, RIA methods have been developed for screening drugs of abuse using blood samples spotted onto filter paper, dried and punched out into disks, as for example in an assay for benzoylecgonine.[1103]

8.3 IMMUNOASSAYS USING A NONISOTOPIC LABEL

Immunoassays were established techniques in biochemistry and clinical chemistry long before the advent of radioimmunoassay, and its enthusiastic adoption for drug analysis. Some of the classical immunoassay techniques have found their way into drug immunoassay such as the use of the latex nephelometric immunoassay of theophylline in human serum,[1104] and in screening for drugs of abuse using the Roche ONTRAK latex agglutination system.[1105] However the main advances in immunoassay have focused on new methods of labeling molecules so that expensive equipment or hazardous radiolabels are not required.

8.3.1 FLUORESCENCE LABEL

In this technique, a labeled drug molecule is prepared by attaching a fluorescing compound to it. The most popular compound for this procedure was originally fluorescein and derivatives were prepared for tobramycin,[1106] propranolol,[1107] morphine,[1108] digoxin,[1109] digitoxin,[1110] phenytoin,[1111] gentamicin,[1112] and procainamide.[1113]

The immunoassay is performed in the usual way, separating bound drug from free and measuring the fluorescence in the free fraction. Thus, for the assay for gentamicin the double antibody method was used. For a number of drugs the absorption onto magnetizable particles has been described and for tobramycin[1106] the technique was to absorb the drug onto a dipstick.

The dau-TRAK Eclair System consists of competitive homogeneous radioimmunoassays which are available for several different drugs of abuse (benzodiazepines, cannabinoids) and which can be applied to urine with sensitivity values of 100 ng ml^{-1}.[1114]

8.3.2   FLUORESCENCE POLARIZATION IMMUNOASSAY

The simple fluorescent labeling technique has been almost completely supplanted by a variation proposed by Danliker et al.[1115] If a molecule is excited using polarized light, then the emitted light will be polarized in the same plane as the excited light, provided the molecule does not rotate in the time between excitation and emission. In solution, the free rotation of the molecule reduces the polarization of the emitted light and hence solutions do not usually exhibit fluorescence polarization. When bound to macromolecules, however, the rotation is slowed and the fluorescence remains polarized. When used in immunoassay, the fluorescent-labeled molecules that are displaced from an antibody have reduced polarization fluorescence. Thus, this method can be used without separating free and bound fractions, making the technique much simpler to use or to automate. Such a technique is generally termed a homogeneous immunoassay as opposed to heterogeneous assays where a separation step is required. A review of the technique described typical procedures for gentamycin, phenytoin, phenobarbital, and theophylline.[1116]

The fluorescence polarization immunoassay technique has been shown to be reliable and specific for screening urine samples for drugs of abuse. In particular, it was claimed that herbal drinks could interfere with such assays, but this was refuted by the analysis of 50 such preparations which showed no interference with either the screening method using fluorescence polarization immunoassay or the confirmatory TLC method;[1117] however, this study did not appear to investigate the urine of herbal tea drinkers. Table 8.3 lists a number of drugs for which fluorescence polarization immunoassay techniques have been developed.

In the initial stages of development and introduction of nonisotopic immunoassays, the advantages promised by circumventing the use of radioactivity were to be offset by the need to have relatively expensive specialized equipment instead of the ubiquitous scintillation counter. Recently, however, laboratories carrying out large numbers of assays have tended to commit their instruments to specific tasks, with the result that dedicated instruments were designed and marketed.

8.3.3   ENZYME LABEL
8.3.3.1   General Description

The enzyme-multiplied immunoassay technique commercially marketed by Syva Corporation as EMIT kits has proved a very popular method for the semiquantitative analysis of drugs of abuse,[1140] and has been extended to quantitative therapeutic drug

TABLE 8.3
Analysis of Drugs in Biological Fluids by
Fluorescence Polarisation Immunoassay

Drug	Reference
Amikacin	1118, 1119
Amphetamines	1120
Amphetamines	1121
Benzodiazepines	1122
Carbamazepine	1119, 1123, 1124
Cocaine	1125
Cortisol	1119
Cyclosporin	1126–1132
Digitoxin	1119
Digoxin	1119, 1133
Gentamicin	1118, 1119
Methotrexate	1134
Monoethylglycinexylidide	1135
Phenobarbitone	1119, 1124
Phenytoin	1136
Phenytoin	1119, 1124
Primidone	1119, 1124
Quinidine	1118
Teicoplanin	1137
Theophylline	1118, 1119, 1138
Thevetin B	1139
Tobramycin	1119
Valproic acid	1119, 1124
Vancomycin	1119

monitoring. In this procedure, the drug molecule is covalently linked to a stable enzyme such as glucose-6-phosphate dehydrogenase to provide the tracer. When the immunoassay procedure is carried out, the unbound tracer remains active, whereas the bound tracer loses its enzyme activity. Hence, rather than a separation step, the enzyme activity is measured after adding a suitable substrate. The sensitivity of the standard enzyme labeled assays is of the order of $\mu g\ ml^{-1}$ rather than $ng\ ml^{-1}$. For enzyme-labeled assays, the special reagents are the drug-enzyme complex and the antiserum; for specific drugs, these and other standards are supplied in kit form. Table 8.4 lists some of the procedures and their evaluation by comparison with other analytical techniques. One great advantage of enzyme-labeled assays is that all the steps of the assay can be performed in the same vessel, making it a homogeneous assay. Thus the technique is very suitable for conversion into ready-made kits for rapid use by nonlaboratory personnel.[1147-1149]

Although some EMIT procedures have been developed to a high degree of sophistication, the use of commercial kits may often be thought to be expensive and the laboratory with economies in mind may often find it worthwhile to adapt the kits, or to prepare their own bulk reagents to make the assay more economic to run. Thus, a Scottish laboratory developed cost-effective immunoassays for amphetamines,

TABLE 8.4
Some Enzyme-Multiplied Assays

Drug	Enzyme	Sensitivity	Reference
Antiepileptic drugs	Glucose-6-phosphate dehydrogenase	—	1141, 1142
Clenbuterol	Alkaline phosphatase	—	1143
Clenbuterol	Horseradish peroxidase	0.2 ppb	1144
Taxol	Alkaline phosphatase	0.2 ng ml^{-1}	1145
Vinpocetine	Alkaline phosphatase	0.1 ng ml^{-1}	1146

benzodiazepines, and methadone in urine by combining Syva EMIT reagents with a centrifugal analyzer to make a 100-test kit stretch to 2470 samples[1150] and a London laboratory managed to cut the costs of assays to a few pennies by using in-house reagents and alternative antibodies.[1151-1153]

The development of monoclonal antibodies has already been mentioned in the development of radioimmunoassays, and such reagents are equally applicable to nonisotopic labels.[1154] The Du Pont discrete clinical analyzer was used to assay cyclosporin rapidly and specifically using a monoclonal antibody to cyclosporin.[1155] Sabate et al. have carried out a comparison of RIA, EMIT, and other procedures for cyclosporin,[1156] including use of monoclonal antibodies. Poklis et al.[1157] described the use of EMIT and monoclonal antibodies in the analysis of amphetamine and methamphetamine. Some aspects of comparisons of EMIT methods with others are shown in Table 8.5. Sometimes commercial kits may be improved if the laboratory is prepared to commit the development time and effort necessary,[1158] although this would seem to defeat the purpose of using the commercially available materials.

Although most EMIT assays are claimed to be extremely sensitive and specific, where they are being used for the detection of drugs of abuse, rather than in pharmacokinetic studies, it is always wise to ensure back-up confirmatory tests are available for positive samples. For example, Olsen et al.[1199] showed that metabolites of chlorpromazine and brompheniramine may cause false positives in the urinary assay of amphetamine using an EMIT kit.

A variation on the enzyme-labeled assay is to label the substrate as in the substrate-labeled fluorescent immunoassay, a technique introduced by Burd et al.[1200] The drug is covalently linked to umbelliferyl-β-D-galactoside. When this complex is attacked by β-galactosidase, it produces the fluorescing entity, umbelliferone-drug complex (Figure 8.5). In the assay procedure, the umbelliferyl-β-D-galactoside-drug molecule competes with unlabeled drug for binding sites on the antibody so that the amount available for attack by β-galactosidase is proportional to the amount of the unlabeled drug present. The rate of increase of fluorescence is simply measured to determine the amount of drug in the sample. Substrate-labeled fluorescent immunoassay procedures have been described for gentamicin,[1200] tobramycin,[1201] phenytoin,[1202] theophylline,[1203] phenobarbitone,[1204] and amikacin,[1205] with commercial kits usually available from Ames Laboratories.

TABLE 8.5
Characteristics of EMIT Kits and Comparison With
Other Analytical Techniques

Analyte	Comparative method	Comment on EMIT	Reference
Benzodiazepines	TLC	Equivalent for screening	1140
	GC	Frequently overestimates oxazepam and nordiazepam	1159, 1160
	GC-MS	Excellent agreement	1161
	HPLC	Good correlation	1160
Cannabinoids	RIA, GC-MS	Detects metabolites and is preferred	1162
	GC-MS	Good agreement	1163, 1164
Carbamazepine	GC	Good agreement although epoxide may interfere; no clinical significance	1165–1168
	HPLC	Epoxide may interfere but no clinically significant differences	1169–1171
Clenbuterol	RIA	Good agreement	1171
Cyclosporin A	HPLC	Good agreement	1173
	HPLC	Good agreement	1174
	FPI, RIA	Good agreement	1175
Disopyramide	GC	Good agreeement, r = 0.97	1176, 1177
Ethosuximide	GC	Interchangeable	1178
Gentamicin	RIA	r > 0.95	1179
	Microbiological	r = 0.95	1180
α-Hydroxytriazolam	GC-MS	More sensitive	1181
Methotrexate	RIA	Good correlation	1182
	HPLC	Good correlation	1182
Morphine	GC	Similar sensitivity but less precise and accurate	1183
Phenobarbitone	HPLC	Several drugs cross-react, r = 0.83	1184, 1185
	Difference spectroscopy	Correlated well	1186
	GC	Good agreement, r = 0.95, methods generally interchangeable, no clinically significant difference	1178, 1187–1189
Phenytoin	HPLC	r = 0.94, p-hydroxy metabolite may interfere	1184
	GC	r > 0.95, methods generally inter-changeable although values may be falsely high in renal insufficiency	1178, 1187–1189
Primidone	GC	Good agreement with no clinically significant differences	1178, 1187, 1188
Procainamide	Spectrofluorimetry	Correlated well in therapeutic range	1190
	HPLC	Correlated well in therapeutic range, r = 0.98	1177, 1191
Quinidine	HPLC	Values 25% higher due to cross-reacting metabolites	1192–1194
Theophylline	HPLC	r = 0.985	1195
Valproic acid	GC with FID	Falsely elevated values	1196
	GC	Good agreement	1197
	HPLC	Satisfactory correlation	1198

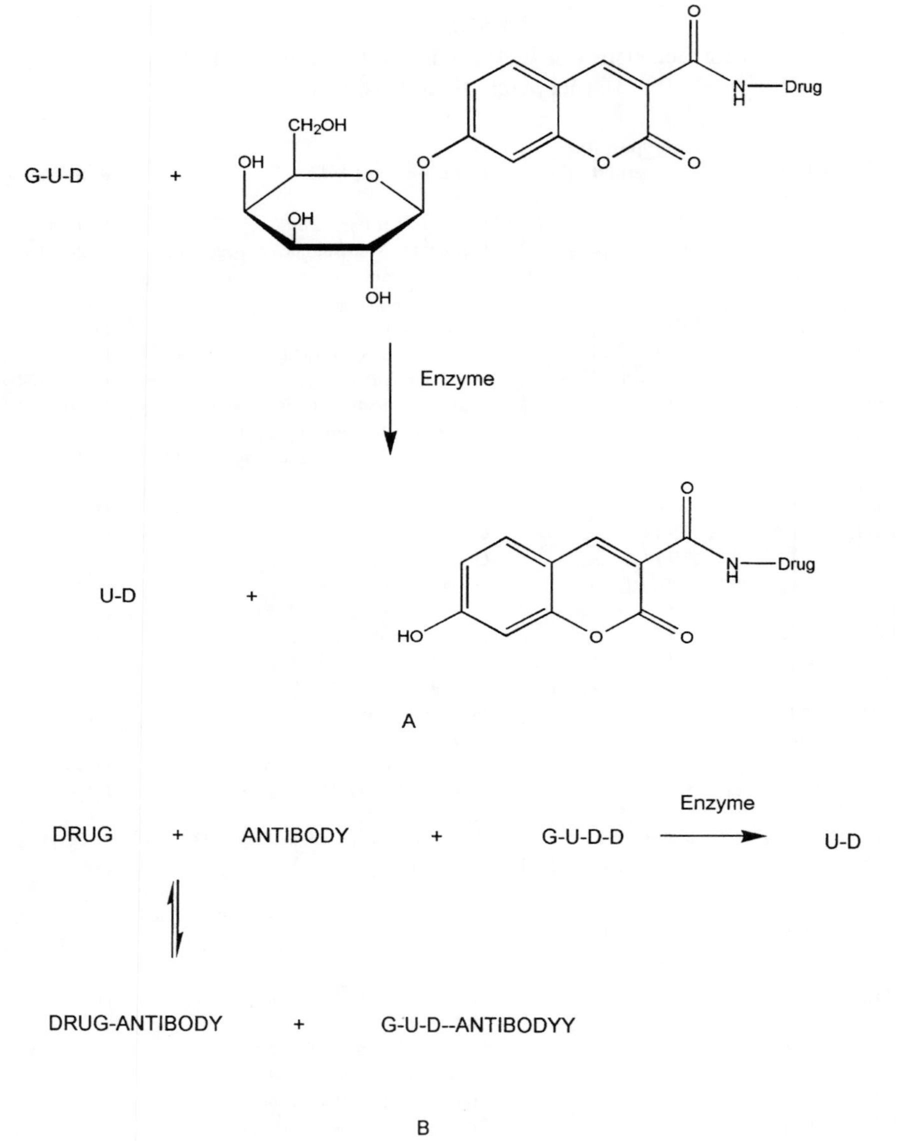

FIGURE 8.5   Reactions involved in substrate-labeled fluorescence immunoassays. (A) Conversion of nonfluorescent galactosyl-umbelliferone-drug (G-U-D) reagent to fluorescent umbelliferone-drug (U-D) by β-galactosidase; (B) competitive reactions in an assay system.

8.3.3.2   Solid-Phase Enzyme Immunoassay

A development of the enzyme labeled technique was to immobilize the enzyme on a solid support in the reaction vessel so that the complete assay can be carried out by monitoring the solution. The method has been most highly developed in the range

FIGURE 8.6   Free radical assay technique. (A) Signal obtained from a radical bound to immunoglobulin; (B) signal obtained from a freely rotating radical.

of ELISA (enzyme-linked immunosorbent assays) kits, such as for the screening with GC-MS confirmation of the tranquilizer chlorprothixene administered in subtherapeutic doses to horses.[1206] This assay actually used a commercially generic promazine kit with cross-reactivity for chlorprothixene. ELISA kits have also been available for delta-9 tetrahydrocannabinol,[1207] papaverine,[1208] isometamidium,[1209] tacrolimus,[1210] medroxyprogesterone,[1211] cephalexin,[1212] and dimethindene.[1213] ELISA procedures using monoclonal antibodies have been developed for theophylline,[1214,1215] and for thiabendazole.[1216]

As well as using the enzyme assays to assay pharmaceutical material, the method has also been used to investigate the location of specific binding sites on the albumin molecule, such as in the solid-phase competitive enzyme immunoassay using acetyl-cholinesterase as the label to determine benzylpenicilloyl groups derived from penicillin G by cleavage of the β-lactam ring.[1217]

ELISA kits have been used to determine the presence of β-agonists, clenbuterol, and salbutamol (after hydroysis of the conjugate with β-glucuronidase) in bovine urine. The ELISA kits comprised a microtiter plate coated with antirabbit IgG, anti-β-agonist antibody, and salbutamol-horseradish peroxidase conjugate.[1218]

An enzyme immunoabsorbent assay has been used for clenbuterol in bovine liver and urine, using hydroxyclenbuterol-alkaline phosphatase conjugate after solid-phase extraction of urine on C_{18} columns.[1143]

8.3.4   SPIN LABEL

This technique was introduced by Leute et al.[1219] in 1971 for morphine and was developed commercially by Syva as FRAT (free radical assay technique). A free radical such as nitroxide in solution when examined in an electron spin resonance spectrometer showed a simple three-line spectrum. This spectrum is seen even when the radical is linked to a drug molecule. However, when the drug-radical complex is bound to a large molecule such as immunoglobulin in a typical immunological reaction, its rotation in solution is slowed and the spectrum becomes considerably flattened. Thus, in the absence of added drug, all the drug-radical complex is bound and no sharp ESR signal is seen. With increasing amounts of added drug, the complex is displaced from the antibody and the sharp signal becomes apparent (Figure 8.6). For this technique, the major disadvantage was the need for an electron spin resonance spectrometer, whereas most other displacement methods were able to use standard spectrometers, scintillation counters, or fluorimeters. The technique does not appear to have been further developed, possibly because it offers no essential advantages over the more highly developed enzyme or radioassays.

8.3.5 COLLOIDAL GOLD

The commercially available TRIAGE immunoassay system uses a competitive reaction between a drug in a urine sample and the drug bound to colloidal gold for a specific monoclonal antibody. Unreacted drug was recovered using a membrane-bound antibody and examined for presence of a violet-colored band. In the absence of unlabeled drug in the sample, all the colloidal gold labeled drug is bound and there is no violet band. Thus the method is a rapid qualitative assay for drugs of abuse and has been applied to the simultaneous screening of phencyclidine, benzodiazepines, cocaine, amphetamines, tetrahydrocannabinol, opiates, and barbiturates in urine.[1220,1221]

8.4 IMMUNOASSAYS WITH OTHER METHODS

Immunoassays have been combined with other separation techniques with mixed success. Hochhaus et al.[1222] developed a selective HPLC-RIA for dexamethasone and its prodrug dexamethasone-21-sulfobenzoate sodium in biological fluids; fractions were collected then assayed by RIA with a detection limit of 0.5 ng ml^{-1}. Detection of cocaine using a flow immunosensor has been described.[1223] A column packed with Sepharose 4B coated with antibenzoylecgonine monoclonal antibody was used. The sample in phosphate-buffered saline was passed through the column and the displaced fluorophore labeled antigen was monitored at 520 nm (490-nm excitation) with a claimed limit of detection of 5 ng ml^{-1}.

Flow-injection electrochemical enzyme immunoassay was reported for theophylline using a protein A immunoreactor and p-aminophenyl phosphate-p-aminophenol as the detection system.[1224] Ethinyloestradiol and norethindrone were determined by radioimmunoassay following Sephadex LH-20 column chromatography;[1225] eluted fractions were assayed by standard RIA procedures with a sensitivity of 10 pg ml^{-1}. Immunoaffinity chromatography combined with gas chromatography negative-ion chemical-ionization mass spectrometry has been described for the confirmation of flumethasone abuse in horses[1226] with picogram amounts being detectable.

On-line immunochemical detection in liquid chromatography has been described using fluorescein-labeled antibodies.[1227] The eluate is mixed with polyclonal antidigoxigenin Fab labeled with fluorescein, next a column of immobilized digoxin removes free antibodies, and then measurement is by fluorescence. Sensitivities of 200 fmol digoxin and 50 fmol digoxigenin were reported.

8.5 RADIORECEPTOR ASSAYS FOR DRUGS

The previous sections have described saturation analysis where the binding protein is biosynthesized directly as a specific reagent for a drug. Radioreceptor assays are, in principle, identical with radioimmunoassays, the only difference is that the proteins are derived from existing natural sources. One advantage of this situation is that standard procedures can be developed for the preparation of the receptors, and

TABLE 8.6
Radioreceptor Assays for Drugs

Drug	Class	Tracer	Source	Reference
Nuvenzepine	Muscarinic	[³H]Pirenzepine	Rat cerebral cortex	1237
Timolol	β-antagonist	[³H](–)-GCP-12177	Rat reticulocyte membrane	1238
FK-506	Immunosuppressant	[³H]Dihydro-FK-506	Calf thymus	1239
Flunitrazepam	Benzdiazepine	[³H]Flumitrazepam	Rat brain	1240
Mepirodine	Calcium-channel antagonists	[³H]PN 200-110	Porcine coronary artery	1241, 1242
Clobazam	Benzodiazepines	[³H]Clobazam	Mammalian brain	1243
Mifepristone	Progesterone	[³H]16_-ethyl-21-hydroxy-19-norprogesterone	Human myometrium	1244
Deoxycortone	Corticosteroid	[³H]Prednisolone	Human liver cytosol	1245

reproducibility from laboratory to laboratory is better. The first assays of this type were the protein-binding assays described by Ekins[1228] for thyroxine, and by Murphy[987] for corticosteroids using proteins isolated from plasma. Specific proteins have also been described for steroids,[1229] adrenocorticotrophic hormone,[1230] opioid peptides,[1231] cyclic AMP,[1232] and γ-aminobutyric acid.[1233] Such receptors will also bind drugs of closely related structure and assays have been developed using some of these receptors. What was initially surprising was the discovery of receptors that appeared to bind specifically to synthetic compounds that appeared to have no counterpart in natural systems, such as the receptors for diazepam,[1234] neuroleptics,[1235] and β-agonists and antagonists.[1236] However, as more and more receptors are discovered by the pharmacologists, it is evident that drugs which bind to the receptors, whether as agonists or antagonists, do not necessarily have to resemble natural compounds in their basic structure. One advantage of this line of research into new drugs is that those very compounds that bind strongly to receptors can be readily assayed by receptor methods, although it is essential, as always, to verify the specificity of such assays before they can be used reliably for applications where specificity is a prerequisite. Table 8.6 lists some of the radioreceptor assays that have been developed based on a number of recently characterized receptor preparations.

8.5.1 BENZODIAZEPINE RECEPTOR ASSAYS

The benzodiazepine receptor was first identified by Squires and Braestrup[1246] and by Mohler and Okada,[1247] who showed that [³H]diazepam will bind to a specific fraction isolated from mammalian brain tissues. As pharmacological actions are assumed to occur at the molecular level, the sensitivity of assays using such receptors should be high, and the specificity lies in the fact that drugs from other classes and nonactive metabolites should not bind to the receptors. Hunt[1243] evaluated the binding of a number of benzodiazepines and their main metabolites to the membrane fraction of rat cerebral cortex. The principal metabolites of diazepam (*N*-desmethyldiazepam and oxazepam) and clobazam (*N*-desmethylclobazam) exhibit similar binding to the parent compound (Table 8.7). As the metabolites are themselves active, then it is

TABLE 8.7
**Displacement of [3H]Diazepam from Rat Brain Receptors by
Metabolites of Diazepam and Other Benzodiazepines[1247]**

Compound	Concentration Required to Displace 50% Label Under Standard Conditions (μg ml^{-1})
Diazepam	7.1
N-Desmethyldiazepam	8.8
Oxazepam	20
Nitrazepam	10
Loprazolam	7.2
Clobazam	170
N-Desmethylclobazam	210

claimed that the assay will be useful in evaluating total benzodiazepine activity in biological fluids. Mohler et al.[1248] and Braestrup et al.[1249] have reported that the ability of benzodiazepines to displace [³H]diazepam correlates well with their clinically recommended doses. However, in a review of published data, Mennini and Garattini[1250] concluded that no simple correlation exists between benzodiazepine activity and *in vivo* occupancy and the complex pharmacological spectrum of activities that are evaluated. This is not surprising as binding to the receptor may have agonist or antagonist properties, so that the activity can be rated positive or negative for similar concentrations of slightly different structures. Nevertheless, the sensitivity of the assay is a useful addition to the range of methods available to the analyst.

Hunt developed the radioreceptor for benzodiazepines to provide a useful general method for active benzodiazepines and used it to evaluate the pharmacokinetics of diazepam, nitrazepam, clobazam, and loprazolam in humans. For loprazolam, the results were compared with a specific chromatographic method of similar sensitivity, both to obtain a single dose profile (Figure 8.7) and a multiple dose profile (Figure 8.8). During the absorption phase, both techniques give closely similar serum concentrations with a peak of 8 to 9 ng ml^{-1} between 2 and 3 h after a 2-mg oral dose. At later times, after about 5 h, as presumably active metabolites appear, the radioreceptor method gives higher concentrations than the chromatographic method. This difference is maintained on multiple dosing over 8 days. The results indicate that the presence of metabolites active in the receptor assay can make an important contribution to the overall pharmacokinetic profile. Jochemsen et al.[1251] carried out a similar comparison of the radioreceptor assay for nitrazepam and triazolam with a gas chromatographic method and reported good correlation.

Just as radioimmunoassay methods have their counterparts in nonisotopic methods, so have equivalent methods been used for receptor assays. Thus, Takeuchi described a method for the assay of diazepam using the receptor isolated from cow brain and a biotinylated conjugate.[1252] The same laboratory described an assay where immobilized biotin and free biotin compete for the binding sites on avidin, permitting the determination of benzodiazepines labeled with biotin; a well-defined dose–response curve was obtained over the range 1 pM to 1 mM.[1253] The determination of

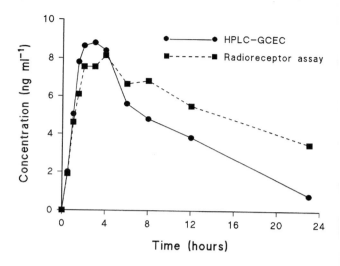

FIGURE 8.7   Plasma concentrations of loprazolam in plasma of a subject following a single dose (2 mg) of the drug, as measured by a radioreceptor assay and a specific HPLC-GC method.

benzodiazepines in biological fluids using receptor assays has recently been reviewed.[1254]

8.5.2   NEUROLEPTIC RECEPTOR ASSAYS

The neuroleptics consist primarily of three classes of synthetic compounds: phenothiazines, butyrophenones, and thioxanthenes (Figure 8.9).

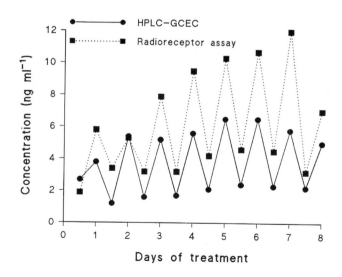

FIGURE 8.8   Plasma concentrations of loprazolam in plasma of a subject following once a day oral doses (2 mg) of the drug, as measured by a radioreceptor assay and a specific HPLC-GC method.

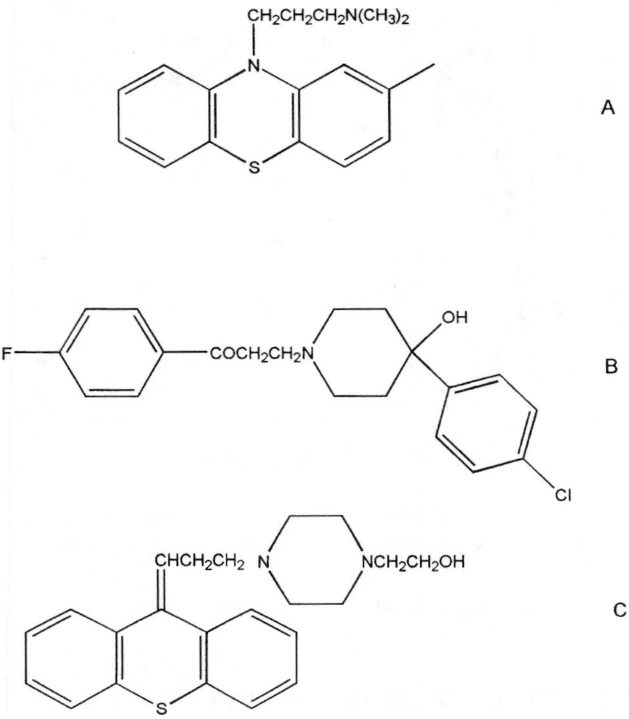

FIGURE 8.9   Structures of principal classes of neuroleptics. (A) Phenothiazine; (B) butyrophenone; (C) thioxanthene.

In clinical use, neuroleptics are thought to exert their therapeutic actions by blocking brain dopamine receptors and indeed [³H]haloperidol and [³H]spiroperidol bind selectively to dopamine receptors in mammalian brain.[1255,1256] The bound labeled butyrophenones can be displaced by other neuroleptics in parallel with their clinical potencies.[1256,1257] Creese and Snyder[1235] described an assay procedure for neuroleptics based on their ability to compete with tritiated butyrophenone (haloperidol or spiroperidol) for binding sites on membranes of the corpus striatum of rat. Assays were carried out by standard procedures, with the free and bound being separated by filtration, and then counting the label trapped on the filter paper. Using as little as 15 to 150 μl serum, the assay was suitable for serum concentrations of 10 to 240 μM haloperidol, fluphenazine, trifluoperazine, chlorpromazine, and thioridazine. Results obtained for haloperidol compared well with those using radioimmunoassay.

8.5.3   PROGESTERONE RECEPTOR

A human myometrium progesterone receptor has been used for determination of the abortifacient mifepristone. The method utilizes competitive replacement of [³H]16α-ethyl-21-hydroxy-19-norprogesterone, over a useful range of 10 to 120 pmol with a detection limit of 8.7 pmol.[1243]

8.5.4   GLUCOCORTICOIDS

A partially purified receptor was isolated from human liver cytosol. In the assay for glucocorticoids, the receptor was incubated with sample and [³H]prednisolone as the tracer to be displaced. Separation was achieved using activated charcoal. Hydrocortisone, deoxycortone, 4-pregnene-17α,21-diol-3,20-dione, 17α-hydroxyprogesterone, corticosterone and β-hydroxyprogesterone all competed for the binding sites.[1245]

8.5.5   ACE INHIBITORS

Because angiotensin converting enzyme (ACE) inhibitors exert their effect by binding reversibly to the enzyme, they can also be measured using the displacement binding techniques. Fyhrquist et al.[1258] developed such a method using a radioiodinated inhibitor as the label and measuring its displacement from its binding to the enzyme when analyte or sample was added. The method has been reported for the assay of enalapril, lisinopril, cilazapril, and the active form of benazepril.[1259]

9 Miscellaneous Methods of Analysis

CONTENTS

9.1 INTRODUCTION

The numbers of analyses of drugs in biological fluids carried out in laboratories throughout the world run into millions per year. The vast majority of these will almost certainly fall into one of the major categories discussed in the preceding chapters (i.e., optical spectrometry, chromatography, or saturation analysis). Occasionally, however, a particular technique, not applied to a wide range of compounds, will nevertheless find its own particular applications. Some of these less-used methods may be old established, classical, analytical procedures that have been overshadowed by more modern methods and may yet be revived by modern supporting technology, and some may themselves be relatively new developments not yet accepted as widely applicable techniques. In this chapter, these less popular methods will be briefly described, with an indication of their applicability.

9.2 MICROBIOLOGY

Microbiological methods of drug assay are among the few bioassays for drugs where the biological effect rather than chemical or physicochemical properties of the drug was used to measure its concentration in unknown samples. Antibiotics inhibit the growth of susceptible microorganisms, when the organisms have been inoculated in a suitable growth medium. Their discovery has been often retold in the serendipitious discovery of penicillin by Alexander Fleming, when a mold on some unwashed petri dishes was noticed to have a zone where the mold had ceased to grow. Further experiments with the material causing the inhibition of growth, indicated that a minimum concentration of the agent was required for the effect, and this concept of minimum inhibitory concentration (MIC) is the basis of all microbiological assays.

9.2.1   DIFFUSION OR ZONE OF INHIBITION ASSAY

In the most popular form of microbiological assay, the solution under test is spotted onto a uniform layer of suitable nutrient seeded with a microorganism which is known to be sensitive to the drug being assayed. After an appropriate period of incubation the layer is examined for growth of the organism; if the antibiotic is present at concentrations greater than the MIC, then there will be a zone of inhibition in the form of a clear ring around the spotted sample. By varying the concentration of the spotted sample and comparing the diameters of the zones of inhibition with those produced by a series of calibration standards, the concentration of the antibiotic can be determined.

9.2.2   SERIAL DILUTION METHOD

In the serial dilution method of analysis, the sample is incubated in a liquid nutrient containing viable microorganisms. As the organism multiplies, the solution becomes cloudy and diluted samples of the test sample are used to find the dilution at which this cloudiness is inhibited, and hence the concentration of antibiotic can be determined from its known MIC value.

The microbiological method is, by definition, sensitive enough to detect the lowest concentrations at which the compound is active. Even when novel and extremely active compounds are discovered, the microbiological method will be equal to the challenge, unlike the analysis of drugs by conventional chemical analysis, where each step toward more active drugs results in a further advance being required in sensitivity of measurement. However, microbiological methods do have limitations as far as precision is concerned. The zone diameters in the zone diffusion method are proportional to the logarithm of the antibiotic concentration. Hence a large concentration change will be reflected in only a small change in the diameter, which itself may not have very clearly defined boundaries.

The method can be made specific by the careful choice of organism. For example, the antibiotic cefotaxime has been assayed using the microorganism *Escherichia coli*. This organism is also susceptible to the metabolite desacetylcefotaxime, but a strain of *Proteus morganii* was found that was sensitive to cefotaxime but not to the metabolite and this could be successfully used for the elucidation of the pharmacokinetics of the parent drug.[1260]

Some laboratories favor HPLC methods for microbiological drugs because they are more rapid, and indeed if the result is required in a matter of hours, or even minutes, then the standard microbiological methods will be inappropriate. For re-

TABLE 9.1
Pharmacokinetics of Cephalosporins Determined Using
Microbiological Methods of Assay

Drug	Route	Organism	Peak after 1-g dose (μg ml^{-1})	Half-life (min)	Reference
Cefalexin	oral	*Bacillus subtilis*	40	50	1261
Cefalotin	iv, im	*Bacillus subtilis*	—	87–11	1262
Cefalotin	iv	*Staphylococcus aureus*	—	60	1263
Cefamandole	iv	*Bacillus subtilis*	—	54	1264
Cefatrizine	im	*Bacillus subtilis*	86	—	1261
Cefatrizine	oral	*Bacillus subtilis*	10	160	1261
Cefazoline	iv, im	*Bacillus subtilis*	39–44	—	1262
Cefazoline	iv	*Staphylococcus aureus*	—	96	1263
Cefotaxime	iv	*Klebsiella pneumoniae*	—	75	1265
Cefoxitin	iv	*Staphylococcus aureus*	—	59	1266
Ceftezole	im	*Bacillus subtilis*	45	—	1267
Cephacetrile	iv	*Staphylococcus aureus*	—	54	1263
Cephapirine	im	*Sarcina lutea*	25	—	1268
Cephapirine	iv	*Staphylococcus aureus*	—	72	1263
Cephradine	oral	*Bacillus subtilis*	20–30	50	1269
Moxalactam	iv	*Escherichia coli*	126	—	1270
Temocillin	iv	*Pseudomonas aeruginosa*	200	360	1271
Ticarcillin	iv	*Pseudomonas aeruginosa*	—	—	1271

search and development use, however, elapsed time for a single assay may not be important; it may be more useful to be able to perform very large numbers of simultaneous assays and this can be done using microbiological methods on large multiwell plates, automatic dispensing, and automated readers. Despite the limitations on precision, microbiological assays have been used satisfactorily for bioequivalence studies (Table 9.1).

An important point to note in relation to the analysis of antibiotics using microbiological methods is that plasma and serum are not interchangeable; certain anticoagulants used in the preparation of plasma will interfere with the agar matrix in the diffusion method, and it is recommended to use serum rather than plasma in such assays.[1272]

The analysis of antibiotics in biological fluids was one of the earliest drug level assays of therapeutic value and microbiological assay was the method of choice. The prime reason for chosing such a method is that it measures the concentration of the biologically active species. In fact, it is more correct to say that it measures the potential biological effect, as due to protein binding and other effects, the standard curves are not identical for protein-free matrices such as urine and protein-containing matrices such as serum.[1273] The fact that metabolites that are also microbiologically active will also be included in such measurements, may be considered either an advantage or a disadvantage, depending on the application. On the one hand, if the investigator is trying to demonstrate an effect, for example a long-lived antibiotic action of a particular dosed drug, then the total antimicrobial activity is an appropriate measurement. On the other hand, if the investigator wishes to demonstrate basic pharmacokinetics of a new chemical entity, then, as is now recognized for all valid pharmacokinetic work, a method that measures only a single chemical entity is necessary. It is for this reason that for research purposes, HPLC has virtually replaced microbiological assays.

9.3   ENZYME INHIBITION

The mechanism of action of drugs is now better understood. Many new drugs are designed to have very specific actions at the molecular level, including binding to specific proteins or enzymes.

9.3.1   ANGIOTENSIN CONVERTING ENZYME INHIBITORS

The first successful compound in this class of drugs (ACE inhibitors) was captopril, and this was also the first drug to be measured using its ability to inhibit the enzyme *in vitro*, by a 1:1 stoichiometric binding.[1274] In its simplest form, Reydel-Bax et al.[1275] measured the plasma enzyme activity directly and compared this with suitable controls; the circulating level of inhibitor was then assumed to be inversely proportional to the activity. The method was made highly sensitive by measuring the enzymic activity using radiolabeled substrate.

Most workers, however, measured the ability of the drug to inhibit the enzyme by measuring the production of [3H]hippuric acid from the substrate [3H]hippurylglycylglycine. The hippuric acid can be readily extracted from the incubation system and the enzyme activity determined according to conventional methods.[1274] The method was used for other ACE inhibitors, including enalaprilat and the active form of the prodrug, fosinopril.[1276] Indeed, one feature of these methods is that only the active forms of the drugs are measured.

Because ACE inhibitors exert their effect by binding reversibly to the enzyme, these drugs can also be measured by displacement binding techniques as described in Chapter 8. Fyrquist et al.[1258] developed such a method using a radioiodinated inhibitor as the label and measuring its displacement from its binding to the enzyme when analyte or sample was added. The method has been reported for the assay of enalapril, lisinopril, cilazapril, and the active form of benazepril.[1277] A similar radioinhibitor binding assay for lisinopril gave results comparable with radioimmunoassay methods.[1278]

9.4   POLAROGRAPHY

Polarography is a branch of electrochemistry that is used to measure species, organic or inorganic, which can be oxidized, or more commonly reduced, in solution by an electric current (Table 9.2).

Classical electrochemical procedures do not have sufficient sensitivity for the small concentrations of drugs in biological fluids and cannot normally be used without prior purification and concentration. Methodological advances have, how-

TABLE 9.2
Reducible Structures Used for the
Polarographic Determination of Drugs

Structure	Example	Reference
C=C	Cephalosporins	1278
C–Cl	Hexachlorophane	1279
C=O	Haloperidol	1280
C=N	Benzodiazepines	1281
N–N	Benzhydrylpiperazine	1282
N=N	4-Hydroxyazobenzene	1283
NO_2	Nitrazepam	1281
N–O	Trimethoprim *N*-1-oxide	1284
C=S	Thiobarbiturates	1285
S–O	Chlorpromazine *S*-oxide	1286

ever, enabled the development of polarographs with sufficient sensitivity to allow the direct application of the technique to plasma and urine.

A typical configuration of an electrode assembly of a polarograph is shown in Figure 9.1 and the traces obtained using this assembly for the analysis of the benzodiazepine, chlordiazepoxide, are shown in Figure 9.2. Chlordiazepoxide has three reducible bonds (Figure 9.3) and hence shows three peaks in the polarographic scan, at –0.275, –0.006, and –1.135 V under the conditions reported.[1287] The desmethyl metabolite has the same three reducible double bonds and shows almost the identical three peaks. The second metabolite, demoxepam, has only two reducible double bonds and hence shows only two peaks in the scan. Because of the similar reducible groups in the parent compound and in the metabolites it is not possible to measure the drug directly in the presence of

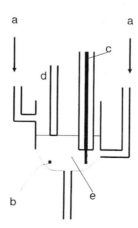

FIGURE 9.1   A typical configuration of a polarographic electrode assembly (a) nitrogen; (b) auxiliary electrode; (c) dropping mercury electrode; (d) saturated calomel electrode; (e) sample in electrolyte. (From Hackman, A. R., Brooks, M. A., De Silva, J. A. F., and Ma, T. S., *Analyt. Chem.*, 46, 1075, 1974. Copyright (1974) American Chemical Society. With permission.)

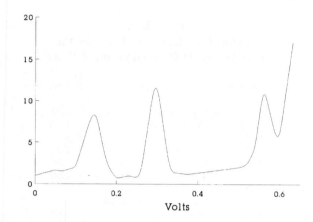

FIGURE 9.2   Differential pulse polarography of chlordiazepoxide. (From Hackman, A. R., Brooks, M. A., De Silva, J. A. F., and Ma, T. S., *Analyt. Chem.*, 46, 1075, 1974. Copyright (1974) American Chemical Society. With permission.)

its metabolites. Following a single dose of chlordiazepoxide, the desmethyl metabolite accounts for a large proportion of the circulating drug, whereas demoxepam is a very minor constituent. However, on chronic administration, demoxepam, because of its extended half-life, becomes an important constituent. Typical values for these three compounds after steady state had been reached following daily 30-mg doses of chlordiazepoxide were: chlordiazepoxide, 1.34 µg ml⁻¹; desmethylchlordiazepoxide, 0.36 µg ml⁻¹; and demoxepam, 0.40 µg ml⁻¹. For satisfactory analysis it was necessary to purify samples by thin-layer chromatography.[1287] Similarly, it was demonstrated that differential pulse polarography could be used to measure levels of diazepam at concentrations of ~10 ng ml⁻¹, although specificity could only be achieved by differential extraction procedures.[1288] Nevertheless, the slight difference in polarographic parameters between parent compounds and their metabolites can be exploited using computer techniques to enable complete analyses to be performed, without this separation.

The original method for the assay of chlordiazepoxide as described by Hackman et al.[1287] was sufficiently sensitive for benzodiazepines administered at dose levels of

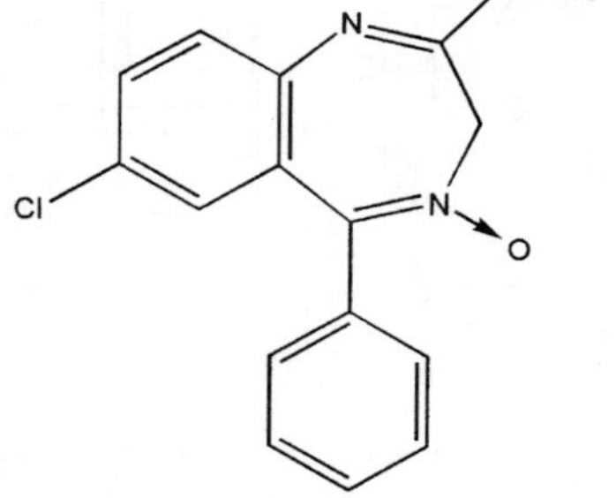

FIGURE 9.3   Structure of chlordiazepoxide showing three reducible double bonds.

TABLE 9.3

Analysis of Drugs in Biolgical Fluids Using Polarography

Drug	Fluid	Sensitivity	Reference
Ceftriaxone	Serum, urine	100 nm	1292
		0.2 nm	1292
Cytarabine, Cytidine	HPLC fractions	—	1293
Daunorubicin	Serum, plasma, urine	0.43 μM	1294
		0.42 μM	1294
Doxyrubicin	Serum, plasma, urine	0.43 μM	1294
		0.42 μM	1294
5-Fluorouracil	Serum	—	1295
Lormetazepam	Urine	400 ng ml^{-1}	1296
Paracetamol	Plasma	—	1297
Buprenorphine	Plasma, urine	40 ng ml^{-1}	1298

10 to 30 mg, and had a sensitivity of the order of 50 ng ml^{-1} using 2 ml plasma. The sensitivity of the general method was extended by using a polarographic microcell with miniaturized electrodes (Princeton Applied Research Corporation) and applied to the analysis of bromazepam in blood following a single oral dose of 3 mg.[1289] No bromazepam metabolites are found in blood to any significant extent,[1290] and hence the method could be used directly without separation of metabolites. The limit of sensitivity appeared to be about 10 ng ml^{-1}, which is comparable with chromatographic detection limits for this compound. In a comparative study with a sensitive and specific gas-chromatographic method, a good agreement was found, confirming the specificity of the polarographic method for bromazepam.

Compounds that do not contain appropriate reducible bonds, as listed in Table 9.2, can be converted to appropriate derivatives. A widely used procedure in this field is nitration of the aromatic nucleus.[1291] Nitration can be simply carried out by dissolving the sample in concentrated nitric acid and removing the acid in a stream of oxygen-free nitrogen at room temperature. The residue is dissolved in the appropriate electrolyte for assay by differential pulse polarography. Specificity normally depends on a prepurification step, but may also by achieved by some degree of selectivity in the nitration reaction itself. Smyth and Smyth[1291] also noted that nitrosation and condensation to form hydrazones as useful methods of derivatization to compounds suited to analysis by polarography.

Table 9.3 lists some of the more recent examples of the application of polarography to the analysis of drugs in biological fluids without recourse to a chromatographic step.

9.5 FLOW INJECTION ANALYSIS

Flow injection analysis is an analytical technique which emerged in the early 1970s and has attracted much theoretical and practical development, particularly in pharmaceutical analysis where there is a demand for the monitoring of materials and where the demands of sensitivity and specificity are not great, such as analysis of caffeine,[1299] corticosteroids,[1300], and codeine[1301] in tablets. The term was apparently coined by Ruzicka and Hansen,[1302] who, with Stewart,[1303] were responsible for much of the early development. Stewart's review of the principles of the technique had the revealing subtitle "New tool for old assays." This gives a good indication of the origins and applicability of the methods, that is, flow injection analysis draws on the well-established methods of chemical analysis, but uses the new tool to automate and miniaturize them for the more up-to-date requirements of speed and quantity. Recently, Ruzicka[1304] has reviewed the history and development of the technique and shown that with the addition of modern computer-controlled procedures, the technique has undreamed of potential.

The technique itself is extremely simple to describe. A stream of reagent is continuously monitored by a suitable detector. The sample is injected into the stream as a single "slug." The analyte reacts with the reagent to form a new species which is passed into the detector downstream of the injection point. For the successful conversion of a standard chemical assay to a flow injection analysis there are three crucial aspects: the injection, the control of dispersion, and the timing.

For injection of the sample, a reproducible, exact technique is needed. The actual volume of injection is relatively unimportant but it must be the same for each injection. For this reason, syringe injection through a septum[1305] soon gave way to rotary or loop valves of the type used in liquid chromatography.[1302,1305,1306] The sample is thus swept forward toward the detector. A certain amount of dispersion will take place in the flow injection technique as opposed to the air-segmentation technique invented by Skeggs[1307] and familiar to clinical chemists as the Technicon AutoAnalyzer, where the sample/reagent mixture is equivalent to a separate test tube or cuvette.

The important difference between the air-segmentation technique and flow injection analysis is that in the latter method the dispersion is allowed but controlled. The theoretical aspects of this dispersion and its control by design of flow injection analysis modules have been the subject of many publications.[1302,1308,1309] With suitably designed apparatus the typical display for flow analysis is a series of sharp, almost symmetrical bell-shaped peaks (Figure 9.4). Analysis rates of greater than 600 samples per hour were reported by Tijssen[1310] who used 20 μm bore coiled tubes to minimize dispersion.

Reproducible timing is the third important consideration in flow injection analysis. The time between injection and detection is solely dependent on the pumping speed, which can be altered to set the residence time as desired for dispersion and reaction. Reproducible residence times can be obtained for long and very short times as desired. This precise timing negates the need for reactions to go to completion, or reach equilibrium. This advantage is beneficial when detection limits are not a problem. In fact, the characteristic of most applications of flow injection analysis is that they are automated versions of "bucket" analysis. Thus the technique has never really been whole-heartedly adopted by bioanalysts concerned with drugs in biological fluids, despite its underdoubted capacity for high-throughput analysis.

Nevertheless, some of the techniques of flow injection analysis can find a useful place in automated analysis in general. A case in point would be the ingenuous device

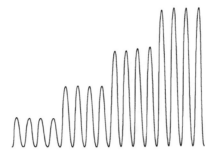

FIGURE 9.4   Typical analytical output for flow injection analysis.

described in Karlberg and Thelander's flow injection method for the determination of caffeine in pharmaceutical preparations.[1299] An aqueous phase containing the caffeine is segmented with a regular pattern of an organic phase. The stream is directed into a reaction coil where the resulting wall drag causes efficient mixing and fast extraction of the caffeine into the organic phase. For separation of the phases, the stream is passed into a T-piece, one leg of which is lined with Teflon fibers; the organic phase passes preferentially down this leg and the aqueous phase runs down the other to waste. Such devices could be applied to biological fluids where extraction of drug from biological fluids is necessary before assay.

Because of the extra demands of some sort of purification step for biological fluids, the dividing line between flow injection analysis and automated liquid chromatography may become somewhat blurred. However, dividing lines in science are to be abhorred, and dividing lines in analytical chemistry are no exception. The analyst should use all available tools to obtain the best possible analysis for the problem at hand. A recent example of this mix of methods was typified in a paper entitled "Flow injection fluorescence immunoassay for serum phenytoin using perfusion chromatography."[1311] Flow injection immunosensors have also been used in the assay of theophylline.[1312]

9.7   ISOTOPE DILUTION ANALYSIS

Isotope dilution analysis is carried out as follows. A radiolabeled form of the analyte is added to the sample and thoroughly mixed. The analyte is then separated and purified by conventional techniques, such as chromatography and sometimes culminating in crystallization of high-purity material. A sample of this material is accurately weighed or quantified by other methods and counted for radioactive content. From the specific activity of this purified material, the total amount in the original sample can be calculated. The main utility of such a method is the accurate measurement of reasonably large amounts of analyte in the original sample. Adequate methods exist for quantitation of the purified material. This would not, therefore appear to be a promising method for determination of drugs in biological fluids at therapeutic concentrations. However, the technique can be useful in the definitive identification of analyte by the process of purification to constant specific activity; if successive purifications result in decreasing specific activity, then this would be an indication that the analyte does not have the same structure as the radiolabeled material. An alternative to the final purification step of crystallization is to use countercurrent extraction apparatus; successive tubes containing analyte and radioactivity should have the same specific activity of analyte to confirm identity. An extension of the use of countercurrent apparatus is to use radiochromatography and check that the mass response is proportional to the radioactive response over the whole of the observed chromatographic peak. The method has a further limitation, however, if there is a significant isotope effect in the physicochemical purification as can occur for tritium labeled drug in chromatographic procedures; thus ^{14}C-labeled drug, with less-likely isotope effects is preferred to tritium-labeled drug.

Although direct isotope dilution as described has limited application in the area of drug analysis in biological fluids, it is extensively used in drug research using radiolabeled drugs, in the form known as reversed isotope dilution. In this procedure, the amount and identity of a radiolabeled entity in the sample is determined by adding a relatively large amount of known compound to the sample and purifying it to constant specific activity. The advantage of this application over the usual techniques for confirming identity of a drug is that losses during the purification procedure are automaticaly accounted for, and losses usually associated with manipulations at the nanogram level will be minimized because relatively large amounts of carrier can be added.

The method has been particularly useful in drug metabolism studies where it is desirable to obtain quantitative values of a particular labeled species from a mixture of radiolabeled drug and metabolites. The great advantage when the method is used in this way is that the only requirement is a liquid scintillation counter, a supply of authentic, crystalline material, and some skill in recrystallization techniques. Usually, the technique would be applied to estimate the parent drug, but it is equally applicable to radiolabeled metabolites as long as authentic crystalline material is available. The technique can be applied simultaneously to the analysis of several compounds as long as adequate methods are available for separation of the components; however, if the amount of sample to be analyzed is not a problem, then it would appear more sensible to carry out separate experiments for each putative analyte. When used in the multiple mode, HPLC is very useful as the separation method, as typified by the determination in biological fluids of [^{14}C]oxprenolol and nine radiolabeled metabolites.[1313] The carriers are ideal internal standards and the method was suitable for determining the disposition and metabolism of oxprenolol in a quantitative manner, which had not been possible by any of the previously published methods.

9.7   ISOTOPE DERIVATIVE ANALYSIS

Just as drugs can be made more sensitive to spectrophotometric analysis (Chapter 4) or to specific types of chromatographic detectors (Chapters 5 to 7), so they can be rendered sensitive to radioactivity detectors by reaction with radiolabeled reagents. For extremely sensitive analyses the label used needs to be of high specific activity because otherwise the amount of label introduced would be too small for accurate counting in a reasonable time. Hence ^{3}H is more usual than ^{14}C for this purpose and a method has been described for the determination of desipramine in plasma by derivatization with [^{3}H]acetic anhydride.[1314] Because the reagent needs to be present in excess, the main problem in preparation of the sample is to remove other radio-labeled components after the reaction. One of the happy characteristics of the isotope derivatization method, however, is that unlabeled carrier derivative can be added to the sample after derivatization is complete and the relatively large amount of material can be purified by the usual techniques (including recrystallization to constant specific activity if necessary). Despite its elegance, however, the demands in modern laboratories for rapid analysis and high sample-throughput have meant that this method is rarely, if ever, used.

9.8   MASS SPECTROMETRY

Occasionally, a chromatographic method will be reported in the literature which claims as proof of its specificity that the chromatogram contains no peaks other than the peak due to the compound of interest. In the author's view, this is an inappropriate use of chromatography as a separation method because it is apparent that the detector is only being used as a measuring device for a purified sample. A technique where this situation becomes apparent is in GC-MS using single ion detection; if the generated chromatogram does indeed show a single peak, then eliminating the chromatographic column may be feasible. A number of authors have described such assays using the extreme specificity of the mass spectrographic system, such as in the analysis of quaternary ammonium compounds in human urine by direct-inlet elec-tron-impact-ionization mass spectrometry,[1315] an isotope-dilution mass spectrometric method for procaine determination[1316] and a screening method for the rapid detection of barbiturates in serum by means of tandem mass spectrometry.[1317]

9.9   NUCLEAR MAGNETIC RESONANCE

Nuclear magnetic resonance spectroscopy is a powerful technique for examining the behavior of suitable atomic nuclei in a magnetic field. In its most familiar form, the technique separates signals due to single hydrogen nuclei depending on their chemical situation in the molecule, and the signal is further split depending on the immediate molecular environment of the hydrogen atom. Thus, Figure 9.5 shows the NMR spectrum of ethanol; three main signals are shown corresponding to the three types of hydrogen atom in the molecule: a hydrogen atom which is part of the methyl group, a hydrogen atom which is part of a methylene group, and a hydrogen atom which is part of a hydroxyl group. In this molecule, the methyl and methylene hydrogens are affected by the proximity of the other group and the signal is further subdivided. The methyl hydrogens are affected by two methylene hydrogens and are therefore split into a characteristic triplet pattern, the signals being in the ratio 1:2:1. The methylene hydrogens are similarly affected by the proximity of three methyl hydrogens and are split into a characteristic quartet, the signals being in the ratio

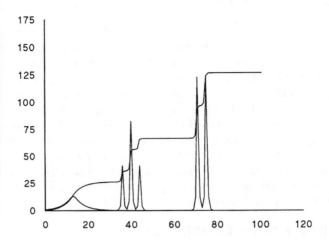

FIGURE 9.5   NMR spectrum of ethanol.

1:3:3:1. Integrating the groups of signals shown gives the the relative number of each type of hydrogen atom.

Most organic molecules contain hydrogen and will have a characteristic spectrum; however, these spectra can be extremely complex and most of the signals will overlap. Even for pure compounds the spectrum is difficult to interpret accurately and for samples of biological origin the situation is even worse. Nevertheless, an organic molecule can exhibit a single NMR signal that is characteristic of the molecule and which can be distinguished from the myriad of other competing signals. This is well illustrated in the work by Nicholson on the detection of endogenous compounds in human urine, in such a way as to be able to diagnose many metabolic conditions hitherto requiring months of conventional separation and spectral techniques.

This use of NMR has only been possible with the use of very high-field spectrometers, by using sophisticated computer techniques to analyze complex patterns, and by using ingenious sequences of magnetic radiation to distinguish the fine detail of the atomic environment, and in some cases to eliminate the effect of the presence of water and other organic solvents that might be present. In a work of this type it is impossible to go into any detail on the technique and the reader should consult some of the original references in this area.

The technique of studying endogenous materials in urine was extended to the detection of drug metabolites, particularly where the drug was given in relatively large doses. Such direct analyzes have been reported for elucidating, or confirming, the metabolic patterns for oxpentifylline,[1318] paracetamol,[1319] N-methylformamide (an antitumor agent),[1319,1320] for cephaloridine,[1321] and hydrazine (an industrial chemical and toxic metabolite of isoniazid).[1322] The technique shows particular promise when coupled with liquid chromatography as was demonstrated for metabolism of antipyrine[835] and paracetamol,[834] using stopped-flow techniques and transfer to the NMR probe. NMR studies are not limited to examining the protons in drug substances; other nuclei can serve as useful probes, as exemplified by the elucidation of the metabolism of fluorine-containing compounds using [19]F[1323-1325] and of nitrogen-containing compounds using[15]N.[1326]

An interesting extension is to use deuterium-labeled compunds such as in the method reported for *N,N*-dimethylformamide.[1327] The use of NMR in drug studies is developing rapidly and some of the most powerful developments will occur in the coupling of the technique to separation methods such as HPLC.

10 Development and Evaluation

CONTENTS

10.1 INTRODUCTION

One of the most important considerations when deciding on the method of analysis of drugs in biological fluids is the use to which the method is put. Chapter 1 detailed the reasons for these investigations and it is apparent that the desirable factors of precision, accuracy, sensitivity, specificity, speed, and ease of use will have different degrees of importance in the different situations. A procedure involving extraction of large volumes of plasma followed by several chromatographic purifications is hardly suitable for emergency application in overdose cases. It will also be valuable to consider the type of information to be obtained in relation to the expenditure involved. When absolute precision and sensitivity is vital to the outcome of a study then an expensive analytical method could well be justified. On the other hand, when the analytical values are of only peripheral interest to the main experiments, then, if the analyses cannot be performed cheaply, one should reconsider the value of including them in the protocol.

This chapter will discuss the selection, development, and evaluation of analytical methods for the measurement of drugs in biological fluids. In order to present a reasonably comprehensive treatment, the discussion will be biased toward methods intended to be used for elucidation of pharmacokinetics of the drug, since this application is probably the most demanding in analytical terms, and the discussion should therefore allow all aspects to be covered.

Every analyst has a favored method. When faced with a new drug for which an analytical method is required, the tendency would be to think in terms of his favorite technique. This approach may not be as narrow as it would first appear. It is unlikely that a single analyst can keep at his fingertips all the the expertise for all the various techniques, particularly where some of the more sophisticated modern procedures such as nuclear magnetic resonance require continual hands-on expertise. Thus, it is more likely, in a small laboratory, thorough excellence in a handful of techniques is more fruitful than a passing acquaintance with a large number. In the larger laboratory it is more feasible to cover a wide range of techniques. When the laboratory head accepts that there is a need to concentrate expertise in particular techniques it is important to select those techniques in the light of the problem to be expected. Thus, despite remarks in this and other chapters on the selection of the most suitable method for particular applications, one should not ignore the fact that the expertise available may dictate the choice of method. Even when a relatively simple assay may be all that is required, the individual laboratory may decide to use a more sophisticated technique that is up and running because this will be quicker, more convenient, and indeed cheaper, than to try to set up a new method with which the analyst is unfamiliar.

The next important consideration which may well dictate the method to be developed is the sensitivity required; that is, the lowest concentrations which are likely to be measured. When beginning the development of a new drug, this is often unknown. Generally speaking, however, there should be some indication from the projected dose. In the early phases of development mentioned in Chapter 1, relatively large doses of drug will be administered to animals and sensitivity requirements may not be so demanding. However, if the drug is unexpectedly extensively metabolized, then even with large doses the plasma and urine levels may be extremely low. If this does turn out to be the case, then there may be a need to reconsider the development program for that particular drug.

A useful diagram on the concentration range for which various techniques are applicable has been developed by de Silva.[1328] An adaptation of this diagram is

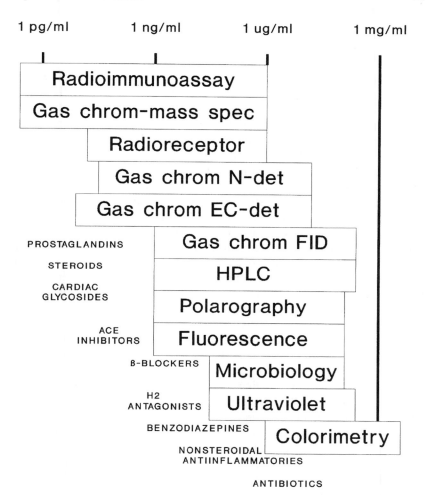

FIGURE 10.1 Applicability of various analytical methods used to analyze drugs in biological fluids.

presented in Figure 10.1, based on the discussions in earlier chapters. In addition, the areas of probable application are also included. Some additional characteristics of this diagram are worth pointing out. It may be hard to imagine now, but in 1950, methods of analysis that measured drugs at the mg ml^{-1} level were considered sensitive. The requirements for sensitivity, and the analyst's response to the challenge have changed by a factor of ten each decade since. As a result, at the beginning of the 1990s, fg ml^{-1} sensitivities were being claimed for some methods.

This drive toward ever-decreasing limits of detection in analytical chemistry has gone from milligrams to micrograms to picograms and on to strange-sounding units apparently named after the lesser known members of the Marx Brothers. Thus the sequence after pico- (10^{-12}) is femto- (10^{-15}), atto- (10^{-18}), ... yocto (10^{-24}). The most sensitive detector yet developed appears to be the laser-induced fluorescence detector for sulforhodamine 101, which, in conjunction with capillary electrophoresis chromatography, can detect 10 yoctomoles of analyte, or six molecules.[1329]

The type of drug has also changed over this time scale, from large-dose antibiotics with a relatively broad application to very small doses of highly specific drugs, much closer to the natural biological molecule; to obtain sensitivities required for such new entities, methods relying on their biological activity have once again become necessary. Nevertheless, these type of compounds are unlikely to replace all new drugs developed and the chemical and physicochemical methods which have served so well in the past will be with us for a while yet.

One of the first practical steps in the development of a new method is to examine the known chemical characteristics of the drug. In the pharmaceutical industry, a useful first move is to to consult with the chemist who first synthesized the compound. Often, the chemist will have considerable structural information on the drug including NMR and mass spectra, which may seem irrelevant to our immediate needs. However, the chemist should also have ultraviolet spectra, which may be useful as an assay in its own right, or may serve as the detection system in some sort of chromatography. The synthetic chemist will certainly have some thin-layer chromatography data and although such data will have been compiled with a view to separating precursors, side products and degradation products from the compound of interest, the same systems may be useful in the separation and detection of metabolites. Apart from the analytical data the chemist may have, more importantly, he should have valuable information on the compound's stability and may even be aware of unusual reactions leading to specific colored or fluorescent degradation products. The chemist will be able to point to specific functional groups with unusual activity, or to steric hindrances preventing what would seem an obvious derivative from forming. An example of this can be seen in the analysis of a series of β-agonists and antagonists. The relatively unhindered amines and hydroxyl groups of propranolol, labetalol, and atenolol are readily derivatized with trifluoroacetic anhydride. However, when a similar method was applied to penbutolol, the derivatives proved extremely unstable and although a method could be carried out with extreme care, it did not have the robustness of the method applicable to other compounds of the series (Figure 10.2).

10.2   ANALYTICAL PROFILE

Having made sure that the potentially rich source of material from the synthetic chemist is not ignored, the next step is to compile a full analytical profile. Although the experienced analytical chemist may be able to predict the particular analytical features that will be useful, from the structure, and hence will select the most likely method at an early stage, it is always useful to take a systematic approach to the preparation of an analytical profile. This should include spectra for the characterization of the compound (NMR, mass spectrometry, infra red) for completion, particularly in these days of strict quality control and attention to good laboratory practice, where the analyst must be sure he is dealing with the correct structure as proposed

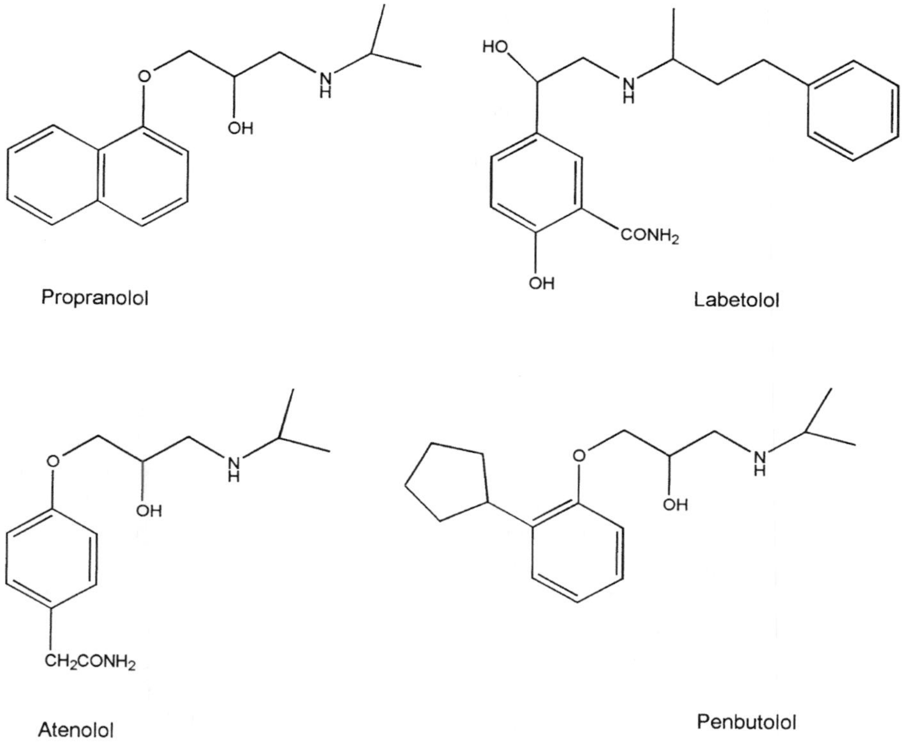

FIGURE 10.2 Structures of β-blockers.

by the originating chemist. The series of publications on analytical profiles of drug substances is a useful guide to the analytical data that should be included in an analytical profile; Table 10.1 shows a typical contents list. This series of publications may also serve as a starting point for the development of methods, although for the most part the series is more concerned with pharmaceutical analysis rather than analysis of biological fluids.

The analytical profile so compiled should not be limited to the unchanged drug but should be developed with attention given to the usual methods used for highly sensitive and selective detection. This can be illustrated by the following considerations for some of the main sections of the profile.

10.2.1 PHYSICOCHEMICAL CHARACTERISTICS

The analyst should have a thorough understanding of the physicochemical characteristics of the drug being studied. These will include solubility in a variety of solvents, pK_a values of ionizable compounds (which will lead to solubility–pH relations), molecular weight and, of increasing importance, whether the drug exists in optically active forms.

TABLE 10.1
Analytical Profile

Physicochemical characteristics
Ultraviolet-visible spectrometry
Fluorescence spectrum
Thin-layer chromatography
High-pressure liquid chromatography
Gas chromatography
Mass spectrometry
Nuclear magnetic resonance
Solubility

10.2.2 ULTRAVIOLET-VISIBLE SPECTROMETRY

Does the compound have inherent high optical absorption? The spectrum should be examined even at wavelengths close to the normal cut-off of 200 nm. Although absorption at such low wavelengths will not be useful in developing a direct method because so many other compounds will also absorb in this region, it may be useful in conjunction with HPLC if a suitably transparent mobile phase is used. Can the spectrum be influenced by a change in pH? Any such change can be useful in increasing the absorption, and hence increasing sensitivity. Can a characteristic spectrum be produced by application of energy such as heat or strong short wave-length ultraviolet irradiation? Such devices may lead to very characteristic spectra, but often the ultraviolet absorbing species may be unknown and specificity may be unproved. In general, absorption spectra are little changed by metabolism (and any change is likely to be an increase in molar absorbtivity rather than elimination of absorption), and hence absorption methods are only specific for unmetabolized drugs, or if a specific extraction step is applied.

10.2.3 FLUORESCENCE SPECTRUM

All compounds which absorb light will have some fluorescence, as explained in Chapter 4; fluorescence can be induced in many drugs by the application of strong shortwave irradiation. Fluorescent derivatives can be made by appropriate coupling with such reagents as dansyl chloride or fluorescamine. The ease of formation of such derivatives and their stability should be considered. Metabolism can markedly alter fluorescence characteristics, and although there are many examples where fluorescence methods have been subsequently shown to be nonspecific because of fluorescent metabolites (spironolactone, triamterene), by understanding the underlying principles, it is possible to devise very specific fluorescence methods.

10.2.4 THIN-LAYER CHROMATOGRAPHY

As mentioned, the chemist who synthesized the drug will have much data on TLC of the drug and closely related compounds, including their detection by a range of spray reagents. The specificity of some of these spray reagents may suggest colorimetric assays in their own right. In the analytical profile being compiled, it is useful to supplement the data of the chemist with R_F values for analogues and potential metabolites, as well as degradation products. Naturally the chromatographic process will confer considerable specificity on a potential method, but quantitative detection systems in TLC have not had the wide acceptability of other forms of chromatography. For plasma analysis, TLC is rarely sensitive enough; however, the sensitivity demands for urine analysis are generally less and the cheapness, versatility, and robustness of some densitometers make TLC attractive for screening urine samples for evidence of drugs of abuse.

10.2.5 HIGH-PRESSURE LIQUID CHROMATOGRAPHY

A variety of HPLC systems should be included in the analytical profile. The best system for analysis of drugs in biological fluids is likely to be a reversed-phase system, with its best chance of sharp, well-resolved peaks. Most metabolites of drugs are more polar than the parent compound and hence will be eluted first from the reversed-phase systems. Hence the conditions can be optimized to obtain short analysis times for the parent drug, without the need to elute late-running metabolites. The potential of utilizing any pH effects should not be missed in considering HPLC systems, nor should the possibility of using ion-pair chromatography. These physicochemical effects may be very useful in solving the problems of difficult separations.

Derivatives are little used in HPLC, except in the special cases of forming highly colored or fluorescent compounds just prior to detection as for example in amino acid analyzers. Thus an exhaustive list of retention times of standard derivatives is not particularly useful for the HPLC section of an analytical profile.

10.2.6 GAS CHROMATOGRAPHY

A simple analytical profile will concentrate on systems which give retention times of 5 to 10 min on at least two phases: (1) a nonpolar phase such as OV-1, and (2) a selective phase such as OV-101. However, for analysis of drugs in biological fluids, as opposed to pharmaceutical analysis, there is no need to limit the profile to the drug itself, and the retention characteristics of easily formed derivatives are useful. This is particularly important where the parent drug is itself not amenable to gas chromatography. The emergence of HPLC as a potent analytical tool, has to a large extent, reduced the need for the analyst to go to extraordinary lengths to make derivatives of nonvolatile or labile compounds that will survive the high temperatures used in gas chromatography. Nevertheless, the commercial development of gas chromatography has produced troublefree instruments which are easily automated, with very high separation efficiencies so that, where appropriate, gas chromatography is still a suitable choice for general methods of analysis of drugs in biological fluids.

The chromatographic step provides most of the specificity in this type of analysis and the analytical profile should reflect this by indicating the separation of similar compounds, or of compounds that are expected to be present in the sample. The detector used will be the primary determinant of absolute sensitivity of the analysis and again it is useful to indicate the limits of sensitivity using the different detectors. If a gas chromatographic system cannot be developed for the drug or a simple derivative, then the bulk of further development is likely to fall on HPLC.

10.2.7 MASS SPECTROMETRY

The mass spectrum of the compound should be established under standard conditions. Mass spectrometry probably made its first appearance in the analysis of drugs in biological fluids as a device for characterizing highly purified metabolites, and then as a very specific detector, only used for evaluating the specificity of simpler, routine procedures. However, the use of selective ion monitoring (even without chromatography) and the use of tandem mass spectrometry has indicated that the specificity of mass spectrometry may be a viable method of analysis. Different types of ionization procedures as well as different derivatives may be used to increase the specific ion yield and if possible the spectra from such methods should be collected into the analytical profile.

10.2.8 NUCLEAR MAGNETIC RESONANCE

Only a few years ago, the use of nuclear magnetic resonance (NMR) as a technique for the analysis of drugs in biological fluids would have been derided. Like mass spectrometry, it may have occasionally been used to characterize metabolites, although the amount of metabolite usually isolated would rarely be sufficient for the instruments available. NMR spectra are notoriously subject to modification by other components of the sample and all these features suggest that NMR would be unsuitable for analysis of biological samples. Nevertheless, as described elsewhere, with modern high-field instruments, computer techniques, and a greater understanding of NMR spectroscopy, the use of NMR is gaining ground and preliminary NMR studies should now be included at this early stage.

10.2.9 SOLUBILITY

The solubility of the drug in a range of solvents should be determined because it will aid in the selection of the best extraction method. The optimum solvent is one that extracts all, or at least a large proportion, of the drug but does not extract endogenous compounds or metabolites. The importance of a high efficiency in the extraction step is not merely to obtain the maximum amount of material for measurement; if the extraction is low, then the variability is high with a consequent poor precision of the method. The pH of the solution from which weak acids and weak bases is extracted is important. By using judicious changes of pH and appropriate solvents, specific extraction sequences can be devised, and it is useful to construct a pH solubility profile (Figure 10.3). The pH solubility profile shows the solubility of the drug, determined by a suitable method, plotted against the pH, usually over the range of 1 to 12. In Figure 10.3, the concentration of drug in an organic phase in equilibrium with a suitable buffer is shown, plotted against the pH of the buffer. At pH values greater than approximately 4.5, most of the drug will be extracted by the organic phase. A suitable procedure would therefore be to buffer the sample to pH 4, wash with extracting solvent, adjust the pH to 5, and then extract the drug with the same solvent.

At this stage of development the availability of a radiolabeled form of the drug may prove invaluable for determining such profiles very rapidly, particularly if the experiments need to be carried out for very low concentrations. However, there are

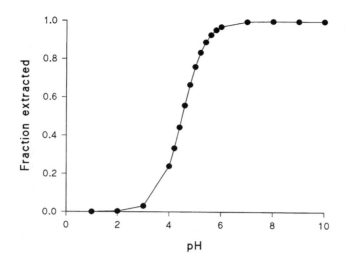

FIGURE 10.3   pH solubility profile of a model drug. The drug is partitioned between an organic solvent and the appropriate buffer; the amount of drug in the organic phase is determined and plotted against pH.

certain dangers in using radiolabel, particularly at extremes of pH, where the drug may be unstable and the observed radioactivity may be due to degraded species.

10.3   SELECTION OF A METHOD

10.3.1   EXTRACTION CONDITIONS

The theory of extraction conditions, based on lipophilicity and the ionization of weak acids and weak bases is well established, and optimum conditions for the extraction of a particular drug from buffered biological fluids can be predicted from these physicochemical properties. However, the analyst is usually presented with the need to exclude endogenous material or metabolites as well as to extract the relevant analyte and the optimum conditions are not readily predictable. Although one approach would be to systematically investigate a host of extraction conditions, a number of workers have devised schemes whereby a limited number of experiments are required to find the best conditions.[1330,1331]

Law and Weir[1332] devised a set of 13 basic solutes with controlled pK_a, logP, and plasma protein binding characteristics and concluded that no one set of extraction conditions could be obtained using reversed phase liquid-solid procedures. Casa et al.[1333] used liquid chromatography using C_{18} phases to predict extraction conditions for the selective isolation of benzodiazepines from biological fluids using C_{18} Sep-Pak cartridges.

From the analytical profile so far reviewed we should have a good idea of a probable separation-end point combination. For example, an HPLC system with fluorescence detection, and a scheme for efficient extraction of drug with minimal interferring coextractants may be indicated. The next step is to apply the combined steps to biological fluids, both native fluids and fluids which have had drug sub-stances added at the appropriate concentration. If the results appear promising (i.e., a chromatogram is produced with no peak corresponding to drug is obtained for the blank samples and a clear single additional peak appears in the samples containing added drug), then the range of metabolites, either expected or known, should be examined. It may be possible to exclude metabolite interference by the extraction step, by the chromatographic separation, or by the detection mode.

At this point we are assuming that the development is running smoothly. How-ever, complications will almost certainly occur. For example, the drug may be unexpectedly difficult to extract from plasma because of very strong protein binding, and denaturation of the sample may may be additionally necessary. This may lead to undesirable absorption of the drug onto the precipitated protein. Endogenous com-pounds may refuse to be separated from the drug either by extraction or by chroma-tography and it may be necessary to reinvestigate the chromatographic systems for the particular application, or to use detectors with greater specificity, such as single-ion mass spectrometric monitoring.

Once a satisfactory system has been established, it is then necessary to look at the quantitation. Even though the most efficient method of extraction would be chosen in the development of the method, it is unlikely that recovery will be 100% at all stages of the process. To properly quantify drug it is necessary to correct for any losses. One procedure would be to calculate the overall efficiency from a parallel blank sample to which has been added a known amount of drug and use the recovery of drug in this sample to correct all values determined in the same batch. The success of this approach depends on a very reproducible technique, a situation which has become viable recently with the use of automated equipment and robotics. Reproduc-ibility of extraction methods can be aided by taking aliquots for transfer operations rather than attempting to transfer total solvent phases. Although the absolute recov-ery of analyte may be reduced in this way, provided the distribution of drug into the required phase is close to 100%, the method will not suffer from variability as would happen if partitioning was low.

10.3.2   THE USE OF AN INTERNAL STANDARD

The use of an internal standard is a powerful technique for ensuring quantitative analysis, particularly for chromatographic methods where reproducibility of sample application is difficult, and where the analyst has a clear picture of the variation from one analysis to the next. There are some dissenting views on the desirability of using internal standards, but on the whole, when properly used, the internal standard method fulfills the claims made.

Although there is sometimes confusion in the literature on the meaning of internal standard, the generally accepted definition is that it is a compound similar in chemi-cal structure to that being analyzed, which is added at a known concentration to the sample prior to analysis; this standard is measured at the end point and its percentage recovery applied to correct the end point value for the drug being analyzed.

For the simplest and least controversial application, the internal standard is added to the final extract immediately prior to the injection of the sample into a chromatograph. This procedure merely allows the analyst to calculate what fraction of the final sample is injected. Thus, the only requirements for this internal standard is that it should be soluble in the injection solvent and should have appropriate chromatographic properties for the system being used. Nevertheless, even when used in this limited way, the properties of the internal standard need to be considered more carefully, including the possibility of differential volatilization of the components of small samples. It is for such reasons that even when used as a purely chromatographic internal standard, the chosen compound should have properties closely matching those of the analyte. Thus, the analyst should beware of using the same internal standard for all his analyses irrespective of the structure and properties of the target compounds. Steroids should not, for example be used as internal standards for benzodiazepines, however attractive the chromatographic parameters may appear.

At the other extreme, the internal standard is added to the original biological sample (blood, plasma, urine, bile) and taken through the entire extraction, transfer, and derivatization procedures as well as the final chromatographic separation and end point measurement. Used in this way, the requirements for an internal standard become more demanding, with the demands increasing as the various steps become more uncertain in their outcome. The internal standard used in this way will almost certainly be a closely related structural analog with physicochemical properties close to those of the analyte. Such properties will include partitioning behavior between aqueous and organic phases at relevant pH values, general stability in all conditions used, extent of conversion to derivatives and extent of absorption to glassware; in particular, similar protein-binding properties may be a crucial factor in considering the appropriateness of an analog used as an internal standard, if the sample is not to be denatured during the analytical procedure.

An internal standard, with its potential to correct for large differences in recovery, does not absolve the analyst from the need to optimize the various steps, but should be used to minimize the inevitable small variations that will occur between analyses. If an internal standard is proposed, it is essential to show its use is valid by carrying out appropriate experiments, and indeed to monitor its use as the method widens in application. For example, the perfect internal standard in volunteer studies may be distorted by comedicated drugs in clinical samples.

Although the principle of the internal standard suggests that both drug and internal standard are quantified and the concentration of the former is determined from the recovery of the latter, in practice the analyst will measure the ratio of the responses of the two compounds in the analytical system and will quantify the analyte by reference to a calibration curve. For chromatographic methods, the measurement of response is most often the peak height, the peak area, or the corresponding electronically integrated signal.

Initial experiments to show the validity of using an internal standard will obviously include the construction of a calibration line where the response ratio is plotted against the amount of analyte added to the calibration samples. The absolute amount of internal standard does not need to be known; instead it is the ratio of response that is important. Further validation experiments should be carried out to establish that

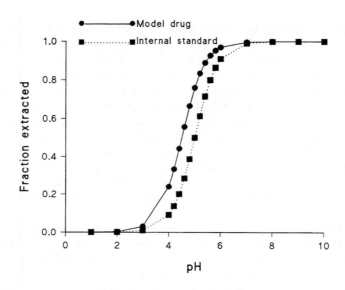

FIGURE 10.4   pH solubility profile of a model drug from Figure 10.3, and its proposed internal standard.

there is no change in relative recovery of the compounds at very high or very low concentrations of analyte.

It is important to be aware of the factors which can cause the recovery ratios to be critically dependent on extraction conditions. One of these is pH. The pH will have been chosen to obtain the best separation of analyte from other compounds and the analyst may have chosen a very precise pH for these conditions, based on the pH-partion profile shown in Figure 10.3. However, in the enthusiasm to pick the best pH, the analyst may chose a pH on the part of the curve where the solubility is most dependent on the pH; if the corresponding curve for the internal standard is only slightly different from that of the analyte (Figure 10.4), then at the pH chosen for extraction, the recovery ratio will depend highly on slight variations in this pH. In this instance, it would be better to choose a pH where a slight change in pH is not so critical. Starkey et al.[1334] have provided a similar illustration of the effect of pH on the extraction of amiodarone, its desethyl metabolite, and an internal standard L 8040 (Figure 10.5). Although these authors note that the internal standard had similar extraction character- istics and similar spectra to the analytes, their published illustration indicates that the internal standard was extracted more efficiently than the analytes at pH values greater than 6, and less efficiently at pH values less than 6; thus the chosen pH is very critical if the internal standard is to perform its function of correcting for extraction losses.

If a chemical transformation is to be performed as part of the analytical procedure, the internal standard should undergo the same transformation, so that the similarities between analyte and standard are maintained throughout the analysis. However, if the reaction itself is not quantitative, then the problems of reproducibility as men- tioned for quantitative extraction need to be solved; the change in rate for one

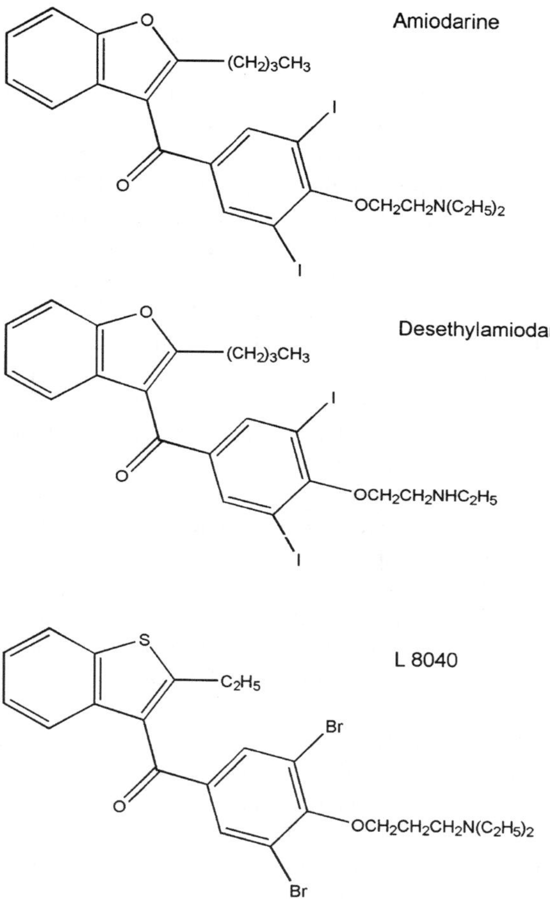

FIGURE 10.5 Structures of amiodarone, its desethyl metabolite, and a proposed internal standard L 8040.

compound might not be the same change in rate for the other, when conditions are slightly different. A good example exists in a general method proposed for benzodiazepines. In this method, the benzodiazepine is converted by acid hydrolysis to the corresponding benzophenone, a simpler, more stable compound, readily analyzed by gas chromatography. For an internal standard, one of the other plentiful benzodiazepines may be chosen (provided it does not hydrolyze to the same benzophenone), but both hydrolyses should be highly efficient under the same conditions for successful use of the internal standard.

Derivatization is perhaps just another form of chemical transformation and such derivatizations should be checked just as rigorously as for chemical transformations. The unquestioned acceptance of many standard derivatization reagents may not be justified, particularly when steric hindrance can begin to play a part in the reaction, such as the poor derivatization that was met with for penbutolol compared with other similar β-antagonists.

The internal standard should reflect changes in sensitivity of the end point. If the chromatographic column deteriorates with use and poorer peak shapes result, then both compounds should be affected in the same way, and if the detector loses sensitivity then both compounds should be equally affected. This is not because the quantification will be subject to change — the daily recalibration should take care of any day to day corrections needed — but because such changes that become necessary may not be using the optimum conditions for the assay. For example, the amount of internal standard may have to be increased beyond the concentrations used during evaluation of the method, to maintain comparable detector signals as the sensitivity of the internal standard falls.

Apart from these considerations on the relative behavior of the drug and its internal standard, several points need to be considered regarding the selection of the internal standard itself. In the chromatographic step the internal standard should achieve base-line separation from the drug and from endogenous compounds and metabolites that may survive the extraction procedure. In developing the method this aspect can be taken into consideration; however, the problems are more likely to arise in the application of the method to real samples, especially pathological samples when compounds may be present which do not occur in normal subjects. If the existence of these compounds results in extra peaks in the chromatogram, but this is not recognized, there may be an overestimate of the internal standard and consequent underestimate of the analyte, even though the chromatographic characteristics of the analyte are unaffected. In addition, the presence of the internal standard may prevent useful observations to be made on the presence of unexpected metabolites.

In the pharmaceutical industry when an analytical method is being developed for a new chemical entity the choice of internal standard is eased by the availability of suitable analogs and homologs prepared in the same series of compounds for pharmacological evaluation. Thus, once again the medicinal chemist should be consulted on the availability of such compounds and their properties. The chemist will not only have several suitable compounds already made, but may even be willing to synthesize new structures to order.

The structure should be chosen with care. The previous remarks relate to the structure which will give a suitable internal standard provided there are no interfering factors in the chromatogram. However, some apparently suitable structures may be metabolites of the drug in question and should be avoided. Such structures would include desmethyl compounds, amines replacing nitro groups, and phenols. Dehalogenated aromatic compounds should also be avoided. The possibility that the internal standard itself may degrade to the analyte should also be considered. As previously stated, the exact concentration of the internal standard need not be known, provided the same amount is added to all calibration samples and experimental samples; however, the discipline of adding known amounts on all occasions enables proper interassay comparisons over extended periods and ensures that the analyst remains in control.

10.4   EVALUATION OF A METHOD

10.4.1   INTRODUCTION

Evaluation must be done for the species being studied in toxicology and metabolism studies because of the potential of other interfering compounds in some species, and the differences in metabolism, which can be quite dramatic.

Thus far, one should have a good idea of the final method, and a detailed standard operating procedure can be written for the analysis. It is now necessary to evaluate the method according to the well-established criteria of precision, accuracy, sensitivity, and specificity. Due to space limitations and a desire to avoid possible repetition, a published method for the analysis of a new drug in biological fluids will seldom have detailed evaluations. However, this is not to say that the evaluation should be limited to the apparent standards reported in the literature. For internal use it is essential that there is a thorough understanding by the analyst of all factors of the analysis and this can only be obtained by a detailed evaluation; when using a literature method, the analyst needs some indication of the performance characteristics of a method, particularly if a choice has to be made among a number of literature methods. If sufficient information is given in the published or in-house literature, then valid choices can be made, without the expensive exercise of setting up and characterizing the methods in one's own laboratory. Unfortunately, however, there is no complete agreement in the literature on the meaning and definition of some of the terms used to characterize a method. Even in the same laboratory, different analysts may have different views on how to express these characteristics. The following discussion is an attempt to describe the philosophy of the necessary evaluations; for more rigorous treatment of these subjects, the reader should consult the literature referred to below.

The terms precision and accuracy are often lumped together in the literature followed by a discussion of standard deviations, means, regression lines, and so forth, without actually defining either term. The novice would be forgiven for assuming the two terms are synonymous, or at least too closely related to discuss separately. It is as well to clarify the distinction between them before embarking on further discussion. De Ridder[1335] defines accuracy as the closeness of the measured value to the true value, whereas precision is the the variation within replicate measurements. More vividly the difference can be illustrated by a robot on a production line that consistently drills holes in exactly the same place on a piece of work (precision) but does it consistently in the wrong place (precision without accuracy).

10.4.2   PRECISION

The precision of a method describes the agreement between replicate measurements. It has been suggested that the numerical value to describe this characteristic of a method should be referred to as imprecision, so that a high value would reflect high imprecision (bad) and a low value would reflect low imprecision (good). The imprecision then would be measured as the standard deviation or coefficient of variation of the results in a set of replicate measurements. The mean value and number of replicates must be stated as well as the design of the study, that is whether we are talking about within-day or between-day imprecision. The most meaningful value is the between-day precision because it is most relevant to the routine laboratory. The author recommends setting up a series of six concentrations of the drug in the biological fluid over the expected concentration range. Samples of each concentration should then be analyzed on six separate occasions (in effect constructing daily calibration lines). The standard deviation (SD), or imprecision, is then calculated at each concentration as:

$$SD = \sqrt{\frac{\Sigma(x - x)^2}{n - 1}} \qquad (1)$$

where x is the found value, x is the mean value, and n is the number of replicates — in this case, six. If the imprecision is expressed as a percentage (that is, the coefficient of variation) then the imprecision can be directly compared for the different concentrations. In many cases, the coefficient of variation can be correlated with the mean value; the coefficient of variation becoming larger as the mean concentration falls. It will then be useful to divide the imprecision into two components: one independent of the concentration and one dependent on the concentration. Thus, the SD value can be expressed in the form of

$$SD = (a + bx) \text{ ng ml}^{-1} \qquad (2)$$

where a and b are constants. When x is a high concentration, then a has little effect on the value of SD; when x is small, then a becomes the main factor in determining the SD value. If a is close to x, then the coefficient of variation will be unacceptable, which is one way to assess the limit of detection as discussed below.

Imprecision is not an absolute characteristic of a method, but rather depends on the skill and expertise of the individual analyst, as well as the reliability of the equipment. As more and more laboratories become fully automated, the imprecision of many procedures will be better and better. Despite the best endeavors of evaluation and advisory groups, the term "imprecision" has not really caught on among analysts and the use of "precision" is still very much the norm.

10.4.3   ACCURACY

The previous discussion on imprecision, although using samples prepared with differing concentrations of drug, does not use the known values, but only the found values in the calculation of imprecision. Accuracy is the agreement between the best estimate of a quantity and its true value. For the reasons stated above in the discussion on imprecision, the numerical value of the parameter is, strictly speaking, the inaccuracy. Thus any method should aim at an inaccuracy of zero. The inaccuracy can be calculated from the comparison of the mean values found in the exercise described above with the known values. This value is often referred to as the bias of the method and, if found to be constant, should be investigated and eliminated. Strictly speaking the estimate of inaccuracy can only be made if the true value is known. Some workers define a true value as one obtained by adding weighed amounts of material to control biological fluids; others maintain that an absolute reference method must exist which can be applied to the same samples. On consideration it will be realized that the use of samples containing added drugs in known amounts is just another calibration line, and one is forced to conclude that we are merely comparing two such calibration lines; there is no guarantee that the analyst making up the samples for accuracy is any more skilled in this activity than the analyst preparing his own calibration samples. Accordingly, the author has concluded that the accuracy should be stated as that assured by the use of a calibration line constructed by taking samples containing known amount of drug through the entire process.

10.4.4 SPECIFICITY

A factor which may indeed affect the accuracy when measuring real samples is the interference by endogenous compounds or metabolites. This comes under the special heading of specificity. Specificity is the ability of the analytical method to determine solely the component it purports to measure and to remain unaffected by the presence of other substances. Taken in isolation, this could mean that the analytical method should be tested in the presence of suspected interfering factors. As a first step this means checking that there is zero response in all analyses where drug is known not to be present. For chromatographic systems this means there is no peak at the retention time of the drug being tested in the system chosen. When the method is used in early volunteer trials, the extent and nature of metabolism should be considered in assessing any interference by metabolites. In such studies, metabolites will be well separated from parent drug, and indeed this separation is often the only criterion used as evidence of metabolism. Accordingly, in those relatively few instances when metabolite and drug have the same retention time, it is good to have a separate check on the identity of the component giving the response. For GC assays this can be done relatively easily using mass spectrometry of the eluted component, provided that there is sufficient material, and by comparing the spectrum both quantitatively and qualitatively with an authentic standard. This approach is also now feasible for HPLC, even for the most-used chromatographic systems containing aqueous mobile phases.

When the analytical method is used in clinical trials, where concomitant therapy and even diet is less stringently controlled than in volunteer studies, then there is a potential for interference from other drugs (and their metabolites) and unexpected dietary components. It is an impossible task to test in advance for all such possible interfering substances, and it is wiser to restrict these tests to a limited number of drugs which may be expected to be prescribed, and then to keep a careful watch on possible reasons for unexpected analytical results.

10.4.5 LIMIT OF DETECTION

Because most drugs developed today are administered in very small doses, the analyst is continually being challenged to devise methods of detection and measure-

TABLE 10.2
Determination of a Drug Following Its Addition in Known Amounts to Human Plasma

Drug added (ng ml^{-1})	Drug found ± SD (ng ml^{-1}, n=6)	Coefficient of variation (%)
0.5	0.9 ± 0.6	67
1.0	1.5 ± 1.0	67
2.6	3.5 ± 1.7	50
5.1	4.9 ± 0.9	19
10.3	9.8 ± 1.4	14
25.6	24.0 ± 3.3	14
51.4	50.1 ± 4.9	9.7
103	104 ± 3	3.1
258	271 ± 32	12
508	525 ± 50	9.6
1023	1109 ± 111	10

ment of very small concentrations of drug. As explained in Chapter 1, for pharma-cokinetic studies, it is at the low level of drug that the quantitation becomes extremely important. Thus the limit of detection obtained using a particular method is of prime importance in such studies. Many reports use the term sensitivity for this parameter; however, sensitivity is also used to describe the slope of a concentration–response line (i.e., a measure of how sensitive the end point is to the change in concentration). Thus the author recommends that the term limit of detection is more appropriate in this context. Of all the parameters used to characterize a method, it is probably the one that has the most varied treatment from analysts, so that a statement that "the sensitivity is x ng ml^{-1}," should only be accepted in the context of the definition given by the author of that particular paper or its description.

A popular definition is that the sensitivity of the method is the smallest single result which with a stated probability (usually 95%) can be distinguished from a suitable blank. The limit may be a concentration or an amount, and defines the point at which the quantitation becomes just feasible. If no blank exists, it is defined as the concentration or amount for which the estimated imprecision generally becomes greater than a given limit. Using this latter approach, the limit of detection can be assessed from the values obtained in assessing the imprecision. This is illustrated in Table 10.2. The values in this table are from a real evaluation and serve to illustrate that experimental error is not an intrinsic property of a method with the coefficient of variation changing smoothly with concentration. If a coefficient of variation of 20% is the maximum acceptable level of imprecision, then the limit of detection can be simply stated as 5 ng ml^{-1} in this particular example. There are other, more elaborate, ways to calculate the limit of detection, but in practice the value decided on is always a rounded up number and the methods described here are adequate.

The limit of detection described here is the lowest level which can be quantified with adequate imprecision; levels below this may well be distinguishable from zero but would be reported as not detectable using the above definitions.

10.5 APPLICATION AND DOCUMENTATION

10.5.1 APPLICATION

The acid test of any analytical method comes when it is applied to real samples. No method can claim to have been developed and evaluated until it has been applied. This is to say that the performance criteria achieved are in fact acceptable for the purpose for which they are developed. In applying a method in a real study, one is usually faced with analyzing more than just a handful of samples and often with a limited time scale and insufficient material to have two or more attempts at the analysis. At this stage we become aware of the robustness of the method, and this will determine the acceptance of the method. Allied to the robustness of the method is the sheer practicality. Considerable ingenuity may go into developing a specific method with a low limit of detection, but if the manipulations place too great of demands on the analyst then there may be strong resistance to its acceptance as a routine analytical method. In my experience, when a large number of samples needs to be analyzed, the robustness of a method should be actively considered at the development stages.

10.5.2 DOCUMENTATION

Finally, any method should be fully documented. Not only may important decisions be made on the results of an analysis, but registration of new drugs may require full documentation of the points already raised in this chapter on development and evaluation. Thus there are generally two sorts of documents required. In one document, the practical details of the final method should be presented in such a way that a laboratory technician can carry out the analysis. It is not necessary to give the rationale of the method, or the instructions in greater detail than would normally be found in the scientific literature. The second document, which may incorporate the first in some detail, should be the definitive method report to which a subsequent researcher can refer to answer any queries on the method. Such a detailed report will probably exist only in the documentation of a method developed for pharmaceutical research but should be made available to outside researchers. A typical layout for documentation of such a method is detailed in Table 10.3.

10.5.3 PUBLICATION OF AN ANALYTICAL METHOD

One indication of acceptance of a new analytical method is that it should be published in a reputable scientific journal. Not all methods are published this way, however, with in-house developments of new methods for investigational drugs often confined to the archives of drug companies or to the official documents submitted to regulatory bodies. Thus perusal of the analytical literature may reveal few methods originating from industrial laboratories and many more from academic departments. This is rather paradoxical as it is probable that the methods in the industrial laboratories are most likely to be those most thoroughly evaluated, whereas many from academic departments may be published through the "publish or perish" imperative and are often methods developed from equipment or expertise available rather than from the most appropriate resources. When a candidate drug reaches advanced stages

TABLE 10.3
Analytical Method Report

Introduction and general statement of problem to include sensitivity expected
Detailed description of proposed method
Development of method
    Extraction conditions
    Stability of drug in solvents
    Selection of chromatography systems
    Selection of detection system
Evaluation of method
    Recovery during method
    Precision (imprecision)
    Accuracy (inaccuracy)
    Specificity
    Sensitivity
Application

of development or when it is in clinical use and needs therapeutic monitoring, then it is in the drug company's own interest to ensure that the most appropriate method is made available to researchers investigating that drug. At this stage publication in a widely circulated and readily available learned journal becomes desirable.

The first consideration in writing a paper for publication is that the author must have something new to say. Having something new is not always obvious for analytical methods of new drugs as the method may merely be an application of recognized principles without any new principles being put forward or tested. Thus one should chose a journal with this in mind.

The next most important consideration in writing a paper is to consider the style and requirements of the journal to which it will be submitted. If the author would give one piece of advice on successfully submitting research work for publication, that advice would be to read the section "Instructions to Authors." Not only will you please the editor, but, providing the content is acceptable, then the passage of the paper through the publishing process will be smoother, with fewer mistakes being introduced as the editorial staff attempts to bring the paper into the general style of the journal.

For analytical method papers, the guidelines given above for documentation are generally useful. There is one very important section that must be included for publishing methods in most journals and that is that there must be a strong indication that the method has been successfully used. The literature is, unfortunately, full of details of methods purporting to be for the determination of drugs in biological fluids, which give no applications, and which may never appear again in the scientific literature as having been applied. Some of these methods may have been developed in isolation using standard drugs added to biological fluids, or even to buffered solutions, and the real problems (required sensitivity may be underestimated, metabolites may interfere) had not been appreciated until the experimental samples were received. The method is then modified so that the final applied method is no longer the one originally published.

11 Quality Control

CONTENTS

11.1 INTRODUCTION

Quality control in the analysis of drugs in biological fluids can cover a wide range of activities, or may be merely concerned with the analytical procedure in the measuring laboratory. In the extreme case of the former view, Scales[1336] stated that quality control begins with the administration of the compound and does not end until the report is signed. Thus the analyst should be involved in such aspects as study design, integrity of administration, state of animals or volunteers, methods of sample collection, and preparation, storage, and dispatch of samples. Scales gives an entertaining, if at times worrying, description of the many mistakes that can occur even before the analyst receives the samples. In addition, he advocates that the analyst should only develop beautiful methods and not waste time on analysis of samples that are so suspect as to make any interpretation meaningless. Though a tongue-in-cheek conclusion, the serious point is that the most elegant analytical work and sophistication of pharmacokinetic interpretation is indeed valueless if the connection between the result and the original experiment is suspect. It is an unfortunate corollary that where a series of analytical results are uninterpretable (or perhaps do not conform to the kineticist's preconceived ideas), the blame will inevitably fall on the unfortunate analyst whose best efforts will be disregarded.

There was a time when the analyst could devote all his quality control efforts on the more immediate aspects of his work, which are discussed below under the headings of internal and external quality control. However, the realization of the effect of extraneous factors on the concentrations of drugs in biological fluids coupled with increasing demands of regulatory authorities has made the analyst much more aware and involved in these other factors, as discussed under the heading of good laboratory practice.

11.2   INTERNAL QUALITY CONTROL

Quality control, in the sense of controlling the quality of the analysis, begins in the setting up and evaluation of the method to ensure that the method itself is designed to give precise and accurate measurements. Internal quality control has been defined as the long-term and continuing assessment of imprecision of an assay with the object of minimizing intralaboratory variation. In the initial development of a method, a rough guide to the imprecision can be expected to be obtained, although generally the number of samples assayed and the timespan allowable means that it is only a rough guide. A reasonable approach is to analyze a range of samples (up to six different concentrations) on six separate occasions. In a perfect world, the standard deviations obtained will be absolutely related to each concentration and will be reproduced each time the experiment is repeated. In reality, the values quoted are those found for those particular analyses carried out on a particular day; the hope is that they will be representative of any future analysis, but at this stage they should only be used to indicate that the method achieves acceptable precision. The continuing quality control aspect is to ensure that the imprecision does not fall outside these acceptable limits.

The first step in ensuring the quality of an analytical method is to establish that any standards — either for calibration or for quality control — are correctly made up. It is particularly useful to run an ultraviolet spectrum of stock solutions of drugs. This guards against a weighing or diluting error in the first stages of standard preparation. For quality control purposes, these validated stock solutions are added to suitable large pools of plasma or urine to prepare pools of at least three different concentrations to cover the expected range of measured concentrations.

The quality control samples are then supplied to the analyst carrying out the relevant study and at least one of each concentration included in each batch of analyses. It is not necessary, or even practical, to "blind" such quality control samples because the analyst would quickly discover which level each sample represents and the nominal value has to be revealed for his own quality control records; new quality control samples for each batch of analyses would place an unnecessary burden on the laboratory's resources for making up so many different quality control samples.

The practice of adding known amounts of drug to blank biological fluid as described above to obtain quality controls may be criticized on the grounds that such samples are unlike real samples in that they contain pure drug and no metabolites. However, the author believes this criticism is unwarranted and arises from an idealistic point of view of analytical work. In real samples, the ratio of drug to metabolites is infinitely variable over the time course of a typical pharmacokinetic experiment and there will be no sample that could be considered appropriate for such ideal quality control purposes. As far as interference by metabolites is concerned, this should have been evaluated in the initial development of the method. An overriding advantage in using samples containing added amounts of drug is that the true value (assuming competent weighing and dispensing) is known and hence the control analyses give information on accuracy as well as precision.

When a new analyst is starting a particular assay, these control samples can be used as a practice ground to ensure that the analyst can produce acceptable precise and accurate analyses, before embarking on the analysis of valuable sample from an expensive trial. A good analyst will need only one of these trials to establish his credentials. Less experienced analysts may be unable to obtain adequate precision or accuracy when meeting a particular procedure for the first time. In such a case, where the method had previously worked acceptably, then the analytical supervisor should beware of embarking on experiments to "find out what is wrong with the method"; it could be that the less experienced analyst is on the steep portion of a learning curve and repeated runs of the same procedure will be more fruitful in improving the results than a myriad of different procedures — none of which will reinforce experience with the officially accepted method.

In using quality control data, the values obtained are used to decide whether the analysis of a particular batch can be accepted or rejected. This means that the criteria for acceptance or rejection must be decided in advance. One approach, and that favored by clinical chemists, is to evaluate each set of quality control results in terms of standard deviations. This necessitates knowledge of the standard deviation of the method. A typical rationale would be to reject the analyses for a particular batch if all the quality control values were more than two standard deviations from the mean in the same direction. However, the author believes that such an approach has two main drawbacks: (1) the required precision may not be as great as that demanded by the quality control criteria and perfectly acceptable results *for the purpose of the analysis* may be rejected; and (2) the established standard deviation is not an intrinsic characteristic of the method, but is a characteristic of the analyst. A logical approach would follow that each analyst would set up his own standard deviation for the method. This would mean that analysts of different ability working on the same project would, by definition, reject the same proportion of their analyses, with results that would be rejected by the better analyst being accepted by the less experienced analyst. This situation would be unacceptable, and it would be better to set acceptance criteria on a percentage basis and in relation to the needs of the assay, rather than on a standard deviation. The fact that some analysts of exceptional ability may routinely return precisions much better than the acceptable limits should be a cause for celebration rather than concern.

A typical quality control chart is shown in Figure 11.1. The values found at the specified levels are plotted for each batch and inspected for acceptance or rejection of each batch of results. In the chart shown, limits of ± 10% of the known quality control values are used. One quality control sample at each concentration is analyzed per batch. Results of any batch of analyses are rejected if three or more quality control results are outside the 10% limit in the same direction, or two or more quality control results are outside the 20% limit in the same direction.

Results need not be rejected, but corrective measures to the procedure must be taken if two successive values of the same quality control specimen are outside the 20% limit in the same direction, or seven successive values for the same quality control specimen are above, or are below, or are on, the mean value. A thorough discussion of this type of quality control chart has been presented by Mullins.[1337]

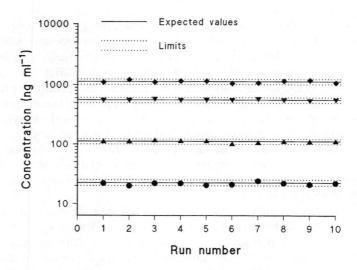

FIGURE 11.1   Quality control results for the analysis of 10 successive batches for the analysis of a drug in human plasma.

An alternative sometimes used as an internal quality control procedure is to provide the analyst with duplicate samples of some of the experimental samples as a separate, coded, series. The results from these samples are then compared with the reported results for their counterparts. This procedure cannot be used on its own as a quality control device. This is because evaluation must wait until the code is broken and by then it is often too late to repeat batches of analyses, and procedures which identify serious faults in analyses early on must be used. The analysis of such duplicate samples has its greatest value in what may be termed quality reassurance; when controversial values are reported, good agreement between paired samples following breaking of the code will have more effect in reassuring the sponsor that the values are real — never mind the skill and reputation of the analyst. This is rather like the instant video replays that will show that the cricket or baseball umpire is more often right than wrong!

The complete quality control data, including data from rejected batches of analyses should be maintained for subsequent evaluation. Over a sufficient period an accurate figure for between-day precision can be obtained from these records. Table 11.1 is a detailed analysis of such records from a number of analysts carrying out the same assays. The values obtained by one analyst show a much narrower distribution about the mean than those obtained by the others. In addition, the means obtained by two analysts were different and also statistically different. The interesting point is that all the series of results were perfectly acceptable in analytical terms yet the differences were detectable. Less experienced analysts would display wider distributions and the differences would not be detectable; it is ironic that in this instance of very competent analysts, the detected differences may even cast doubt on the analytical values, whereas values from less competent analysts would be indistinguishable and there-fore more readily accepted.

TABLE 11.1
Quality Control Results From a Number of Analysts Analyzing the Same Compound in Plasma

Study	Analyst	Mean, standard deviation, and number of analyses			
		QC1	QC2	QC3	QC4
I	A	40.6 ± 4.6 (21)	254 ± 14.5 (25)	511 ± 26.7 (18)	1026 ± 52.6 (23)
	B	38.1 ± 7.7 (35)	242 ± 16.0 (31)	510 ± 40.8 (31)	1015 ± 56.7 (30)
II	C	38.1 ± 1.5 (3)	257 ± 10.3 (3)	547 ± 27.0 (3)	997 ± 27.4 (3)
III	C	39.5 ± 2.5 (12)	266 ± 15.5 (13)	526 ± 29.9 (13)	1062 ± 83.0 (11)
IV	D	40.1 ± 4.3 (11)	257 ± 17.1 (11)	533 ± 33.5 (11)	1121 ± 119 (13)
	E	41.7 ± 4.4 (13)	260 ± 14.9 (13)	529 ± 36.5 (13)	1103 ± 78.3 (13)
V	E	40.4 ± 3.3 (13)	261 ± 12.5 (13)	512 ± 22.8 (13)	1055 ± 109 (13)
VI	F	40.1 ± 7.3 (2)	249 ± 9.5 (2)	532 ± 7.4 (2)	1047 ± 16.3 (2)
	G	43.8 ± 4.9 (19)	279 ± 21.1 (17)	552 ± 24.5 (14)	1093 ± 83.3 (17)
Nominal value		42.4	265	530	1060

It is often of interest to the analyst, and to those researchers making use of the analysts' services, to know what level of precision is usually achieved, and is usually acceptable to the scientific community at large. The acceptable figure is not the value claimed by the analyst in his pristine publication, but rather the value obtained in routine application. The figure is not the one stipulated by the clinician or pharmacokineticist as the one that he needs to attain significance in his particular experiment but rather the one obtained in practice, and any imprecision or inaccuracy must be accepted into the study design just like any other variable in a scientific experiment. In fact, although the imprecision of the analysis is sometimes much greater than professional analysts may be prepared to admit, this imprecision may be insignificant compared with the biological observations with which the analytical parameters are being correlated. In a detailed discussion of quality control in drug assays, Burnett and Ayers[1338] published a number of interesting tables which indicated that the coefficient of variation for routine analyses of a number of anticonvulsants is approximately 10%.

11.3 EXTERNAL QUALITY CONTROL

Burnett and Ayers[1338] recommend that any laboratory engaged in the analysis of drugs, whether as a routine service or as part of a research project, join an external assessment scheme in addition to establishing its own internal quality control procedures. For research projects with new chemical entities the opportunity to join an external assessment scheme is limited, but it is a wise precaution for the originating laboratory to be aware of other laboratories which may become involved in such analyses; the originating laboratory should take the initiative in developing such a scheme. Contract laboratories, particularly academic laboratories earning extra funds by taking on collaborative trials, may wish to develop their own methods and often considerable effort is expended in sorting out the tangles of discrepant results. It is certainly true that where the results of ring trials have been published, there may be considerable differences among laboratories analyzing the same samples.[1339] For antiepileptic drugs,[1340-1343] the wide variation in reported results for the same sample prompted the establishment of several external assessment schemes for these drugs, and schemes have also been established for other classes of drugs where the therapeutic level is crucial in controlled therapy.[1339,1344,1345]

Unlike internal quality control schemes, external schemes cannot be used for deciding on acceptance or rejection of results because there is usually a substantial time interval before the results are known. In general, participants in such a scheme should have the opportunity to assess their performance in relation to others rather than to use the scheme to control their own results. Thus the reports to the participating laboratories should include the number of results used in the calculation, the mean, the standard deviation, the true value as originally prepared, and the participant's own values.

To some extent, participants do not need to be told if their position in the results charts is pleasing or gives cause for concern. However, a number of methods have been outlined for giving performance indexes. The BARTSCONTROL index for antiepileptic drugs has been described in some detail.[1339] The index is based on ranking the reported results in terms of the differences of individual results from the mean value reported. However, this type of index compares the individual performance with the overall performance of the group, so that a poor overall performance will tend to increase the acceptability of individual poor performers and rejection of the single laboratory with excellent results. This index and other methods of assessing performance are further discussed by Burnett and Ayers,[1338] and recently by Wilson et al.[1346]

A special case of external quality control is when the analytical method has been developed in-house by the pharmaceutical company and the analysis of routine samples is carried out by a contract laboratory. Simmonds and Wood[1347] have detailed the special problems involved in this situation and have emphasized the importance of a troublefree, robust assay being developed in house, while at the same time recognizing that the performance of the assay need only meet the standards of precision and accuracy demanded by the experimental protocol, and not by the potential of the method. The purist might argue that the assay should be as precise and accurate as possible in all circumstances, but the purist does not have a drug to develop; that is, the purpose of the exercise, not the production of beautiful analytical results. However, once this philosophy has been accepted then the clinical pharmacologists and pharmacokineticists must not use such results for other purposes, and then when finding them wanting, declare them null and void.

The conclusions of Simmonds and Wood on their experience in obtaining the best results from outside contract laboratories are shown in Table 11.2. Some of these conclusions reflect the points the author has already made in the preceding discussion.

11.4   GOOD LABORATORY PRACTICE

The early 1970s were a time of upheaval in the American politics and in the American drug industry. However, whereas the sensational events of the Watergate

TABLE 11.2
Obtaining the Best Results From Outside Contract Laboratories[1347]

1. Contract out only those methods for which there is experience in the principal laboratory
2. Simplify, sacrificing sensitivity for robustness if necessary
3. Consider the work a collaborative effort
4. Establish agreed protocol, ensure lines of communication
5. Set specifications but leave room for scientific judgment
6. Define what is meant by valid data
7. Check quality control data as generated and look for trends, not being overconcerned with random variations
8. Setting up the method is of prime importance and needs full evaluation
9. Deliver good laboratory practice with good science, not slavish attention to administration
10. Remember the aim of quality control is to back up scientific judgment and therefore need only meet a reasonable quality standard.

affair have receded into history as a rather bizarre episode, the ramifications of the upheaval in the drug industry were more permanent and the resulting legislation and regulatory measures are still found in laboratories throughout the world. Just like the Watergate affair, the series of events leading to the initiation of good Laboratory Practice regulations began with a relatively minor event. This event was when the American pharmaceutical company, G. D. Searle, reported to the Food and Drug Administration (FDA) some discrepancies in one of their toxicity studies in relation to their drug metronidazole. This drug for certain protozoal infections had been on the market for many years. Searle maintained that the changes in their reports did not affect the overall estimate of the safety of the drug and that they were merely putting their house in order. Indeed, nothing since has suggested any particular toxicity problem with this drug. The FDA is the government-appointed body in the United States which is responsible for the regulation of drugs. The discrepancies reported by Searle were not automatically accepted by the authorities and a thorough investigation was carried out by the FDA into Searle's records in their Chicago laboratories. In addition to investigating the records of metronidazole, the FDA took the opportunity to investigate all aspects of Searle's toxicological operations. The first public events in the Searle saga occurred on July 10th 1975, when the Kennedy hearings into the Searle laboratories were launched. To the American public, it must have seemed like another example of the amoral state of big business in general and the pharmaceutical industry in particular.

Without underestimating the seriousness of the findings against Searle, however, the more discerning and less hysterical observer would recognize that many of the errors coming to light were the result of carelessness rather than any attempt at fraud or malpractice on behalf of the company or its employees. This can be demonstrated by the type of errors that were uncovered. For example, in some toxicology studies involving long-term dosing of rats, some animals were noted to have died during the study but the same number apparently completed the study as had started it. This would seem to indicate carelessness on the part of the scientist or technician compiling the report, and also on the part of management reading and checking it. Similar discrepancies, such as observations of tumors in some animals on some days, but nonobservation subsequently, again is indicative of carelessness or uninformed observation. The FDA investigations into Searle were eventually published in November 1976.[1348] The opening statement of the report reads:

> At the heart of the FDA's regulatory process is its ability to rely on the integrity of the basic safety data submitted by sponsors of regulated products. Our investigation clearly demonstrated that in the G. D. Searle company we now have no basis for such reliance.

The conclusion was even more damning in its particulars:

> There was total disregard by Searle technical personnel of a number of important aspects of their work. Too little importance was placed on the significance of the studies they were conducting, they did not adhere to the research protocol and there was not enough accurate observations of the appropriate parameters made. The observations that were made, failed in their documentation and there were no signatures or dates on the records of these observations. The need to assure the accuracy of the data transcribed from original documents to final reports and the need to assure proper and accurate administration of the product under test, along with observations of proper laboratory work, animal husbandry, and data management procedures was not carried out.

The investigations and findings did not begin and end with Searle. The FDA made the assumption that if these findings applied to Searle, then they could, and probably would, also apply to any other organization involved in the same sort of work, and especially those laboratories carrying out scientific testing for profit. Thus, enormous resources were devoted to monitoring programs for all such testing facilities in the United States where safety of drugs was at issue. The FDA's worst fears were apparently realized when two contract research establishments were shown to be so deficient in their adherence to the truth that their cases were sent to the Grand Jury. The sins of these contract laboratories, however, were quite different from those of Searle. Whereas Searle was doubtless guilty of much careless and sloppy work, the contract laboratories were apparently seeking shortcuts to completing lucrative contracts for pharmaceutical companies. One of the contract laboratories was unable to produce anything other than summary reports on the work they claimed to have done and in fact were sued by the sponsoring laboratory, Syntex, for damage done to their anti-inflammatory drug, naprosyn, when it was revealed that the work carried out was invalid, if not fictitious. A detailed investigation by the FDA into this case revealed that 30 out of 30 rodent toxicity tests showed sufficient flaws to be declared invalid. Another contract laboratory was indicted for allegedly deceiving its clients over at least an 18-month period. This affair concluded with jail sentences for a number of the contracting laboratory's senior personnel.

All this upheaval in the field of toxicological testing designed to prove the safety of drugs caused great consternation both in the industry and in the offices of the regulatory authorities. As far as the industry was concerned, the individual companies were forced to look at their own operations to ensure they were not guilty of sloppy procedures in investigating their own drugs and those companies with studies contracted to the convicted contract laboratories found much of their safety data useless. Other contract testing facilities also came under close scrutiny, but the lesson for the pharmaceutical industry should have been to tighten up their own procedures and to supervise contract work. The FDA, however, was not content to allow the industry to look to its own efficiency and safety, but perhaps elated by its success in uncovering the Searle deficiencies and the criminal activities at the contract houses, concluded that the safety of drugs was too important to be left to the industry; the FDA maintained that the deficiencies and criminal activities so far discovered were just the tip of an iceberg, and proclaimed the principles of Good Laboratory Practice (GLP). These principles were subsequently developed as regulations but later modified as guidelines. The force of the principles of GLP, however, whether regulations or guidelines, was that any studies submitted to the FDA to support safety of new pharmaceuticals that did not comply with GLP would not be accepted. GLP, therefore, did not just apply to the United States, as because of the size of the United States market, the FDA was virtually able to dictate the conduct of toxicological and other studies around the world.

The application of GLP to all aspects of safety testing of drugs means the analyst comes under scrutiny in several important areas. The most obvious aspect is in the analysis of samples from animals in toxicity studies. For valid toxicology studies, it is essential that animals indeed receive the drug, and, as discussed in Chapter 1, one way of checking that animals are receiving and absorbing the test drug, is to analyze plasma samples for its presence. Hence the analyst will receive samples from such studies and the analytical procedures will themselves come under scrutiny.

It has so far been a matter of considerable debate whether drug metabolism and animal pharmacokinetic studies come under GLP guidelines. It is probably true that most pharmaceutical companies are apprehensive at any extension of GLP bureaucracy into aspects of research and development other than those for which they were originally defined, and indeed the view that preclinical pharmacokinetic and metabolism studies come under GLP has not gone unchallenged; at the 1990 Drug Information Symposium on "International Harmonization on Nonclinical Drug Safety Requirements" mixed opinions were voiced.[1349,1350]

Many companies looked to the FDA for a ruling on whether such studies should be included. At first, the FDA seemed to indicate that they would not be included, but it is hard to reconcile this interpretation with the statement that all aspects of safety testing should be included in the GLP guidelines. It is certain that such studies are indeed part of the safety testing procedure. Certainly one of the stated objectives of drug metabolism studies is to compare different test species with humans in order to help select the appropriate species for further toxicity testing. Apart from the comparison studies, the study of metabolism and pharmacokinetics in animals has safety implications in its own right. For example, the metabolism of a drug to a known toxic agent in one animal species will surely halt its development just as would a toxic effect itself. An unacceptably long half-life may also terminate development, as the potential for accumulation of the drug would certainly be undesirable.

Animal studies are used not merely in this negative sense to weed out potentially dangerous drugs. They are also used to define the conditions of safe usage of the drug in humans. Such studies include bioavailability (extent of absorption, dose-dependence, absorption-time profile, intersubject variation, food effects, formulation effects), effect of multiple dosing (enzyme induction, accumulation), interactions with coadministered drugs (kinetics and metabolic), effects of disease (renal or hepatic impairment) or age and sex. Some of these studies are frequently performed in humans but obviously come under the heading of safety testing in animals when animals are used as test models. These considerations indicate quite clearly that drug metabolism and pharmacokinetics studies in animals are part of safety testing and hence any work in this area should be carried out according to GLP — and this includes analytical work — if the studies are to be presented to the FDA or other regulatory bodies in support of the registration of a drug.

So far, the discussion has been as applied to preclinical safety studies. However, it would be difficult to operate two different codes of behavior in a laboratory that carries out analyses for human studies and animal studies. The promulgation of Good Clinical Practice, and the FDA's Bioavailability and Bioequivalence Guidelines (Table 11.3) have effectively brought about the encompassing of all analytical studies on biological materials into the GLP system. Thus, like it or not, GLP is now an integral and necessary consideration in any laboratory carrying out analyses of drugs in biological fluids and the procedures, methods, quality control, and documentation of such a laboratory must be closely monitored.

Good laboratory practice (without the capital letters) is of course the natural working pattern of all conscientious laboratory workers, and one approach that could have been taken would have been to have a system of accredited laboratories; the laboratory would be established as an accepted laboratory based on its facilities and the professional qualifications of its staff. This approach was considered by the FDA but rejected in favor of a system whereby laboratories would have to establish systems in line with GLP concepts laid down by the FDA; all studies intended for submission to these authorities would have to comply with these laid down procedures and moreover the laboratory would have to establish procedures to show that the studies had been carried out in accordance with Good Laboratory Practice. Thus rather than have a system depending on professional integrity, the authority preferred one based on procedural control. Nevertheless, it must be said that the implementation of GLP has certainly brought benefits in helping point out deficiencies in some working practices.

Although the GLP Regulations, or Guidelines, have been published, the details are more appropriate to toxicological testing laboratories, and do not transfer easily to an analytical laboratory. A more useful document to follow, to enable the analytical laboratory to comply with the FDA's thinking is the manual used by the inspectors involved in the Biopharmaceutical Analytical Testing Compliance Program. This manual, available through the Freedom of Information Act, is in the form of instructions or questions which the inspecting officer would consider in his report. The first edition of this book[1351] contained a detailed description of the relevant parts of the manual, and a regulatory view covering the same ground has also been published.[1352]

TABLE 11.3
Recent Food and Drug Administration Bioavailabilty and Bioequivalence Guidelines

Draft Guidance for the Performance of a Bioequivalence Study for Antifungal Products	1990
Bioavailability Policies and Guidelines	1993
Guidance for Carbidopa and Levodopa Tablets *In Vivo* Bioequivalence Testing and *In Vitro* Dissolution Testing	1992
Guidance for Cimetidine Tablets *In Vivo* Bioequivalence Studies and *In Vitro* Dissolution Testing	1992
Guidance for Conjugated Estrogens Tablets *In Vivo* Bioequivalence Studies and *In Vitro* Dissolution Testing	1191
Various Comments Regarding Bioequivalence Study Guidance for Conjugated Estrogens	1990
Interim Guidance: Topical Corticosteroids *In Vivo* Bioequivalency and *In Vitro* Release Methods	1992
Guidance for Diflunisal Tablets *In Vivo* Bioequivalence Studies and *In Vitro* Dissolution Testing	1992
Guidance for Diltiazem HCl Tablets *In Vivo* Bioequivalence Studies and *In Vitro* Dissolution Testing	1992
Guidance for Indapamide Tablets *In Vivo* Bioequivalence Studies and *In Vitro* Dissolution Testing	1993
Guidance for Metoprolol Tartrate Tablets *In Vivo* Bioequivalence Studies and *In Vitro* Dissolution Testing	1992
Guidance for Nadolol Tablets *In Vivo* Bioequivalence Studies and *In Vitro* Dissolution Testing	1992
Guidance for Naproxen Tablets *In Vivo* Bioequivalence Studies and *In Vitro* Dissolution Testing	1992
Guidance for Nortryptiline HCl Tablets *In Vivo* Bioequivalence Studies and *In Vitro* Dissolution Testing	1992
Guidance for Piroxicam Tablets *In Vivo* Bioequivalence Studies and *In Vitro* Dissolution Testing	1992
Guidance for Ranitidine Hydrochloride Tablets *In Vivo* Bioequivalence Studies and *In Vitro* Dissolution Testing	1993
Guidance on Statistical Procedures for Bioequivalence Studies Using a Standard Two-Treatment Crossover Design	1992
Guidance for Terfenadine Tablets *In Vivo* Bioequivalence Studies and *In Vitro* Dissolution Testing	1992
Guidance on Triazolam *In Vivo* Bioequivalence Studies and *In Vitro* Dissolution Testing	1992

Analytical chemistry has long been recognized as a very skilled science, and analytical chemists have justly taken great pride in their reputation. However, the foregoing discussion has indicated that a high reputation is not sufficient either to ensure the correctness of the results or to provide acceptable records to the regulatory authorities. The correct application of quality control procedures is now recognized as an integral and necessary part of the analytical function. No self-respecting analyst would now issue results without the procedures outlined in this chapter. The quality control aspects described here are procedures that have been developed and are applied to improve the quality of the assays undertaken. Good Laboratory Practice, on the other hand, is a system imposed from the outside, and while it has brought attention to the possible shortcomings of laboratory work, has meant greater bureaucracy which does little to add to the quality of the analysis. In my own experience, a particular standard operating procedure was called into question by an inspector because the written procedure did not state that the tops were placed on the extraction tubes after the solvent was added and before they were shaken in a a rotary shaker. Recently, an instrument manufacturer declared it would no longer be providing servicing for a particular chromatographic model which they no longer made, and that this would mean assays performed on such an instrument would be invalid as far as GLP was concerned.

Despite my sadness at this turn of events, we have to accept that the discipline of GLP as a bureaucratic necessity to further the cause of drug development or even in monitoring of drugs of abuse[1353] is here to stay. This is well recognized by pharmaceutical companies, who prefer to follow the trend as the cheaper alternative to presenting a rational argument, much as many toxicological tests are still carried out, even when they are known to be of doubtful scientific worth. The discussions in this chapter result in a suggested scheme probably now followed in one way or another by all those developing and applying methods for analysis of drugs in biological fluids.

The method is devised by a development scientist and evaluated for precision, accuracy, sensitivity, and specificity. The method is written up in full detail and this is the official analytical method. A standard operating procedure is written, which gives the full practical details on how to carry out the analysis. A series of quality control samples is prepared, preferably by a skilled person outside the routine applications laboratory. Any new analyst taking up the method is provided with the standard operating procedure and should carry out a number of assays using quality control specimens to establish his ability to perform the analysis.

On satisfactory completion of the trial runs, the analyst is provided with the appropriate study protocol for the samples to be analyzed. The analyses are then carried out exactly according to the standard operating procedure and the protocol, including appropriate quality control, and calibration samples with each batch. All raw data and notebooks are checked and signed by the analyst and supervisor where appropriate. At the conclusion of the study all these data are archived and the analytical report issued.

The full documentation of the analytical studies should be contained in only three officially reported documents: the method report, the standard operating procedure, and the analytical report. These documents must contain sufficient references so all statements can be traced back to the properly archived raw data.

If these procedures are followed, then the analytical laboratory can be satisfied that it has provided an efficient and professional contribution to the particular study.

12 Pitfalls and Practical Solutions

CONTENTS

12.1 INTRODUCTION

The analysis of drugs in biological fluids as described in previous chapters in this book has discussed the reasons for such analyses, the available techniques for

determining drugs, particularly at low concentrations, and some of the good laboratory practices that are necessary for ensuring the validity of the data collected.
However, most analysts recognize that the analytical process itself is a very practical
operation, as illustrated in a recent survey on inter- and intralaboratory sources of
variation for a variety of analytical techniques.[1354] Despite the increasing amount of
standard laboratory equipment and instrumentation that can now be installed in an
analytical laboratory, the importance of individual expertise and preferences is
continually underlined by the almost inevitable failure of one laboratory to replicate
exactly the analytical method developed in another. In principle, it should be possible
to write standard operating procedures and instrument specifications in such detail
that such replication is obtained, yet there will still be steps that analysts perform,
almost by instinct, that may not be written into the procedures. This chapter will
attempt to gather together many of the tricks of the trade that may sometimes appear
as part of a method description, but are more often applied to solve a particular
problem with a particular sample. Alternatively, they may even be found to be
necessary in one laboratory but not in another, and as such should be considered as
useful, but not essential, elaborations in particular circumstances. Some of the
procedures described below will be "generally known," and some will be attributable
to the ingenuity of individuals. The author has given references where possible, but
apologizes to those anonymous innovators whom he has not been able to credit. A
rich source of material, however, is often found in the discussion sections of reports
on analytical symposia or meetings. One such source is the series of books on the
Analytical Forum held at Surrey University biennially since 1975.[1355-1364] The author
firmly believes that every bioanalyst should read these volumes, including even the
early ones from cover to cover; he may be dismayed at the problems encountered, but
he will also be considerably enlightened. Attending the Forum is even better. The
author has unashamedly plundered these volumes, but has used his own conference
notes to help compile the rest of this chapter.

   The author must emphasize that many of the procedures mentioned should be
regarded as hints to solve a particular problem and should not be considered recommendations in general. Rather like Rubik's Cube, you may find that a small
operation to correct one problem will play havoc with parts of the assay apparently
well divorced from the proposed adjustment. If your development time is unlimited,
you may eventually be able to try all possible combinations to achieve the perfect
method.

12.2   STORAGE OF SAMPLES

   The traditional view of an analyst is that of one who takes your sample and then
tells you what is in it and how much. The history of the sample may be considered
irrelevant to the analysis. However, the recognition that the analyst is usually an
expert in the make-up of complex matrices, means that it is most important that he
is aware of any unusual conditions under which a sample is taken and stored. The
analyst must be aware of any potential for prelaboratory error,[1365] a point reinforced
by the so-called Good Laboratory Practice guidelines promulgated by a number of

regulatory authorities, and should be in a position to minimize these errors by giving appropriate advice. Thus when a laboratory offers a particular assay, it should issue instructions for collection, storage, and transport, and the development of the method should include appropriate investigations. In laboratories which are close to the source of samples, such as in the bioanalytical laboratory engaged in pharmacokinetic studies in the pharmaceutical industry, tight control over these aspects can be achieved. However, as the pharmaceutical research departments are less keen to commit skilled analytical staff to routine analyses more of these studies are being contracted out; in such cases the company needs to have well-developed guidelines and quality control procedures[1347] to ensure the contract laboratory performs the analyses as competently as the in-house analytical laboratory. Such matters as stability at various temperatures and in various fluids are readily assessed but it is the less obvious pitfalls that are to be discussed here.

Methodological problems already arising during the collection of blood for assay of metriphonate and dichlorvos have been pointed out.[1366,1367] The former rapidly breaks down to the latter in blood samples but this can be prevented by direct acidification with phosphoric acid. Another possibly unrecognized problem in collection of samples may arise when the concentration of drug is rapidly changing and the results are to be used in pharmacokinetic analysis as exemplified for the analysis of lidocaine in sheep blood following a bolus injection.[72]

One immediate problem with ester-type drugs is the presence of serum esterases (Chapter 2). If these enzymes are active against the drug the presence of sodium fluoride in the sample tube as an anticoagulant will also inhibit the esterases, as for example in an assay for erythromycin propionate.[1368] Adam[1369] recommends a fluoride concentration of 5 mg ml^{-1} for this purpose rather than the usually supplied 1 mg ml^{-1}, when intended as an anticoagulant alone. Brogan et al.[1370] have suggested that sodium fluoride is essential in inhibiting the degradation of cocaine in blood samples. Similarly, failure to consider the effects of pseudoesterases has confused research on the metabolism of cocaine in humans.[1371,1372]

The decomposition of diamorphine (heroin) in plasma has been studied in some detail. The drug is rapidly deacetylated in aqueous solution at alkaline pH to form 6-monoacetylmorphine and further deacetylated to form morphine.[1373,1374] Barrett et al.[1375] suggested that the degradation by serum esterases takes place even at 4°C and recommended immediate freezing of collected plasma and liquid-liquid extraction to prevent misleading analyses in the study of heroin pharmacokinetics.

The most usual procedure for ensuring the stability of drugs on storage is to deep freeze samples as soon as possible after collection. This is not without problems, however. The problems can begin as soon as freezing starts, as pointed out by Scales[1376] and described as the apple jack effect; that is, the drug is concentrated in the first few drops to freeze and hence will remain at the surface and be susceptible to oxidation. Thus it might be necessary to stipulate that freezing should be carried out while gently swirling the sample tube.

Once frozen, the drug may still not be safe from decomposition. Recently, Edmonds and Nierenberg[1377] reanalyzed samples that had been stored at −70°C for 5 years for the retinol, d-α-tocopherol and β-carotene and concluded that in samples

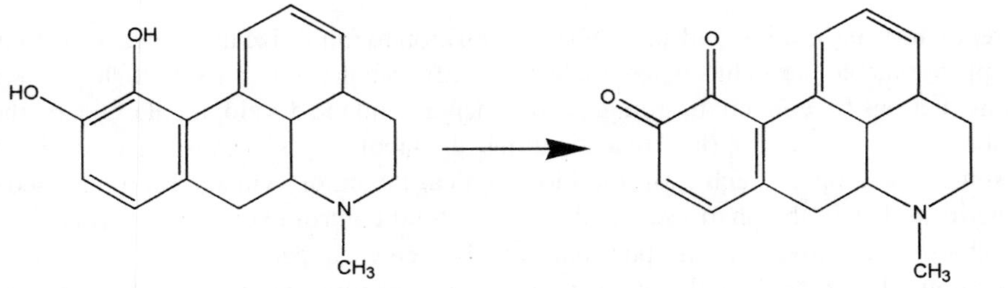

FIGURE 12.1 Oxidation of apomorphine to a quinone.

from some patients, clinically important changes in concentration had occurred, although after 5 years this would hardly matter to the patient. Apomorphine, like most catechols (Figure 12.1), is prone to oxidation to quinones, even when stored at −15°C unless ascorbic acid is added as an antioxidant.[1378] The presence of antioxidants can have the reverse effect, however, and there have been reports on the reduction of N-oxide metabolites of chlorpromazine reverting to the parent compound on storage of samples containing antioxidants.[1379,1380] Indomethacin has also been reported to be unstable in the deep freeze.[1381] One consequence of freezing samples is that the protein in plasma or serum samples may become denatured to an extent which depends on the temperature;[1382] this may have more subtle consequences in the subsequent analysis, for example, by releasing bound drug in assay methods where the free drug only is measured.

Some authors report apparent instability of drugs in deep frozen conditions; however, instability is not always distinguishable from loss of sample by other routes and it is wise to establish that there are confirmatory breakdown products present before assuming that the deep freeze conditions are inappropriate for the safe storage of samples containing a particular drug. Thus, de Silva[1383] prefers to confirm the claim of instability by listing the appearance of unnamed drug and the appearance of a decomposition product with the passage of time.

An interesting phenomenon has been described in experiments to determine the stability of hydralazine on storage at −20°C; hydralazine added to plasma and stored at this temperature was apparently very unstable probably being converted to the pyruvic acid hydrazone, yet plasma samples in pharmacokinetic studies could be repeatedly frozen and thawed and still return the same apparent hydralazine content.[1384] This discrepancy was put down to the presence of metabolites which were readily hydrolyzed to hydralazine in the assay, but were not degraded on storage.[1385]

Problems can also arise when the sample is thawed prior to analysis. As the sample is thawed, there may be localized concentration effects and the sample should be shaken gently to minimize such effects.[1382] The sample must then be confirmed to be homogeneous before taking aliquots for further analysis. This is a notorious problem, particularly for large urine samples, when impatient and careless laboratory workers may take samples of semithawed urine, with the consequent error in the final result. On thawing, enzymes that have been dormant during storage may be reactivated and hence thawed samples should be processed as soon as possible on thawing. However, the enthusiasm for rapid thawing may mean warming the sample to too high a temperature with the danger of both decomposition of the analyte or even raising the temperature for optimum activity of any enzymes that may be present.

12.3 EFFECTS OF PRESERVATIVES

Sodium metabisulfite is a commonly used preservative for urine samples; however, it may also have an inhibitory effect on enzymic hydrolysis of conjugates as has been demonstrated for its effect on sulfatases used in the enzymic hydrolysis of 4-hydroxyantipyrine sulfoconjugate.[1386]

12.4 EXTRACTION

For partial purification of biological fluids containing drugs, the most widely used procedure is still extraction into an organic solvent. It is not the purpose of this chapter to discuss the theory of the best choice of extractions, the details of which can be found elsewhere; however, the general points to be observed should be noted. The drug needs to be in the un-ionized form to be extractable by the organic solvent from aqueous solution. Hence the pH of the aqueous phase needs to be adjusted to ensure that this is so for ionizable drugs. A standard scheme for extraction of drugs for toxicological screening,[1387] although based on extraction from pH-adjusted aqueous phases, is inappropriate when considering analysis for known drugs,[1388,1389] and it is desirable to optimize the pH for a specific drug. This may be done by measuring the pK_a and selecting the pH on theoretical grounds. The second important parameter for extraction of drugs from aqueous phases is the nature of the organic solvent. Highly lipophilic compounds will be readily extracted by nonpolar solvents such as hexane (see Chapter 2 for the table of solvent polarity), whereas drugs containing, for example, hydroxyl groups, will require correspondingly more polar solvents.

Thus, on paper, the optimum conditions for drug extraction can be selected, ensuring efficient extraction of drug consistent with low extraction of other materials. However, in practice, other factors will determine the choice of conditions in a particular laboratory as discussed below. These factors mostly relate to ease of handling, although the chemistry of the solvents is also important, and the use of internal standards may also be a consideration.

In the extraction step, the density of the extracting solvent with respect to the aqueous phase may be a consideration. Thus, if traditional separating funnels are used, then the extracting solvent may be chosen to be heavier than water, so that the phase containing the drug can be run off first without contamination from the upper phase; if the separation of phases is carried out in centrifuge tubes — as is more often the case for batch operation of large numbers of samples — then it may be more convenient to have the organic phase as the upper layer which can then be removed by aspiration without disturbing the aqueous layer. Of course, these arguments can be reversed, depending on which phase is to be discarded or retained for further processing. There is no firm rule on this matter; some individuals may prefer to carry out all their washing and extraction operations in a single tube, while others prefer to transfer phases to clean tubes for the subsequent steps.

A solvent may be selected for its volatility, so that it is easily evaporated in a subsequent concentration step. The most obvious contender for this rule would be diethyl ether, but too high a volatility can be a problem if the sample is not kept cold, for example, during centrifugation, and ether is not favored for quantitative work without an internal standard. The rule of thumb that the extracting solvent should be just polar enough to extract the drug of interest is usually followed, the choice being made by trial and error. Whelpton and Curry[1390] described a systematic separation of fluphenazine from its metabolites using the principle of least polarity combined with pH considerations. The more polar solvents will, it should be noted, be partially miscible with water and therefore the apparent organic phase will contain a large amount of water and hence water-soluble material. The consequences of this is that it is possible for water-soluble conjugates of drugs to be transferred into the organic phase. If they are subsequently hydrolyzed this will lead to erroneously high levels of parent drug. Similarly, protein may be transferred and this could have undesirable consequences for chromatographic columns. To some extent this can be obviated by using a small amount of a very polar solvent such as amyl alcohol bulked up with hexane. The extraction properties of the drug are thus conserved, but with little carryover of the aqueous phase. It is interesting to note the perceived role of amyl alcohol in different laboratories using it as an additive with organic solvents of low polarity. Some workers use it merely to prevent absorption of drugs onto glassware and assume the extracting property resides totally in the main organic solvent. The importance of the alcohol as an extractant in its own right was illustrated by Curry and Whelpton[1391] on the one hand and Bailey and Guba[1392] on the other. Reid,[1393] however, has pointed out that in the United Kingdom, "amyl alcohol" is mainly 3-methylbutan-1-ol with some 2-methylbutan-1-ol, whereas U.S. laboratories would describe this solvent as isoamyl alcohol.

The chemical properties of extracting fluids can give rise to unwanted effects. These may be of two types: the chemical reactions of the solvent itself and the insiduous breakdown products they may contain. Some chemical properties may be fairly obvious, but just the same, they are often overlooked. Primary amines such as ethylamine will react with ketones and aldehydes to form Schiff's bases, and conversely ketones used as extracting solvents, such as methylbutylketone, will react with drugs containing primary amines. In some instances the reaction may go unrecognized and the drug may be unknowingly (but still acceptably) assayed as the derivative. This phenomenon may often be at the root of puzzling chromatographic results when the nature of the extracting solvent is altered. This is well illustrated in Curry's[1391] procedure for chlorpromazine. Isoamyl alcohol used for extraction of chlorpromazine contains isovaleraldehyde as an impurity, which reacts spontaneously with primary amines such as didemethylchlorpromazine; this product then separates very efficiently from chlorpromazine and the monodesmethyl metabolite. In the absence of isovaleraldehyde, the separation of the two metabolites is very poor. As Curry put it: "We used bad isoamyl alcohol without knowing it." Less obvious is the formation of hemiketals between ketones and alcohols, especially if traces of acids are present to catalyze the reaction.

Ethyl acetate is a popular extracting solvent because of its high polarity, cheapness, and volatility. However, at pH values greater than 9.0 and at normal laboratory temperatures it may be hydrolyzed to acetic acid and ethanol. This has a dual consequence; the pH of the sample may unwittingly be lower than needed and the alcohol will modify the polarity of the extracting solvent. Scales[1394] has noted that if ethyl acetate must be used as an extracting solvent at elevated pH values then low temperatures (2 to 10°C) and short extraction times may be successful.

Apart from impurities in solvents such as amyl alcohol mentioned above, impurities may arise from decomposition of the solvent. On exposure to air, chloroform and dichloromethane degrade to the extremely reactive phosgene. It is for this reason that commercially supplied chloroform contains a stabilizer such as ethanol. This may lead to the presence of ethylchloroformate which has been shown to be instrumental in the conversion of norcodeine to norcodeine carbamate.[1395] This proved a parallel to Curry's experience with chlorpromazine, norcodeine only being measurable in the gas chromatographic method because of its conversion to the carbamate. The occurrence of carbamate derivatives has also been noted when tricyclic antidepressants are extracted with chloroform not containing ethanol as a preservative.[1396]

Phosgene can be removed from chloroform before use by the simple expedient of washing the solvent with water, and ethanol can be removed if desired by passing the solvent through a column of alumina, a device that can also be used to remove peroxides from ether. The latter procedure is particularly recommended because of the safety hazard when large volumes of ether are to be reduced in volume, and also to prevent oxidation of extracted drugs. Needless to say, such purified solvents should be used immediately before the decomposition from which the ethanol is affording protection sets in.

The occurrence of cyanogen chloride as an impurity in dichloromethane has also been reported to lead to artifactual "metabolites" in drug studies.[1397] A metabolite of Wy 23699, on extraction with this solvent containing cyanogen chloride, readily formed the cyano derivative (Figure 12.2).

In order to encourage partitioning of ionic compounds into organic phases, some workers saturate the aqueous phase with various salts, usually ammonium sulfate. Other desirable effects of doing this are to reduce the occurrence of emulsions, to ensure that otherwise miscible solvents (acetonitrile[81] or 2-propanol[1398]) form separate layers, or to convert a heavier than water organic phase into a floating one as described by Leppard and Reid[1399] for an ether-chloroform mixture.

In working with large numbers of samples, the analyst is attracted by any device that will enable him to reduce the volumes of solvent necessary for extraction. Classical extraction procedures are generally recognized as inappropriate for methods designed for individual drugs. Thus Campbell[1389] and Reid[1388] both suggest that Jackson's[1400] figure for organic volumes of ten times those of the aqueous sample is unnecessary. Extraction ratios of as low as 1:50 and 1:10 are claimed to be adequate for efficient extraction.[1389,1401]

Similarly, the times required for extraction are often unnecessarily long[1402] and "appear to coincide strongly with the gustatorial habits of a particular laboratory." Campbell[1389] reported a 300-s extraction of fenfluramine (90% after 5 sec) from pH-adjusted plasma with chloroform. Apparently, short, not too vigorous, extractions

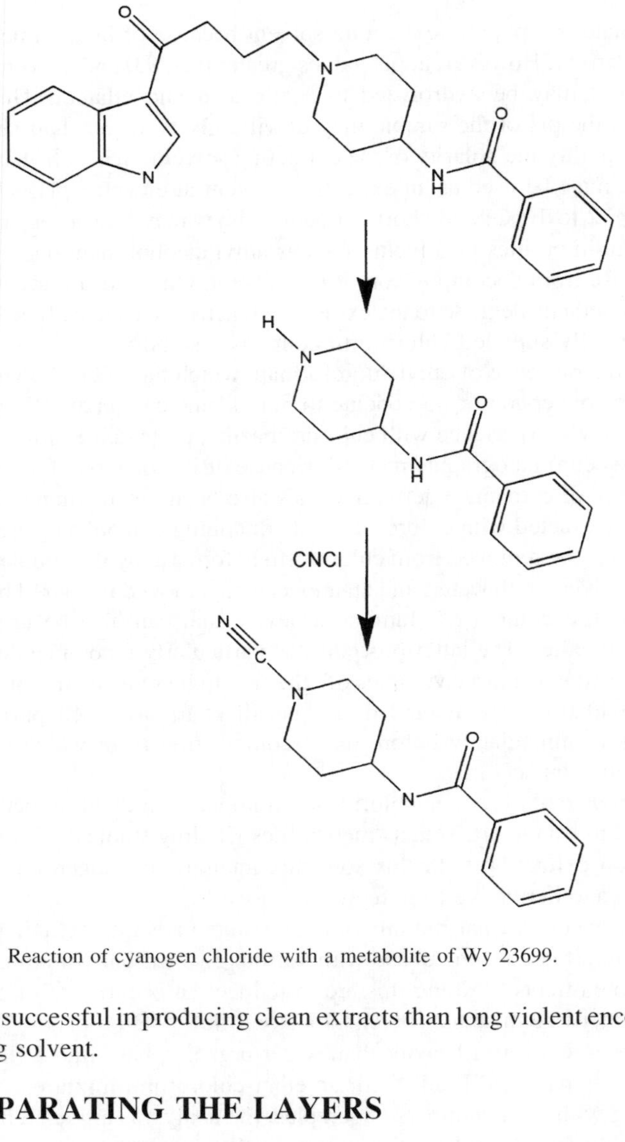

FIGURE 12.2   Reaction of cyanogen chloride with a metabolite of Wy 23699.

may be more successful in producing clean extracts than long violent encounters with the extracting solvent.

12.5   SEPARATING THE LAYERS

Because of the general necessity (and desirability) of working with small volumes of solvent, it is not usual to use classical separating funnels for separating the layers, most operations being carried out in small test tubes or centrifuge tubes. As mentioned previously, the analyst can to some extent control the situation so that he is discarding either the upper or lower phase according to taste. If the layers are separated cleanly and he wishes to discard the top layer and continue working with the lower layer, then the simplest procedure is to remove the top layer with a Pasteur pipet with a very fine tip. To minimize disturbance of the interface, a curved tip can

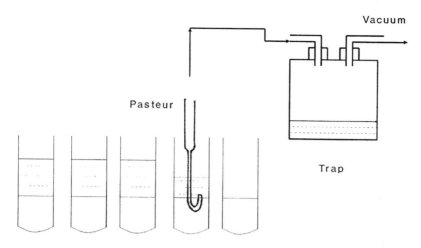

FIGURE 12.3 Use of a vacuum line for rapid removal of an upper solvent layer.

be used as shown in Figure 12.3. In dealing with large numbers of samples, the Pasteur pipet can be connected to a vacuum line, with a suitable trap and the top layers in batches of tubes removed rapidly in sequence. A batch of 20, for instance, could be processed in a few minutes. Removing the top layer in this way should cause no carryover from one sample to the next. If the top layer is to be retained, then the Pasteur pipet connected to a vacuum line is not so convenient and the bottom layer may need to be removed by hand pipetting. However, the complete removal of a lower layer is difficult by this method and it would be better to transfer the upper layer (or an aliquot of the upper layer) to a clean tube for further processing. A quantitative transfer can be achieved if the top phase is sufficiently mobile and the bottom phase is sufficiently viscous, by simple decantation. This can be aided when the lower phase is an aqueous phase, by standing the tubes in an acetone-dry ice bath; the lower layer freezes and the top layer is readily poured off. If this technique is used, the usual evaluation of the complete method must be carried out to ensure that the partition characteristics remain favorable at the lower temperature. An interfacial emulsion, usually undesirable in separation science, may actually be of practical help in restraining a lower layer, when the top layer is being poured off, provided quantitative aspects are otherwise provided for; that is, internal standards are used, or fixed volumes of sample are taken.

On occasion it may be necessary for the lower layer to be transferred to a fresh tube — for example, when the top layer is also to be retained for further processing — and the Pasteur pipet method is used for transferring the lower layer in the presence of the upper layer. The experienced analyst will know that the bulb of the Pasteur pipet need not be fully squeezed until the pipet is in place, to expel the small amount of upper phase which will have entered the lower few millimeters of the pipet as it penetrates the upper layer; the even more experienced analyst will know how to exert just the amount of pressure to prevent ingress in the first place.

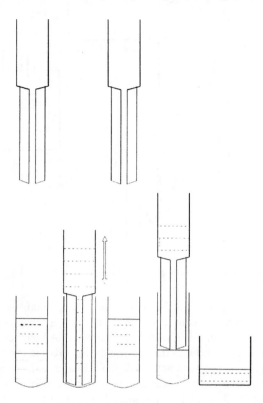

FIGURE 12.4 Novel device for the rapid separation of layers in solvent extraction procedures. (A) sovents placed in lower unit; (B) solvents forced through channel and mixed; (C) upper unit withdrawn and layers allowed to separate; (D) upper layer taken into upper unit; (E) upper layer poured off, retaining lower layer.

An interesting device has been described as an alternative to separating flasks. The device is shown schematically in Figure 12.4, and basically consists of a separating tube fitted with a hollow plunger connected to the main chamber via a narrow channel. The phases to be mixed and separated are placed in the test tube. As the plunger is depressed, both liquids are forced upward through the channel into the plunger, being thoroughly mixed in the process. The plunger is then slowly withdrawn until the interface is exactly at the top of the connecting channel; the upper layer can then be poured off — the capillary effect ensuring that the lower phase remains in place.

Although it is possible to extract drugs from aqueous solution with small amounts of organic solvent, this does give rise to difficulties in handling such small volumes. De Bree[1403] demonstrated how 50 μl of iso-octane can be used to extract analytes from 2 ml of plasma. As illustrated in Figure 12.5, a simple manipulation using a cone-terminated capillary, made from a Pasteur pipet, enlarges the organic layer thickness from 2 to about 20 mm. Thereby even 20 μl may be taken easily from a 50

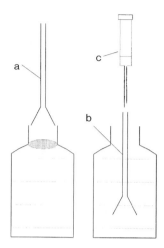

FIGURE 12.5 Simple device for transferring entire upper phase when only a small amount of the solvent is required for extraction. (a) Cut-off Pasteur pipet; (b) pipet filled with upper phase; (c)syringe for withdrawal of upper phase.

μl sample extract. The technique can be used for solvents heavier than water, such as chloroform, by saturating the aqueous phase with potassium carbonate.

Phase separation papers have been suggested, but appear to have been little used in drug analysis, although they can be very convenient. The mixture is simply poured through the paper in a traditional filter funnel, the aqueous phase being barred from filtration by the hydrophobic paper. The paper will eventually become wetted and therefore should not be left in position too long. Such papers may be more suitable for removing the last traces of an aqueous layer, although it should be noted that ordinary cellulose paper may form the same function. However, falsely negative urine drug assay results have been shown to be due to filtration by the loss of drugs by adsorption to the filters (polyvinylidene fluoride), polysulfone, cellulose acetate and nylon filters, and centrifugation is recommended instead.[1404]

The discussion so far has assumed that the layers have already separated into two clean layers. Natural gravity and the immiscibility of the solvents will, it is hoped, ensure that this happens, some systems settling more quickly than others. Some workers accelerate the natural processes by centrifugation as a routine step. Even if there is an apparent rapid separation of phases on standing, it is still often worthwhile to introduce this centrifugation step, as microdroplets of organic phase may still remain in the aqueous phase, being seen as a slight opalescence.

The most common difficulty in the separation of phases is the formation of persistent emulsions. Emulsions are caused when small globules of one phase in a two-phase mixture are allowed to form; vigorous shaking will break the phase into small globules and if the surface tension is low, the globules will persist. Thus, persistent emulsions occur where vigorous extraction is performed in the presence of proteins, soaps, or detergents. Centrifugation, as mentioned above, is often the most successful method of breaking emulsions, but a number of other techniques have been proposed, including successive freezing and thawing, filtration, touching with a glass rod, saturating the aqueous layer with salt, and the addition of a small amount of alcohol. Addition of alcohol to some samples that form emulsions is undesirable if reproducibility of extraction is to be maintained. Most workers would agree that, rather than introduce stages to break up emulsions, it is better to design methods where the emulsions are prevented from occurring. Thus it is now well recognized that gently stirring, shaking, tumbling, or rolling of samples is better than vigorous extraction procedures; in my experience, extraction tubes gently tumbled at 20 rotations per minute does not cause intractable emulsions when buffered plasma samples are extracted with most organic solvents.

12.6   DRYING THE ORGANIC PHASE

Once separated from the organic phase, it may be necessary to ensure that the organic phase is free of water. The presence of water in this phase can be critical, but has received little attention. If the solvent is to be evaporated to dryness, the last droplets of water will entail more forcing conditions than necessary for the solvent alone; at the end of the drying step the drug will be out of solution and, if supplementary heat is being used, decomposition will be likely. Some workers will add a few drops of ethanol to help in this final evaporation of the last traces of water, but this may provide perfect conditions for the formation of ketals as described above.

The final residue may also contain mineral acids or alkalis and these can have even more disastrous consequences for the drug. Direct injection of organic phases into HPLC columns may produce anomolous results due to transient changes in mobile phase composition if the organic phase contains unknown amounts of water. In gas chromatography, the injection of salts or water-soluble proteins in the organic phase may lead to undesirable deposits at the top of the column, or the water itself may contribute to the stripping of the stationary phase.

It is sometimes desirable to carry out chemical derivatization in the organic extract, for example the use of acylation of amines or alcohols prior to gas chromatography. The different rates of acylation in different solvents for the heptafluorobutyrylation of the sterically hindered amine terodiline is probably partially due to different water contents of the organic solvent.[1405]

Traces of water can be removed from organic phase by the adding a pinch of anhydrous sodium sulfate. Even visible amounts of water can be mopped up by this reagent and the dried solvent can be readily poured off, the hydrated salt sticking in clumps to the sides of the vessel. Alternatively the organic extract can be simply filtered through anhydrous sodium; if there are only trace amounts of water, simply passing the phase through dry filter paper will be adequate with less potential of losing drug by adsorption to the solid salt. For example, calcium chloride added to the extract was shown to cause loss of oxprenolol in an assay for this compound.[1406] Immediate storage of dried-down samples in a desiccator is recommended to remove last traces of water or to preserve the anhydrous condition of samples. Reid[1407] suggests that silica gel, rather than phosphorous pentoxide, should be used as a desiccant; the latter reagent will overdry the atmosphere and cause volatilization of drug from the sample. A similar effect may also be noted in freeze-drying procedures; during the process of freeze-drying the sample will remain cold, but if the process is allowed to continue after the water has been removed, the temperature of the sample will rise to ambient and loss of sample may occur.[1408]

12.7   EVAPORATING TO DRYNESS

In searching for methods of measuring drugs at the limit of sensitivity, it is often necessary to evaporate extracts to dryness in order to reconstitute them into small volumes for total transfer to chromatography columns or plates. It is the evaporation to dryness stage that can be the cause of loss of sample, the most common routes being bumping of the solvent, adsorption onto glassware, and volatilization of the drug. The resulting variations in analytical recovery will be reflected in poor calibration lines and poor reproducibility of results as detailed for tricyclic antidepressants in particular.[1409-1411] Sonsalla et al.[1412] described conditions necessary to reduce this variability by using silanized glassware, evaporating solvents at 40°C, removing drug residues promptly and using two internal standards for losses that occur prior to chromatography. The use of internal standards is a well-respected technique for accounting for small variations in recovery of drug through an extraction procedure, but to rely on internal standards to correct for wide variations would not be good analytical practice. Wide variations must be investigated and appropriate action taken to improve recoveries.

On the small scale usually required for drug extractions, it is unlikely that rotary evaporators will be utilized. Almost all laboratories will evaporate solvent from a warmed test tube or centrifuge tube in a stream of dry nitrogen. For warming the tubes, a water bath, roughly thermostatted, may be used, although a heated aluminum block with appropriately drilled holes for reception of test tubes is very convenient. The now less commonly used hot sand bath does have the great merit that tubes can be placed to any depth depending on the degree of heating required, and supporting apparatus is not required. Evaporation with any of these heating arrangements is usually with the aid of disposable Pasteur pipets or stainless steel needles connected to a gas cylinder via a manifold for batches of tubes; the analyst must beware of contamination from one batch to the next if the nitrogen jets (pipets or needles) are not replaced or thoroughly cleaned between batches.

Loss of sample or part of sample by sudden bumping of the solvent is something most analysts will be reluctant to admit to, although the occasional annotation "lost sample" is often due to this cause. Low heat and a gentle stream of nitrogen may not be sufficient to stop bumping of solvent, particularly if the extract is free of particulate matter — and hopefully it should be at this stage! It may be useful to add a single antibumping stone to each tube, accepting some risk of adsorption of analyte onto the stone. The gentle stream of nitrogen will help maintain the reflux effect of condensing solvent on the upper, cooled part of the tubes washing analyte down the sides and concentrating the residue in the bottom of the tubes. This is in contrast to the enthusiastic blasting off of solvent with high flow rates and consequent spreading of the residue over a larger area.

Careful handling of apparatus, however, will not solve the problem of loss of small amounts of drug in analytical procedures. This is the phenomenon of adsorption (and hence loss) onto surfaces, particularly glass surfaces. This problem can arise when the drug is in dilute solution in a nonpolar solvent but is even more serious at the evaporation to dryness stage. The simplest way to overcome this adsorption is to include a polar solvent in the procedure, either as an additive to the extracting solvent or immediately prior to evaporation. The most common reagents for this purpose are amyl alcohols or ethanol. Reid[1413] suggests adding 2% ethanol. James and Wilson[1414] describe the addition of 0.1 ml of 5% propylene glycol in the evaporation of extracts containing aldosterone. In this case the propylene glycol remains in the tube after extracting solvent (dichloromethane-methanol) has been evaporated and holds the aldosterone in solution, preventing adsorption onto the glassware.

Adsorption losses can be prevented by pretreatment of extraction tubes with ethanol or by coating them with a thin layer of Teflon or by silanization. Reid[1413] requests that authors distinguish between silanization and siliconization, the former being formation of chemically bonded silyl groups, the latter being physical adsorption of silicone compounds onto active sites and hence reducing adsorption of drugs. A typical siliconization is carried out merely by rinsing the glassware in a 1% solution of trimethylchlorosilane in toluene, followed by drying at 100°C for 30 min. Silanization on the other hand uses an active reagent with an appropriate catalyst. Thus in a typical silanization, the glassware is treated with an aqueous solution of 1% siliclad and two drops of concentrate ammonium hydroxide for 3 min, followed by copious washing with water.[1415] Silanization provides greater resistance to hydrolysis than mere siliconization.[1416]

Not all authors agree that silanization is the best treatment for glassware. For example, Schill[1417] has noted a greater loss of dimethylprotriptyline when silanized tubes are used compared with acid-washed glassware. Reid also suggests that chromic acid washing of unscratched glassware may be quite adequate for the analysis of chlorpromazine metabolites.[1398]

If coating of the glassware, in any form, is considered undesirable, then adsorbed drug can always be washed off with hot methanol as described in an assay for surfactants in blood.[1418] In a TLC assay for muzolimine, the extract was spotted directly onto the TLC plate rather than risk adsorption onto glassware in the usual drying down steps.[1419] Adsorption of some drugs can be minimized by adding large amounts of other drugs as carriers; for example, amitriptyline and its metabolites were successfully analyzed by adding maprotiline as a carrier to minimize adsorption onto glassware.[1415]

Natural volatility of small molecules with low boiling points, such as ethambutol,[1420] valproic acid,[1421] and salicylic acid,[1422] can be a cause of loss in the evaporation stage due to volatilization. The problem was solved in the case of salicylic acid by using extra-long tubes (15 cm long in fact) when loss of compound was reduced to less than 2%, even when the samples were left for as long as 30 min after they were dry.[1422] The volatilization of ethambutol was prevented by adding a few drops of 1 M HCl prior to evaporation of the chloroform extracts; the compound in the form of the hydrochloride salt is nonvolatile.[1420] The corresponding situation with valproic acid was tackled by the conversion of this acid to the potassium salt prior to the evaporation of the solvent.[1421] The conversion of organic bases or acids to the nonvolatile salts is a general method to avoid loss of volatile compounds — acetic acid being a useful milder reagent for organic bases, and piperidine being suitable for organic acids.[1423] On the other hand, derivatization of involatile drugs to more volatile derivatives is often performed to make them more amenable to gas chromatography or mass spectrometry; in purifying these derivatives, this very volatility can cause loss of analyte.[1424]

12.8 SAMPLE PREPARATION

Some of the procedures mentioned above are applicable on the normal laboratory analytical scale. One of the features of ultramicro analysis is the need to keep the addition of any chemicals or solvents to a minimum and many workers have endeavored to devise microsystems of analysis. These procedures also have the intrinsic merit of being relatively rapid. For example Bonato and Lanchote devised a procedure for differential extraction of drugs and metabolites from apolar compounds (thought to be detrimental to reverse-phase columns) using partitioning between hexane and the mobile phase before analysis and claimed longer column life as a result.[1425] Salting out solvent extraction for preconcentration of benzalkonium chloride prior to high-performance liquid chromatography has also been claimed to increase concentration tenfold.[1426] Van der Vlis et al. also claimed on-line trace enrichment of doxorubicin on iron(III)-loaded 8-hydroxyquinoline-bonded silica.[1427]

12.9 EXTRACTION CARTRIDGES

Many of the problems associated with solvent extraction and subsequent removal of large amounts of organic solvent can be minimized by using solid phase extraction techniques. One of the earliest versions of such a technique was simply to fill a Pasteur pipet with silica gel over a glass-wool plug and add the sample to the top of this mini-column. Unwanted components were eluted with a solvent of low polarity and the material of interest was eluted with a small amount of the most appropriate solvent. Homemade versions of this apparatus would suffer from irreproducibility, mainly derived from the amount and activity of the silica gel; additionally, for variable samples such as urine, the sample itself could affect the properties of the column. With the rise of HPLC and the production of stationary phases with tailor-made properties, these phases were also utilized as the material for these extraction columns (Table 12.1). It was only a short step then to the design of complete systems of ready-made columns and solvent and buffer sequences to optimize the purification and quantitative extraction of analytes. Some columns have been developed for specific applications, such as the phenylboronic acid phases for catecholamine separations. However, apparently similar columns from different manufacturers may well have different properties toward a particular analyte.[1428]

The use of such commercial cartridges can be extremely time-saving compared with traditional liquid-liquid extraction, but the sample matrix may have unlooked

TABLE 12.1

**Examples of the Use of Pre-Extraction Cartridges in the
Analysis of Drugs in Biological Fluids**

Drug	Sample	Cartridge	Method of Analysis	References
Amiloride	Plasma	Bond Elut C_8	HPLC	1433
Amiodarone	Serum	C_2 column	HPLC	1434
Antipyrine	Urine	C_{18}	HPLC	1435
Cefaclor	Serum	Bond Elut C_{18}	Capillary HPLC	819
Chlorzoxazone	Plasma	C_{18}	HPLC	1436
Cocaine	Brain	X-TrackT	HPLC	1437
Codeine	Plasma	Bond Elut Certify	DB-5 GC	1438
Dextromethorphan	Urine	Bond Elut Phenyl	HPLC	1439
Diazepam	Plasma	Bond Elut C_{18}	HPLC	1440
Doxorubicin	Serum	Bond Elut C_8	HPLC	1441
Epirubicin	Serum	Bond Elut C_8	HPLC	1441
Ethacrynic acid	Urine	Extra-Sep C_{18}	HPLC	1442
Flecainide	Serum	Varian C_{18} membrane	HPLC	1443
Fluconazole	Serum	Bond Elut C_{18}	HPLC	1444
Fluticasone	Plasma	C_{18}	RIA	1445
Hydrocodone	Plasma	Bond Elut Certify	DB-5 GC	1438
Hydromorphone	Plasma	Bond Elut Certify	DB-5 GC	1438
Indomethacin	Plasma	Bond Elut Phenyl	HPLC	1446
Isosorbide dinitrate	Plasma	Bakerbond C-18	OV-17/QF-1 GC	1447
Meprobamate	Serum, plasma	Extra-Sep C_8	3% OV-17 GC	1448
Methadone	Plasma	C_{18} Sep Pak	HPLC	1449
Mexiletine	Serum	Varian C_{18} membrane	HPLC	1443
Morphine	Plasma	Bond Elut Certify	DB-5 GC	1438
Morphine	Plasma	Bond Elut C_{18}	HPLC	1450
Oxazepam	Urine	Bond Elut C_2	HPLC	767
Oxcarbazepine	Plasma	Bond Elut C_{18}	HPLC	1451
Oxiracetam	Plasma	Phenyboronic acid	HPLC	1452
Oxycodone	Plasma	Bond Elut Certify	DB-5 GC	1438
Oxyphenbutazone	Plasma	Bond Elut Phenyl	HPLC	1446
Pentoxifylline	Plasma	SPE	HPLC	1453
Pentoxifylline	Serum	Separcol SI C_{18}	HP1 GC	1454
Phenylbutazone	Plasma	Bond Elut Phenyl	HPLC	1446
Probenicid	Urine	Bond Elut C_8	HPLC	1455
Propranolol	Plasma	Bond Elut PBA	—	1456
Suxibuzone	Plasma	Bond Elut Phenyl	HPLC	1446

for effects on their performance. Urine would usually be expected to be the simplest
fluid for this type of extraction, but because of the larger amounts of sample usually
necessary, the extraction profile may be poorer, with poorer recoveries. Whelpton
and Hurst,[1429] in a discussion of their experience with Bond Elut columns, claim that
whole blood can be analyzed successfully if the cells are disrupted first. Nevertheless, other authors suggest that serum samples are better than plasma samples, as the
fibrin particle remaining in the plasma may block the cartridges, altering flow rates
and reproducibility.[1430]

Because of the greater variety of phases that can be used for solid phase extraction, and the variety of solvents, buffers and other solutions that can be used in the elution sequences, the technique of solid-phase extraction is considerably more versatile than traditional methods, but the corollary of this is that the analyst must be well informed on the mechanism of all the interactions and secondary interactions if he is to maintain proper control of the separations. However, because of the ready analysis of drugs by HPLC, it is possible to use HPLC retention data to predict the behavior of drugs on extraction cartridges which use the same stationary phase, as described by Casas et al.[1431,1432] for benzodiazepines. Those authors suggested that in general, C_2 phases are best for benzodiazepines in extracting fewer extraneous peaks from urine and plasma than C_8 or C_{18}.[1431]

The logical development to the use of chromatographic-type columns as concentrating devices was to include such columns in the chromatographic system and to incorporate the complete elution sequence into an automated system using appropriate column switching techniques. Examples of such commercially available systems are the Advanced Automation Sample Preparation (AASP) from Waters Associates, used for example in a method for the determination of mitoxantrone in serum of cancer patients,[1457] and the Automatic Sample Preparation with Extraction Columns (ASPEC) from Gilson, used for example, to analyze carbamazepine in plasma.[1458] Chen et al.[1459] have carried out a thorough evaluation of the ASPEC system for drug screening of plasma or urine in combination with Clean Screen DAU columns followed by gas chromatographic analysis. The use of mini-columns combined with detailed study of elution schemes has been reported with successful schemes developed for a number of drugs and their metabolites in serum, plasma, saliva, and urine.[1460]

Although not strictly an extraction cartridge, dialysis units have been used for sample pretreatment and included in-line with liquid chromatography systems. For example, Van de Merbel et al. described such a system for the analysis of benzodiazepines in plasma. In this method 100 µl plasma was dialyzed for 7.6 min against a chosen acceptor at a flow rate of 3 ml min^{-1}, then concentrated in a more conventional precolumn before analysis by HPLC. Andresen et al.[1461] also used a predialysis step for the HPLC analysis of pholcodine in human plasma and in whole blood.

12.10 SOURCES OF CONTAMINANTS

In the context of this discussion, a contaminant is an undesirable substance introduced into the sample in the course of processing it, as opposed to the so-called endogenous contaminants which more properly are part of the analytical problem. As true contaminants are introduced by the analytical process, it will be more fruitful to find their source and take the appropriate steps to eliminate them rather than to incorporate special steps into the analytical method. However, a contaminant in one procedure will not necessarily be a contaminant in another. Thus solvents supplied as spectroscopically pure may well contain nonabsorbing species that will affect chromatographic properties. A particular contaminant eluting from filter paper, tri-iso-butyl phosphate,[1462] is more serious in subsequent gas chromatographic procedures when a thermionic detector is used than when using other types of detectors.

The introduction of contaminants can occur at the very beginning with the specimen collection tubes. For blood samples there will invariably be some type of coagulant. In an analysis of frusemide in plasma by HPLC, the type of anticoagulant used was critical for subsequent analysis. It was found that heparin used as an anticoagulant caused interfering peaks in the chromatogram, which were not observed when EDTA was used. When sodium oxalate was used, interfering peaks were observed with an ultra-violet detector but not with a fluorescence detector. It is apparent that, at least in the case of oxalate, the peaks are not derived directly from the anticoagulant but arise as an indirect consequence, for example, by aiding the extraction of organic matter from containers, stoppers, and tubing used in the extraction apparatus.

Laboratory workers and their habits may be a source of contamination, either by contact with fingers leading to excess amounts of natural products in the sample or even by fibers from paper tissues; fibers from lab coats with optical brighteners may have deleterious effects on fluorescence assays.[1463] Other contaminations from the laboratory will include barrier creams, bench polishes, perfumes and cosmetics, greases, plastics, rubber bungs and tubing, oil from air lines, filter papers, and glass wool.[1464] The occurrence of organic material leached out of plastic ware is not surprising; more surprising, however, is the apparent leaching out of organics from plastic tubing used to carry nitrogen, yet this has been reported by several groups. One such contaminant was identified as tributyl aconitate[1465] which potentially will interfere in both gas chromatographic and HPLC assays. Hooper and Smith[1466] found interferences due to bis-(2-ethylhexyl)phthalate as a spurious peak in their gas chromatographic assay for oxprenolol, but were able to eliminate it by using PTFE tubing instead of PVC in the nitrogen line. On the other hand, Feyerabend and Russell[1467] found an unidentified contaminant due to rubber tubing and eliminated it by switching to PVC.

It may be expected in these days of throwaway laboratory ware that contaminations of samples during processing would be less of a problem. Not so, unfortunately, as the factory-fresh equipment is often the biggest source of contaminants and will often need prewashing even before its first use. Even labor-saving devices such as cartridges for extracting biological samples will often contain elutable phthalates, fatty acids, alcohols, and resins.[1468]

The point regarding most of these contaminants is that they are preventable if the source is known. Mass spectroscopy is a very useful tool for tracking down contaminants of manmade origin. In several laboratories[1399] the mass spectrum of isolated material is checked with standard spectra in relevant collections; the identified compound can then be looked up in reference books such as the *Merck Index*, or the manufacturer can be contacted to obtain lists of products likely to contain the material. This is a more satisfactory procedure than the tedious chore of attempting to find the source by extracting everything in the laboratory.

Phthalate esters and other plasticisers seem to be everywhere, and their complete elimination from the laboratory environment may not be possible because they are still used so extensively. One such source that has been studied and well documented is the stopper of Vacutainer tubes, manufactured by Becton-Dickinson. Much of the documentation in the literature refers to the "Vacutainer effect" and it is as well to clarify the use of this term. The problem was not due merely to leaching out of organic material from stoppers, such as has been described for the analysis of Δ^9-tetrahydrocannabinol[1469] or frusemide;[1470] the phenomenon was more subtle and more interesting than that, hence the bestowal of the special name. The term Vacutainer effect should be reserved for the phenomenon observed by Cotham and Shand[1471] for propranolol. In their report it was shown that spuriously low levels of propranolol were measured in plasma when the blood had been in contact with Vacutainer stoppers. The explanation was that the plasticiser, tris(2-butoxyethyl)phosphate,[1471,1472] leached from the stoppers, causing displacement of drugs from plasma protein, particularly α-1-glycoprotein, and consequent redistribution into erythrocytes: the overall effect was to lower the total fraction in the separated plasma or serum. The effect was noted for propranolol,[1472] alprenolol,[1473] quinidine,[1474] lidocaine,[1475] meperidine,[1476] chlorpromazine,[1477] other phenothiazines[1478] and tricyclic antidepressants.[1479,1480] The effect was mainly noted with Vacutainers as marketed by Becton-Dickinson; however, once these tubes had been implicated, the manufacturers made strenuous efforts to recall tubes containing the offending plasticiser. The subsequent types of Vacutainer were free of this problem.[1481] Hopefully the problem is now of only historic interest, but it is still an effect which should be checked if blood comes into contact with stoppers or similar material prior to separation of plasma.[1482]

The presence of phthalates may also contribute to altered extraction behavior of drugs. Westerlund and Nilsson[1483] showed that mono-2-ethylhexylphthalate, a degradation product of the plasticiser di-(2-ethylhexyl)-phthalate can form ion-pair adducts with norzimelidine. This phenomenon may explain the often puzzling greater efficiency of extraction of drugs from plasma when the plasma has been in contact with plastic containers.[1484]

12.11 FORMATION OF ARTIFACTS

A problem the analyst often has to deal with is the formation of artifacts of drugs which subsequently result in false measurements in biological fluids. This artifactual formation may result in decreased levels of drug, increased levels of drug, or misidentification of metabolites.

The problem of artifactual formation can be greatest in forensic toxicology or in screening methods, where it is necessary to use fairly forcing conditions to take in a wide spectrum of possible compounds. Thus, the conditions routinely used in analytical toxicology have been shown to convert clofibric acid to the amide.[1485] The effects of phosgene in chloroform or dichloromethane used for extraction have already been mentioned as a cause of artifactual formation of carbamates of tricyclic antidepressants.[1397]

Increased levels of drug are most often caused by reversion of metabolites to parent compound during the analytical process. This phenomenon is most common for labile conjugates such as *N*-conjugated glucuronides.[1486] The phenomenon of glucuronide stability has been studied in some detail for the anti-inflammatory drug isoxepac.[179]

The main metabolite of isoxepac in urine is the 1-*O*-acyl glucuronide with only small amounts of free isoxepac. On standing, there is some hydrolysis to free isoxepac as well as conversion of the original conjugate to other conjugates. Thus, the value of an assay of free isoxepac is doubtful due to the unknown degree of artifactual conversion. If the conjugates are converted to the parent compound in an attempt to determine "total" isoxepac, further complications arise because the rearranged conjugates have been found to resist the usual enzymes used for cleaving conjugates and again false and variable levels of total isoxepac in urine would be found. It is recommended that analysis for total isoxepac, and for similar carboxylic acid nonsteroidal anti-inflammatories, should be performed following mild alkaline hydrolysis, which has been found to hydrolyze all conjugates to the parent compound.[179] Similar phenomena have been noted for other ester glucuronides such as ketoprofen, naproxen, and probenicid.[1487] The pH-dependent transformation observed for isoxepac may be similar to the pH-dependent transacylation of the 1-*O*-acyl glucuronide of bilirubin to give 2-, 3-, and 4-*O*-acyl isomers.[1488] In the case of probenicid, 1-, 2-, 3-, and 4-*O*-acylglucuronides have been isolated.[1489] Both transacylation and formation of α- and β-furanoside and α-pyranoside forms of the ester glucuronide of clofibric acid[1490,1491] have been proposed (Figure 12.6).

Samples containing N-oxides and S-oxides of chlorpromazine have been shown to have increased levels of parent compounds on storage, displaying an unusual reduction reaction.[1380] N-oxides in general are easily converted to the parent compounds on gas chromatography and often subsequent GC-MS lends unwarranted weight to the interpretation, since the oxygen function is easily lost in injection ports of GC and MS equipment.[1492] Promethazine sulfoxide shows similar behavior on gas chromatography.[1493]

Metabolites are often misidentified, especially when vigorous conditions are used to isolate them in the analytical procedure. For this reason, carefully controlled collection procedures are of great importance in definitive experiments designed to elucidate the metabolic pathways of foreign compounds. An example is furosemide where the assumed major metabolite, 4-chlorosulfamoyl anthranilic acid, is now accepted to be an artifact formed during the isolation.

A particular type of artifact formation is in the racemization, or inversion of optically active compounds. Some drugs are considered to racemize so easily that it is accepted that nonchiral methods are adequate to measure their levels in biological fluids. However, where the racemate is dosed and it is desired to follow the true concentration of the separate enantiomers, then it is essential to ensure no artifactual racemization occurs, either in the extraction procedure or in the derivatization. Wright and Jamali[1494] studied the potential of derivatization with ethylchlorformate for stereochemical conversion during the process of preparing diastereoisomers of several anti-inflammatory drugs with *R*-(+)-_-phenylethylamine or L-leucinamide. Although they concluded that conversion could occur, the degree of conversion was small enough in the assay conditions not to contribute a significant error to the results.

One of the most common forcing techniques in drug analysis is in the volatilization that is necessary for gas chromatography. This was a particular problem during the early days of development of the technique, where the injection port often acted as a very active catalytic surface. Even recently, a false positive for methamphtamine, using a very specific GC-MS method was reported to be due to conversion of

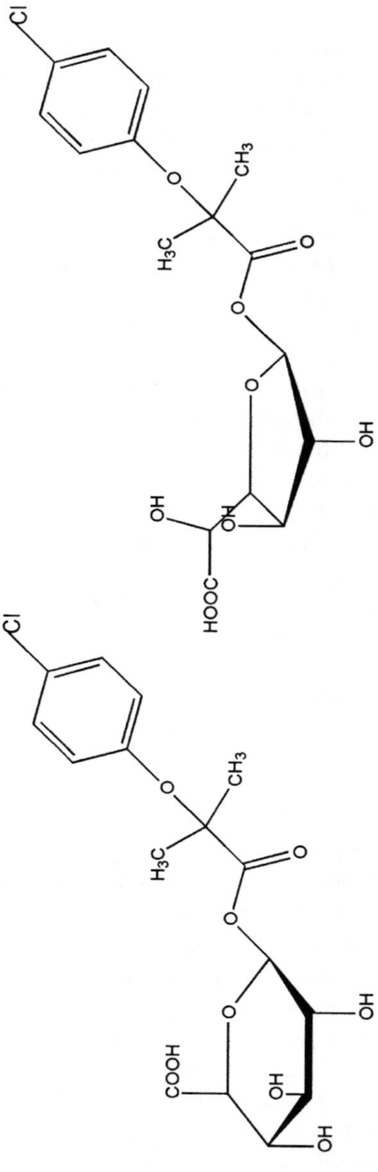

FIGURE 12.6 Different forms of the ester glucuronide of clofibric acid.

ephedrine or pseudoephedrine present in the sample.[1495] The problem can be overcome using prior periodate oxidation to destroy ephedrine or pseudoephedrine.

12.12   QUANTITATION

The formal aspects of evaluation of analytical methods have been discussed in Chapter 10, and for quantitation the use of calibration lines derived from processing blank samples with added analyte is common. However, not all laboratories follow the same procedure for constructing and using calibration curves for their assays. It would appear that once a straight line relationship has been established for a given response to the presence of drug, then the quantitation should follow by reference to the response from a known amount of analyte. However, at the levels of uncertainty that accompany sophisticated trace analyses, it has been customary to construct a calibration line over a concentration range. The problem arises when such a calibration line is recognized as a straight line but the points are scattered about the line. Many laboratories will draw the regression line of the points plotted (and the use of computer packages makes this easy to do) and then assume the straight line relationship for further calculations. The first rule of the use of any calibration line, however, is that the line must not be used for any concentrations outside the range used for constructing the calibration line. Dell[1496] has discussed the problems of trying to use regression lines for quantitation over the wide range of concentrations expected in pharmacokinetic studies and the use of weighting to improve the quantitation. Dell suggested that weighting procedures will always lead to improved quantification for the critical low concentrations. Burrows[1497] has described a computer program which simulates variations in sample and calibration concentrations to illustrate the consequences of various calibration routines and to assist in the selection and optimization of appropriate calibration procedures.

12.13   AUTOMATION

12.13.1   INTRODUCTION
For drug analysis, the preparation step, though no more inherently complex than that used for an individual clinical chemistry parameter, nevertheless needs to be considered separately for each new compound. The analytical step requires more sophisticated equipment such as gas chromatography or HPLC, and the measured value often covers a one hundred- or even a one million-fold dynamic range, necessitating complex calculating and reporting systems. Thus, these considerations have led to the development of automated devices for the preparation of samples, for automatic injection into suitable chromatographic systems, and for computing and reporting integrators. There has also been a burgeoning in total data processing and reporting known as laboratory information management systems (LIMS). The main development has been along the lines of the robot chemist. Some of the following discussion will review the more successful robot chemists as well as the more imaginative development using automated devices that do not reproduce human manipulations and also indicate ways in which the computers are used, not as rapid adding machines, but as unique components of the analytical instruments.

12.13.2 COMPUTERS AND THE ANALYSIS OF DRUGS

In the last decade, computers have had a considerable impact on the analysis of drugs in biological fluids. Some of this impact was predictable once the computer revolution was underway, but some aspects of computing have had their impact in enabling the analyst to do procedures that would not be possible without immense computing power, in particular the newer aspects of nuclear magnetic resonance and the broader field of chemometrics.

In the discussions on the various techniques available for the determination of drugs in biological fluids, the emphasis has been almost entirely on the inherent capabilities of the method; little attention has been given to the requirements of the analytical method in practice. In practice, one of the essential features of any analytical method is the throughput of samples. In analytical chemistry there is a long history of attempts to automate methods. Such attempts have generally been so successful that the calibration and documentation of the results has often outstripped the capacity of a single analyst to cope. An essential development is in automatic printers, and, in the past decade, sophisticated computer programs to keep up with the flow of results and achieve the scientists dream — the conversion of data into information and information into knowledge.

With a few honorable exceptions, the tendency in efforts to automate analytical methods has been to construct a robot chemist; that is, the inventor puts together all the actions of the traditional chemist to replicate the actions of pipetting, diluting, shaking, and pouring, and applying these mechanised devices to carrying out analyses by the traditional methods. The first edition of this book suggested that it would be brave to make predictions in the fast-developing fields of analysis and only touched on the appearance of robots in the analytical chemistry laboratory. The robots envisaged were similar to those used in heavy manufacturing industries, such as automobile construction, scaled down for the more delicate manipulations of test tubes and chromatography syringes. The most successful type of robot has been developed by Zymark.[1498] This system consists of a robot arm placed in a central position with access to a number of stations and its movements controlled by a programmable computer. The entire procedure is designed to mimic the movements of a human operator, with the ability to change the type of tool required from a hand-parking station. The stations in the system can include sets of syringes of different capacities for precise delivery of solvents or reagents, balances, centrifuges, and test tube racks. A complete robotic system requires complete access and custom-designed laboratories ought to be considering circular islands rather than conventional rectangular benches against walls.

12.13.3 AUTOANALYZER

The imaginative analyst should not be constrained to devizing systems that merely mimic the human capabilities. An example of this more imaginative approach is in the development of the AutoAnalyzer™ (the trademark of the Technicon Corporation) by Skeggs.[1307] In this procedure, the samples are set up in a train of tiny reaction vessels, separated by air bubbles. The samples then flow through various mixing, separating, and reaction devices in sequence, each device being designed to process the flowing segments presented to it in rapid succession rather than the separate vessels that would

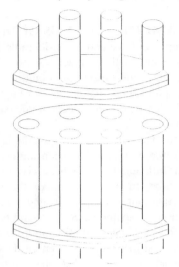

FIGURE 12.7 Column assembly for the automated chromatography of biological fluids. (From Barlow, G. B., in *Trace Organic Sample Handling*, Reid, E., Ed., Ellis Horwood, Chichester, 1981, 227. Copyright (1981) Ellis Horwood Limited. With permission.)

be handled by the robot chemist. The Skeggs system was designed and developed for the clinical chemistry laboratory, and early applications concentrated on the chemistry of the system, generally because most such analyses measured parameters within reasonably narrow limits and the end step was relatively straightforward and constant (UV absorption, flame photometry). The final result was usually read out as a single figure which could be readily interpreted by the physician.

12.13.4 INDIVIDUAL UNITS FOR AUTOMATION
12.13.4.1 Column Chromatography Preparation Units

A system of ion-exchange columns has been reported for the automatic preparation of samples prior to estimation by appropriate end points or as part of a Technicon AutoAnalyzer System.[1499] The essence of the design of this equipment is shown in Figure 12.7. In the six-column version shown here, the column assembly is sandwiched between the feed assembly and waste/collector assembly, the interface consisting of flat discs able to move in close contact. Each column is subjected to various steps in the usual chromatographic procedure, that is, prewash, load with sample, elution with the selected sequence of solvents, and regeneration. The device was successfully used to separate a number of endogenous compounds from urine, including creatinine, xanthurenic acid, kynurenic acid, amino acids, and polyamines. Preparation rates of up to 60 samples per hour were reported and potentially such a system can be used for any small column chromatographic work where the column packing can be regenerated.

Pacha and Eckert[1500] also described the use of absorption columns placed in automatically controlled sequences of solvent elutions and were able to include automatic chemistry for making fluorescent derivatives and subsequent assay by fluorimetry, as in their assay for thioridazine in pig plasma by reaction with $KMnO_4$ to enhance its fluorescence. For the analysis of other drugs, purification was achieved by adsorption onto charcoal or XAD-2 resins, prior to analysis.

12.13.4.2 Solvent Extraction

Solvent extraction devices were included in the Technicon AutoAnalyzer sytem, and can be used to prepare samples for drug analysis. In a typical unit, the liquid sample and the organic extractant are delivered together by a peristaltic pump into a mixing coil, often packed with glass ballotini; in a subsequent step the layers are allowed to stratify and either layer can be resampled for further manipulation. Although the sample-to-extractant ratio can be varied within the limits normally applying to such operations, the maximum concentration factor consistent with good operation is normally about 3:1.[1501] Alternative methods of automatic solvent separators have been developed for AutoAnalyzer systems and for flow injection techniques.

12.13.4.3 Evaporation Units

Another device was developed by Technicon for its Fast-LC system; this was referred to as Evaporation to Dryness Module (EDM). Samples extracted into a volatile solvent are placed on a moving, chemically inert belt which is passed through a vacuum or is subjected to a current of air. The solvent hence evaporates and the sample can be redissolved in a different solvent.

12.13.4.4 TLC Spotters

Analysis of drugs by thin-layer chromatography is potentially the technique which could allow the best simultaneous analyses. However, the spotting of the samples onto the plate is the most difficult to reproduce manually because of the skill required to apply small spots without disturbing the fragile layer. In the usual manual technique, a small portion of the sample is taken up into a fine capillary or a Pasteur pipet and then the tip of the capillary is touched onto the thin-layer surface. The sample is transferred to the absorbent by capillary action and if the sample is warmed at the same time, then the solvent will evaporate rapidly before the sample spreads. The procedure is repeated until all or most of the sample has been applied. This stop-go sequence allows the sample to be applied as a compact spot. The procedure then has to be repeated for all samples in the batch as well as for calibration standards or quality controls, and it can be seen that this would be extremely tedious, especially if large volumes of extract are required for analysis. An automatic TLC applicator has been designed which almost exactly mimics the manual procedure, but has the merit of doing so reproducibly.[1502] In the original version of the instrument,[1503] the sample was picked up by stainless steel loops in precise volumes of 1 μl which then were rotated through an arc to the thin-layer plate to make gentle contact. The movement back to the pick-up position could be interrupted for sufficient time to allow the solvent to evaporate from the plate. In the later version the stainless steel loops were replaced by glass capillaries.

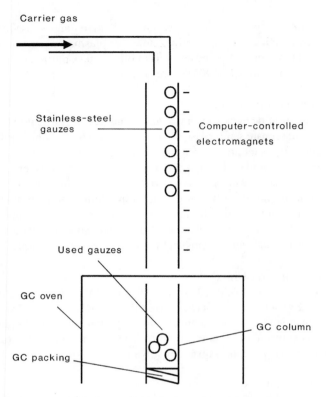

FIGURE 12.8   An automatic sampling device for gas chromatography. Sample is applied in solution to stainless-steel gauze cylinders and the solvent evaporated. Electromagnets keep the individual gauze cylinders in position until released by switching of the individual elecrotromagnets.

12.13.4.5 Application of Samples to HPLC or Gas Chromatography

The popularity of both HPLC and gas chromatography and their resultant commercial development have ensured that many systems have been developed for automatic injection of samples into the chromatograph. One of the earliest automated devices for gas chromatography is shown in Figure 12.8. Samples in solution were applied to stainless steel meshes in shallow Teflon wells and the solvent evaporated. The Teflon repelled the solvent and ensured that the last traces were evaporated from the mesh. The wire meshes were then placed in a glass tube and held in place by electromagnets and the tube was mounted over the head of the chromatographic column. The power to the individual electromagnets was switched off at preset times allowing the meshes to fall onto the top of the column. This device allowed analysis of samples without any solvent interference.

Most of the other early automated devices for chromatography exactly followed the manual procedures; a glass syringe with a stainless steel needle is thoroughly washed by repeated filling and emptying with the solvent, then the sample is taken up in a measured volume and injected through a septum as close as possible to the

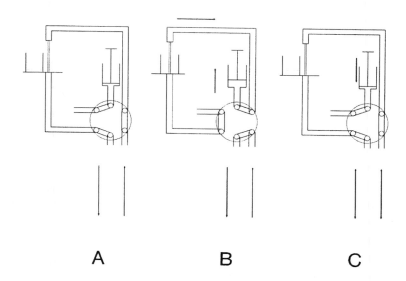

FIGURE 12.9   Loop-injector with 6-port valve. (A) run; (B) load; (C) inject and run.

head of the column. The sequence is then repeated for the next sample after a suitable interval.

An alternative injection system utilizes the loop injector as typified in an HPLC system in Figure 12.9. In this system, in the normal mode, the mobile phase flows through the sampling needle and onto the column. In the load position, the 6-port valve directs the main flow directly onto the column and the injection needle is isolated so that it can perform the loading operation by drawing up a measured sample. In the injection position, the sample is pushed back through the needle and onto the column; during the analytical step, the needle is automatically cleaned for the next injection. The measuring syringe is never in contact with the sample.

Switching valves such as those shown here have been heavily utilized in HPLC for making the maximum use of guard columns or for incorporating solid-phase extraction columns into automated systems. An example is shown in Figure 12.10. The sample, preferably unprocessed plasma or urine, is injected into the extraction cartridge with the solvent flowing through the cartridge to waste. During this phase, the bulk of the endogenous material is eluted from the column while the analytical column is subjected only to pure mobile phase supplied directly from the solvent reservoir. After a preset time the analyte is eluted from the cartridge in the reverse direction onto the analytical column and chromatography proceeds in the normal way. On completion of the analytical phase, the switching valve is reset to the injection mode and the next sample is injected. In this sequence of switching, material that is strongly retained on the column is not eluted onto the analytical column because of the reverse flow technique, although such material will accumulate on the extraction cartridge and may affect subsequent performance.

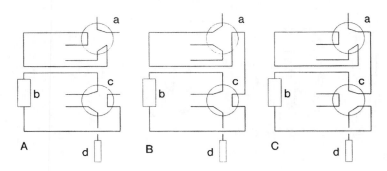

FIGURE 12.10   Switching valve assembly for including an extraction cartridge in an automated HPLC system. (A) Sample load and pre-column wash; (B) sample inject onto precolumn only; (C) elute from precolumn and analysis; (a) injection valve; (b)precolumn for extraction; (c) switching valve; (d) analytical column.

Switching valves may also be used after the chromatographic column, such as in heart-cutting procedures where the most concentrated part of the sample is collected for further processing, as described, for example by Lecaillon et al. for metoprolol and oxiracetam.[1504] These column-switching techniques in turn have benefited considerably by the use of computer control.

12.13.5 CHROMATOGRAPHIC METHODS

Many authors report fully automated chromatographic procedures for drug analysis, but closer investigation may reveal that the automation is limited to injection and data processing; sample preparation, often the most tedious and labor-intensive part of the whole system, is still necessary by manual means. The use of cartridges has been mentioned for the preparation of samples for subsequent analysis, and the logical step of interfacing such devices with chromatographic systems has been successfully applied. Because the sample preparation using cartridges invariable results in the sample as a solution, HPLC has benefitted most by direct interface. The various steps of loading, washing, eluting, and regenerating stages for the cartridge can be carried out by switching valves under computer control, and the same methodology can be used for diverting the appropriate effluent to the analytical column. Methods for a large number of drugs have been developed using these column switching methods to allow direct injection of biological fluids into the integrated system (Table 12.2).

12.13.6 CENTRIFUGAL ANALYZERS

Most of the devices described in this section deal sequentially with samples for analysis. Thus if there are a large number of samples, the complete analysis will take a correspondingly longer time, and systems that could deal with many samples simultaneously were attractive in routine clinical chemistry or drug analysis laboratories.

In the 1960s, Anderson introduced the concept of "fast analyzers" where all operations take place in a very short time preferably processing many samples in

TABLE 12.2
Column Switching Analyses

Drug	Starting material	Reference
Alprazolam	Dried column extract	1505
Ceftibuten	Buffered plasma	1506
Clozapine	Dried column extract	1507
Buprenorphine	Plasma	1298
Methotrexate	Urine	1508
Cefotiam	Plasma	1509
Ivermectin	Bovine plasma	1510
Amitriptyline	Plasma, serum	1511, 1512
Nortriptyline	Plasma, serum	1511, 1512
Furosemide	Plasma	1513
Carboplatin	Plasma ultrafiltrate	1514
Terbutaline	Deproteinized plasma	1515
Bambuterol	Deproteinized plasma	1515

parallel.[1516] This was made possible by the invention of methods for dispensing, transferring, mixing, and measuring the absorbance of a large number of samples over very short time intervals. By including standards in the parallel analyses it was possible to make measurements before reactions had gone to completion. The objectives were realized by incorporating the analytical system into a centrifuge rotor, thereby enabling (1) accurate measurement of volume using centrifugal force to fill, debubble, and level menisci accurately in small vessels; (2) quantitative transfer of fluids using centrifugal force; and (3) rapid measurement of absorbance of a large number of samples by rotating them rapidly past a beam of light. Figure 12.11 illustrates one of the simpler devices for transferring a measured volume of fluid from an unmeasured sample into a cuvette. Anderson has described a number of ingenious devices for other analytical procedures, all of which can be carried out simultaneously on a large number of samples depending on the number of places on the rotor head. In a typical analysis using the centrifugal analyzer the absorbancies of all reception cuvettes are displayed simultaneously on a visual display unit and a computer is used to perform appropriate calculations and report results.

Centrifugal analyzers found most favor in clinical chemistry laboratories where the same routine analyses would be expected to be carried out for the foreseeable future, and the fairly expensive investment could be seen to be worthwhile. As far as drug analyses go, the technique has been used more in monitoring laboratories, for example, for drugs of abuse rather than pharmacokinetic studies where the more traditional methods would suffice for the relatively small number of samples to be analyzed in separate studies.

However, there is some potential for centrifugal analyzers in radioimmunoasays or in enzyme immunoassays where the techniques and reagents may be essentially the same for assays of different drugs. One of the attractive advantages of adapting such methods for immunoassays is that the analyses can be carried out on a small scale and therefore there is an economy in the use of expensive reagents and solvents. Thus, an

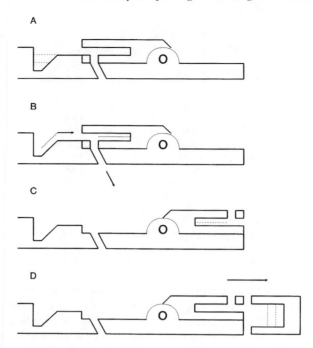

FIGURE 12.11   Diagram of the use of reorienting measuring chamber to measure and transfer a liquid. Centrifugal force is from left to right. (A) Unmeasured sample is placed in depression with rotor at rest; (B) during rotation the liquid flows into the measuring vessel, fills it, and excess fluid is drained off below; (C) at rest, the tube is turned so the open end is positioned next to the cuvette. The tube is small enough so that its contents are retained by the capillary and do not run out in the absence of centrifugal force; (D) during centrifugation the contents are quantitatively transferred into a cuvette held in a cuvette rotor. (From Anderson, N. J., *Am. J. Clin. Pathol.*, 778, 53, 1970. Copyright (1970) J. P. Lippincott Co. With permission.)

adaptation of the Centrifichem analyzer for the assay of theophylline, phenobarbital, phenytoin, carbamazepine, primidone, ethosuximide, and gentamicin used sample volumes of 3 µl and total cuvette volumes of 210 µl. The imprecision was less than 1.5% with drug recoveries of between 90 and 105% over the calibration range used. It was claimed that 600 tests could be carried out with a 100-test kit as normally provided. Table 12.3 summarizes a number of drug assays using centrifugal analyzers.

12.13.7 COMPUTER ANALYSIS OF ROTARY TLC

One of the main features of the centrifugal analyzers is that many assays are performed in parallel. To perform many HPLC or GC analyses in parallel would require arrays of identical columns and possibly the same number of detectors. This approach for achieving high throughput of samples would be extremely costly. Using TLC methods, however, the problem of multiple columns does not arise; usually up to 20 chromatograms can be accommodated on a single plate.

In the first edition of this work, the author speculated on the advantages of adopting the centrifugal approach to TLC analysis. Since those first speculations,

TABLE 12.3
Analysis of Drugs Using Centrifugal Analyzers

Drug	Analyzer	Method	Reference
N-Acetylprocainamide	Multistat	EMIT	1517
N-Acetylprocainamide	Centrifichem	EMIT	1517
Amphetamine	Gemsaec	EIA	1518
Barbiturates	Gemsaec	EIA	1518, 1519
Barbiturates	Centrifichem	EMIT	1517, 1520, 1521
Barbiturates	Mini-CFA	EIA	1522
Barbiturates	Multistat	EMIT	1517
Barbiturates	Rotochem	EMIT	1523
Carbamazepine	Centrifichem	EMIT	1517, 1521
Carbamazepine	Multistat	EMIT	1517
Carbamazepine	Rotochem	EMIT	1523
Dilantin	Multistat	EMIT	1517
Dilantin	Centrifichem	EMIT	1517
Disopyramide	Multistat	EMIT	1517
Disopyramide	Centrifichem	EMIT	1517
Methadone	Gemsaec	EIA	1518
Mysoline	Multistat	EMIT	1517
Mysoline	Centrifichem	EMIT	1517
Opiates	Gemsaec	EIA	1518
Phenytoin	Mini-CFA	EIA	1522
Phenytoin	Rotochem	EMIT	1523
Phenytoin	Centrifichem	EMIT	1520
Primidone	Gemsaec	EIA	1519
Procainamide	Multistat	EMIT	1517
Procainamide	Centrifichem	EMIT	1517
Quinidine	Multistat	EMIT	1517
Quinidine	Centrifichem	EMIT	1517
Theophylline	Centrifichem	EMIT	1517, 1521
Theophylline	Multistat	EMIT	1517
Theophylline	Gemsaec	EIA	1519
Tobramycin	Multistat	EMIT	1517
Tobramycin	Centrifichem	EMIT	1517

such rotary TLC methods have been described, where the centrifugal force is the means of movement of the solvent rather than the usual capillary effect. The other suggestion in this context which was made in the first edition has not been taken up, but I think it is worthwhile to enlarge on the idea here. It is envisaged that 12 samples could be spotted in a ring about the center of the plate and mobile-phase is delivered into the center of the ring so that 12 radial chromatograms are developed on the same plate.

Essentially, the device would include a single detector mounted near the edge of the rotating plate, rather like the playing head of a compact disc player, with a laser light source and a collector. The reflected (or transmitted) light from the plate is sampled once in each radial channel per revolution and digitized for storage in a computer. The stored chromatogram would have the form of an HPLC or GC chromatogram rather than a scanned TLC plate. The 12 chromatograms on each plate could contain suitable standards for quantitation, or it may be possible to carry out a complete pharmacokinetic analysis for a single subject on the one analytical plate, with the computer performing all the necessary calculations, making use of any clinical data that may also be added to the computer file. The scheme is illustrated in Figure 12.12. Development of such a device is probably limited for biological fluids because of the requirements of sensitivity, but it may find a use in stability studies of pharmaceuticals or other materials, where the detection method does not need to be so sensitive.

The use of laser beams in this system would be analogous to the compact disc technology for miniaturizing disc players. It is striking that the revolution in electronics has hardly touched the hardware for analytical tools — a point made by Brinkman et al.[1524] in an excellent review of the possibilities in this area. This paper and a subsequent one from the same laboratory[1525] show the way forward. I would expect that it will be in this sort of miniaturization that the next major advances in practical analytical techniques will be found.

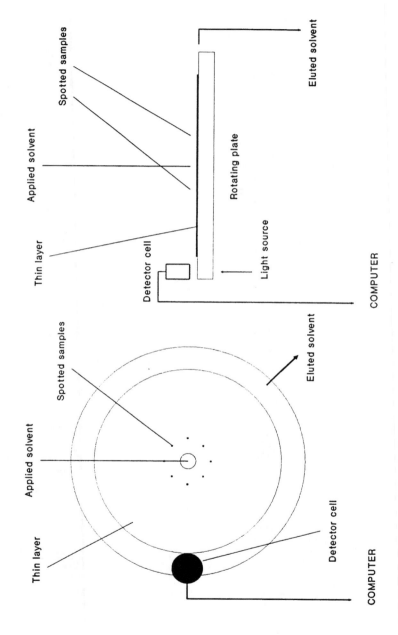

FIGURE 12.12 Proposed scheme for a centrifugal TLC analyzer.

REFERENCES

1. **Finkle, B.,** Forensic toxicology in the 1980's. The role of the analytical chemist, *Analyt. Chem.,* 54, 433A, 1982.
2. **Prescott, L. F.,** Toxicological monitoring, in *Therapeutic Drug Monitoring,* Richens, A. and Marks, V., Eds., Churchill Livingstone, Edinburgh, 1981, 492.
3. **Hansen, A. R.,** Glutethimide poisoning. A metabolite contributes to morbidity and mortality, *N. Engl. J. Med.,* 292, 250, 1975.
4. "Under the Influence, Drugs and the American Workforce," National Research Council, Washington, D.C.
5. **Moss, M. S. and Cowan, D. A.,** Drug abuse in sport, in *Clarke's Isolation and Identification of Drugs, 2nd ed.,* Moffatt, A. C., Ed., The Pharmaceutical Press, London, 1986, 87.
6. **Catlin, D., Cowan, D. A., Donike, M., Fraisse, D., Oftebro, H., and Rendic, S.,** Testing urine for drugs, *Clin. Chim. Acta,* 207, S13, 1992.
7. **Hollister, L.,** *Clinical Pharmacology of Psychotherapeutic Drugs,* Churchill Livingstone, New York, 1978, 44.
8. **Koch-Weser, J.,** Serum drug concentration in clinical perspective, in *Therapeutic Drug Monitoring,* Richens, A. and Marks, V., Eds., Churchill Livingstone, Edinburgh, 1981, 1.
9. **Kuhn, R.,** Über die Behandlung depressiver Zuständer mit einem Imino-dibenzylderivat, *Schweiz. Med. Wochens.,* 87, 1135, 1957.
10. **Anhalt, J. P. and Brown, S. D.,** High-performance liquid-chromatographic assay of aminoglycoside antibiotics in serum, *Clin. Chem.,* 24, 1940, 1978.
11. **DeLuccia, F. J., Arunyanart, M., and Cline Love, L. J.,** Direct serum injection with micellar liquid chromatography for therapeutic drug monitoring, *Analyt. Chem.,* 57, 1564, 1985.
12. **Christofides, J. A. and Fry, D. E.,** Measurement of anticonvulsants in serum by reversed-phase ion-pair liquid chromatography, *Clin. Chem.,* 26, 499, 1980.
13. **Charette, C., McGilveray, I. J., and Mainville, C.,** Simultaneous determination of disopyramide and its mono-*N*-dealkyl metabolite in plasma and urine by high-performance liquid chromatography, *J. Chromatogr.,* 274, 219, 1983.
14. **Dawling, S. and Braithwaite, R. A.,** Simplified method for monitoring tricyclic antidepressant therapy using gas-liquid chromatography and nitrogen detection, *J. Chromatogr.,* 146, 449, 1978.
15. **Broughall, J. M.,** in *Laboratory Methods in Antimicrobial Chemotherapy,* Reeves, D. S. et al, Eds., Churchill Livingstone, Edinburgh, 1978, 194.
16. **Lindberg, R., Salonen, J. S., and Laurikainen, E.,** Improved liquid-chromatographic determination of lidocaine and its desethylated metabolites in serum, *Clin. Chem.,* 29, 1572, 1983.
17. **Soldin, S. J. and Hill, J. G.,** A rapid micromethod for measuringtheophylline in serum by reverse-phase high-performance liquid chromatography, *Clin. Biochem.,* 10, 74, 1977.
18. **Schwertner, H. A.,** Analysis for underivatized theophylline by gas-chromatography on a silicone stationary phase SP-2510-DA, *Clin. Chem.,* 25, 212, 1979.
19. **Sjöqvist, F., Hammer, W., Ideström, C.-M., Lind, M., Tuck, D., and Åsberg, M.,** Plasma levels of monomethylated tricyclic antidepressants and side-effects in man, in *Proceedings of the European Society for the Study of Drug Toxicity, Volume IX. Toxicity and Side-effects of Psychotropic Drugs, Excerpta Medica International Congress Series* No. 145, 1967, 246.
20. **Mahgoub, A., Idle, J. R., Dring, L. G., Lancester, R., and Smith, R. L.,** Polymorphic hydroxylation of debrisoquine in man, *Lancet,* ii, 548, 1977.
21. **Eichelbaum, M., Spannbrucker, N., Steincke, B., and Dengler, J. J.,** Defective N-oxidation of sparteine in man: a new pharmacogenetic defect, *Eur. J. Clin. Pharmacol.,* 16, 183, 1979.

22. **Alvan, G., Bechtel, P., Iselius, L., and Gundert-Remy, U.,** Hydroxylation polymorphisms of debrisoquine and mephentoin in European populations, *Eur. J. Clin. Pharmacol.,* 39. 533, 1990.

23. **Gut, J., Gasser, R., Dayer, P., Kronbach, T., Catin, T., and Meyer, U. A.,** Debrisoquine-type polymorphism of drug oxidation: purification from human liver of a cytochrome P450 isozyme with high activity for bufuralol hydroxylation, *FEBS Lett.,* 173, 287, 1984.

24. **Gut, J., Catin, T., Dayer, P., Kronbach, T., Zanger, U., and Meyer, U. A.,** Debrisoquine/ sparteine-type polymorphism of drug oxidation: purification and characterization of two function- ally different human liver cytochrome P450 isozymes involved in impaired hydroxylation of the prototype substrate bufuralol, *J. Biol. Chem.,* 261, 11734, 1986.

25. **Gut, J., Meier, U. T., Catin, T., and Meyer, U. A.,** Mephenytoin-type polymorphism of drug oxidation: purification and characterization of a human liver cytochrome P450 isozyme catalyzing microsomal mephenytoin hydroxylation, *Biochim. Biophys. Acta,* 884, 435, 1986.

26. **Zanger, U. M., Vilbois, F., Hardwick, J., and Meyer, U. A.,** Absence of hepatic cytochrome P450buf1 causes genetically deficient debrisoquine oxidation in man, *Biochemistry,* 27, 5447, 1988.

27. **Meyer, U. A., Skoda, R. C., Zanger, U. M., Heim, M., and Broly, F.,** The genetic polymorphism of debrisoquine/sparteine metabolism — molecular mechanism, in *Pharmacogenetics of Drug Metabolism,* W. Kalow, Ed., Pergamon Press, New York, Oxford, 1992, 609.

28. **Hammer, W. and Sjöqvist, F.,** Plasma levels of monomethylated tricyclic antidepressants during treatment with imipramine-like compounds, *Life Sci.,* 6, 1895, 1967.

29. **Balant-Gorgia, A. E., Schultz, P., Dayer, P, Balant, L., Kubli, A., Gertsch, C., and Garrone, G.,** Role of oxidation polymorphism on blood and urine concentrations of amitriptyline and its metabolites in man, *Arch. Psychiatr. Nervenkr.,* 232, 215, 1982.

30. **Balant-Gorgia, A. E., Balant, L. P., Genet, C., Dayer, P., Aeschlimann, J. M., and Garrone, G.,** Importance of oxidative polymorphism and levomepromazine treatment on the steady-state blood concentrations of clomipramine and its major metabolites in man, *Eur. J. Clin. Pharmacol.,* 31, 449, 1986.

31. **Potter, W. Z., Bertilsson, L., and Sjöqvist, F.,** Clinical pharmacokinetics of psychtropic drugs — fundamental and practical aspects, in *Handbook of Biological Psychiatry,* van Praag, H. M., Lader, M. H., Rafaelsen, A. J., and Sacher, E. G., Eds., Marcel Dekker, New York, 71, 1981.

32. **Bock, J. L., Nelson, J. C., Gray, S., and Jatlow, P. I.,** Desipramine hydroxylation: variability and effect of antipsychotic drugs, *Clin. Pharmacol. Ther.,* 33, 322, 1983.

33. **Park, Y. H., Kullberg, M. P., and Hinsvark, O. N.,** Quantitative determination of dextromethorphan and three metabolites in urine by reverse-phase high-performance liquid chromatography, *J. Pharm. Sci.,* 73, 24, 1984.

34. **LLerena, A., Dahl, M.-L., Ekqvist, B., and Bertilsson, L.,** Haloperidol disposition is dependent on the debrisoquine hydroxylation phenotype: increased plasma levels of the reduced metabolite of poor metabolizers, *Ther. Drug Monitor.,* 14, 261, 1992.

35. **Gram, L. F., Bjerre, M., Kragh-Sorenson, P., Kvinesdall, B., Molin, J., Pedersen, O. L., and Reisby, N.,** Imipramine metabolites in blood of patients during therapy after overdose, *Clin. Pharmacol. Ther.,* 33, 335, 1983.

36. **Draffan, G. H., Lewis, P. J., Firmin, J. L., Jordan, T. W., and Dollery, C. T.,** Pharmacokinetics of indoramin in man, *Br. J. Clin. Pharmacol.,* 3, 489, 1976.

37. **Kirch, W., Spahn, H., Ohnhour, E. E., Kohler, H., Heinz, U., and Mutschler, E.,** Influence of inflammatory disease on the clinical pharmacokinetics of atenolol and metoprolol, *Biopharm. Drug Dispos.,* 4, 73, 1983.

38. **Dawling, S., Lynn, K, Rosser, R., and Braithwaite, R.,** Nortriptyline metabolism in chronic renal failure: metabolite elimination, *Clin. Pharmacol. Ther.,* 32, 322, 1982.

39. **Sindrup, S. H., Brosen, K., Gram, L. F., Hallas, J., Skjelbo, E., Allen, A., Allen, G. D., Cooper, S. M., Mellows, G., Tasker, T C. G., and Zussman, B. D.,** The relationship between paroxetine and the sparteine oxidation polymorphism, *Clin. Pharmacol. Ther.,* 51, 278, 1992.

40. **Singlas, E., Gonjet, M. A., and Simm. P.,** Pharmacokinetics of perhexiline maleate in anginal patients with and without peripheral neuropathy, *Eur. J. Clin. Pharmacol.,* 14, 195, 201.

41. **Oates, N. S., Shah, R. R., Idle, J. R., and Smith, R. L.,** Influence of oxidation polymorphism on phenformin kinetics and dynamics, *Clin. Pharmacol. Ther.,* 34, 827, 1983.
42. **Schneck, D. W., Pritchard, J. F., Gibson, T. P., Vary, J. E., and Hayes, A. H.,** Effect of dose and uremia on plasma and urine profiles of propranolol metabolites, *Clin. Pharmacol. Ther.,* 27, 744, 1980.
43. **Meyer, J. W., Woggon, B., Baumann, P., and Meyer, U. A.,** Clinical implications of slow sulphoxidation of thioridazine in a poor metabolizer of the debrisoquine type, *Eur. J. Clin. Pharmacol.,* 39, 613, 1990.
44. **Fourtillan, J. B., Courtois, P., Lefebvre, M. A., and Girault, J.,** Pharmacokinetics of oral timolol studied by mass fragmentography, *Eur. J. Clin. Pharmacol.,* 19, 193, 1981.
45. **Yue, Q.-Y. and Sawe, J.,** Interindividual and interethnic differences in codeine metabolism, in *Pharmacogenetics of Drug Metabolism,* Kalow, W., Ed., Pergamon Press, New York, 1992, 721.
46. **Tucker, G. T., Lennard, M. S., Ellis, S. W., Woods, H. F., Cho, A. K., Lin, L. Y., Hiratsuka, A., Schmitz, D. A., and Chu, T. Y. Y.,** The demethylation of methylenedioxymethamphetamine ('Ecstasy') by debrisoquine hydroxylase (CYP2D6), *Biochem. Pharmacol.,* in press, 1994.
47. **Tucker, G. T.,** Clinical implications of genetic polymorphism in drug metabolism, *J. Pharm. Pharmacol.,* 46, 417, 1994.
48. **Küpfer, A., Patwardhan, R., Ward, S., Schenker, S., Preisig, R., and Branch, R. A.,** Stereoselective metabolism and pharmacogenetic control of 5-phenyl-5-ethylhydantoin (Nirvanol) in humans, *J. Pharmacol. Exp. Ther.,* 230, 28, 1984.
49. **Wilkinson, G. R., Guencherich, F. P., and Branch, R. A.,** Genetic polymorphism of S-mephenytoin hydroxylation, in *Pharmacogenetics of Drug Metabolism,* Kalow, W., Ed., Pergamon Press, New York, Oxford, 657, 1992.
50. **Evans, D. A. P.,** N-Acetyltransferase, *Pharmacol. Ther.,* 42, 157, 1989.
51. **Price-Evans, D. A.,** N-Acetyltransferase, in *Pharmacogenetics of Drug Metabolism,* Kalow, W., Ed., Pergamon Press, New York, 95, 1992.
52. **Weber, W. W.,** *The Acetylator Genes and Drug Response,* Oxford University Press, Oxford, 1987.
53. **La Du, B. N.,** Genetic variations in humans and toxicokinetics, in *Drug Toxicokinetics,* Welling, P. G. and de la Iglesia, F. A., Eds., Marcel Dekker, New York, 221, 1993.
54. **Murray, F. T., Santner, S., Samojlik, E., and Santen, R. J.,** Serum aminoglutethimide levels: studies of serum half life, clearance, and patient compliance, *J. Clin. Pharmacol.,* 19, 704, 1979.
55. **Thompson, T. A., Vermuelen, J. D., Wagner, W. E., and Le Sher, A. R.,** Aminoglutethimide bioavailability, pharmacokinetics, and binding to blood constituents, *J. Pharmacol. Sci.,* 70, 1040, 1981.
56. **Blanchard, J. and Sawers, S. J. A.,** The absolute bioavailability of caffeine in man, *Eur. J. Clin. Pharmacol.,* 24, 93, 1983.
57. **Levy, M., Flusser, D., Zylber-Katz, E., and Granit, L.,** Plasma kinetics of dipyrone metabolites in rapid and slow acetylators, *Eur J. Clin. Pharmacol.* 27, 453, 1984.
58. **Shepherd, A. M., Ludden, T. M., McNay, J. L., and Lin, M.-S.,** Hydralazine kinetics after single and repeated doses, *Clin. Pharmacol. Ther.,* 28, 804, 1980.
59. **Männistö, P., Mantyla, R., Klinge, E., Nykanea, S., Koponen, A., and Lamminsiva, U.,** Influence of various diets on the bioavailability of isoniazid, *J. Antimicrob. Chemother.,* 10, 427, 1982.
60. **Scröder, H. and Campbell, D. E. S.,** Absorption, metabolism, and excretion of salicylazosulfapyridine in man, *Clin. Pharmacol. Ther.,* 13, 539, 1972.
61. **Vree, T. B., Hekster, C. A., Baakman, M., Janssen, T., Oosterbaan, M., Termond, E., and Tijhuis, M.,** Pharmacokinetics, acetylation-deacetylation, renal clearance, and protein binding of sulphamerazine, N_4-acetylsulphamerazine, and N_4-trideuteroacetylsulphamerazine in "fast" and "slow" acetylators, *Biopharm. Drug Dispos.,* 4, 271, 1983.
62. **Woolhouse, U. M. and Atu-Taylor, L. C.,** Influence of double genetic polymorphism on response to sulfamethazine, *Clin. Pharmacol. Ther.,* 31, 377, 1982.
63. **Blume, H. E. and Midha, K. K.,** Bio-International 92, Conference on bioavailability, bioequivalence, and pharmacokinetic studies, *J. Pharm. Sci.,* 82, 1186, 1993.
64. **Hamner, C. E.,** Drug Development, 2nd ed., CRC Press, Boca Raton, 119, 1990.

65. **Selby, I. A.,** Analytical chemistry: the hub of pharmaceutical development, *Chem. Br.,* 14, 606, 1978.
66. **Rudofsky, G., Brock, F. E., Urich, M., and Nobbe, F.,** Treatment of patients with occlusive arterial disease with oxpentifylline. Hydrodynamic and ergometric findings, *Med. Klin.,* 74, 1093, 1979.
67. **Hinze, H.-J.,** Zur pharmacokinetic von 3,7-Dimethyl-1-(5-oxohexyl)-xanthin (BL191) am Menschen, *Arzneim. Forsch.,* 22, 1492, 1972.
68. **Di Carlo, F. J.,** Metabolism, pharmacokinetics, and toxicokinetics defined, *Drug Metab. Rev.,* 17, 1, 1982.
69. **Wright, D. S.,** Analytical method considerations for toxicokinetic studies, in *Drug Toxicokinetics,* Welling, P. G. and dela Iglesia, F. A., Eds., Marcel Dekker, New York, 1993, 1.
70. **Garrett, E. R.,** personal communication.
71. **Chamberlain, J.,** *Analysis of Drugs in Biological Fluids,* CRC Press, Boca Raton, Florida, 15.
72. **Huang, F.E., Upton, R. N., Mather, L. E., and Runciman, W. B.,** An assessment of methods for sampling blood to characterize rapidly changing blood drug concentrations, *J. Pharm. Sci.,* 80, 847, 1990.
73. **Houin, G., Lapeyre, C., Rochas, M. A., Tufenkji, A. E., Campistron, G., Coulais, Y., Akbaraly, J. P., Grislain, L., and Beck, H.,** Pharmacokinetic study of fentiazac and its main metabolite hydroxyfentiazac in the elderly, *Arzneim. Forsch.,* 43, 50, 1993.
74. The Control of Drugs for the Elderly. Report on the ninth European Symposium on Clinical Pharmacological evaluation in Drug Control, 18–21 November 1980. Euro Reports and Studies, 50, 1981, WHO Regional Office for Europe, Copenhagen.
75. **Evans, L. and Spelman, M.,** The problem of non-compliance with drug therapy, *Drugs,* 25, 63, 1983.
76. **Maickel, R. P.,** Separation science applied to analyzes on biological samples, in *Drug Determination in Therapeutic and Forensic Contexts,* Reid, E. and Wilson, I. D., Eds., Plenum Press, New York, 1984, 3.
77. **Argenti, D. and D'mello, A.,** Chelation with ferric ions may prevent extraction of 2-hydroxydesiipramine from whole blood, *Clin. Chem.,* 38, 1195, 1992.
78. **Leonidov, N. B., Zorkii, P. M., Masunov, A. E., Gladkikh, O. P., Bel'skii, V. K., Dzyabchenko, A. V., and Ivanov, S. A.,** The structure and biological nonequivalence of methyluracil polymorphs, *Russian J. Phys. Chem.,* 67, 2220, 1993.
79. **Stevens, H. M. and Bunker, V. W.,** Cited by Campbell, in *Assays of Drugs and Other Trace Compounds in Biological Fluids,* Reid, E., Ed., North-Holland, Amsterdam, 1976, 105–106.
80. **Dell, D.,** Sample preparation in HPLC, in *Assays of Drugs and Other Trace Compounds in Biological Fluids,* Reid, E., Ed., North-Holland, Amsterdam, 1976, 131.
81. **Mathies, J. C. and Austin, M. A.,** Modified acetonitrile protein-precipitation method of sample preparation for drug assay by liquid chromatography, *Clin. Chem.,* 26, 1760, 1980.
82. **Osselton, M. D.,** Enzymic liberation of chemically labile or protein-bound drugs from tissues, in *Trace Organic Sample Handling,* Reid, E., Ed., Ellis Horwood Limited, Chichester, 1981, 101.
83. **Bogusz, M., Bialka, J., and Gierz, J.,** Enzymic digestion of bio-samples as method of sample pretreatment before XAD-2 extraction, *Z. Rechtsmed.,* 87, 287, 1981.
84. **Holzbecher, M. and Ellenberger, H. A.,** Simultaneous determination of diphenhydramine, methaqualone, diazepam and chlorpromazine in liver by use of enzyme digestion: comparison of digestion procedures, *J. Analyt. Toxicol.,* 5, 62, 1981.
85. **Callum, G. I., Ferguson, M M., and Lenihan, J. M. A.,** Determination of methylmercury in tissue using enzyme proteolysis, *Analyst,* 106, 1009, 1981.
86. **Campbell, D. B.,** Separation of drugs from body fluids, in *Assays of Drugs and Other Trace Compounds in Biological Fluids,* Reid, E., Ed., North-Holland, Amsterdam, 1976, 105.
87. **Collins, C., Muto, J., and Spiehler, V.,** Whole blood deproteinization for drug screening using automatic pipettors, *J. Analyt. Toxicol.,* 16, 340, 1992.
88. **Ings, R. M. J., McFadzean, J. A., and Ormerod, W. E.,** The fate of metronidazole and its implications in chemotherapy, *Xenobiotica,* 5, 223, 1975.
89. **Chu, S. Y., Olivieras, L., and Deyasi, S.,** Extraction procedure for measuring anticonvulsant drugs by liquid chromatography, *Clin. Chem.,* 26, 521, 1980.

90. **Da Silva, J. A. F.,** Sample preparation for trace drug analysis, in *Trace Organic Sample Handling*, Reid, E., Ed., Ellis Horwood Limited, Chichester, 1981, 192.

91. **Stafford, B. E., Kabra, P. M., and Marton, L. J.,** Some comments on anticonvulsant drug monitoring by liquid chromatography, *Clin. Chem.*, 26, 1366, 1980.

92. **Sallustio, B. C. and Morris, R. G.,** Unbound plasma phenytoin concentrations measured using enzyme immunoassay technique on the Cobas MIRA analyzer — *in vivo* effect of valproic acid, *Ther. Drug Monitor.*, 14, 9, 1992.

93. **Ekblom, M., Hammarlund-Udenaes, M., Lundqvist, T., and Sjöberg, P.,** Potential use of microdialysis in pharmacokinetics: a protein binding study, *Pharm. Res.*, 9, 155, 1992.

94. **MacGregor, T. R. and Sardi, E. D.,** In vitro protein binding behavior of dipyridamole, *J. Pharm. Sci.*, 80, 119, 1990.

95. **Tse, F. L. S., Nickerson, D. F., and Yardley, W. S.,** Binding of fluvastatin to blood cells and plasma proteins, *J. Pharm. Sci.*, 82, 942, 1993.

96. **Piekoszewski, W. and Jusko, W. J.,** Plasma protein binding of tacrolimus in humans, *J. Pharm. Sci.*, 82, 340, 1993.

97. **Jung, D., Mayersohn, M., and Perrier, D.,** The 'ultra-free' ultrafiltration technique compared with equilibrium dialysis for determination of unbound thiopental (thiopentone) concentrations in serum, *Clin. Chem.*, 27, 166, 1981.

98. **Joern, W. A.,** Gas-chromatographic assay of free phenytoin in ultrafiltrates of plasma: test of a new filtration apparatus and specimen stability, *Clin. Chem.*, 27, 417, 1981.

99. **Cramer, J. A. and Mattson, R. H.,** Simple method to obtain unbound antiepileptic drugs in blood, *Ther. Drug Monitor.*, 4, 107, 1982.

100. **Jeevanandam, M., Novic, B., Savich, R., and Wagman, E.,** Serum acetamidophen assay using activated charcoal adsorption and gas chromatography without derivatisation, *J. Analyt. Toxicol.*, 4, 124, 1980.

101. **Plavsic, F.,** Charcoal extraction of acid, neutral and basic drugs from human serum, *Period. Biol.*, 82, 289, 1980.

102. **DiCorcia, A., Ripani, L., and Samperi, R.,** Chromatographic microprocedure for trace determination of phenobarbitone in blood serum, *J. Chromatogr. Sci.*, 18, 365, 1982.

103. **Kwong, T. C., Martinez, R., and Keller, J. M.,** Bonded phase extraction of plasma tricyclic antidepressant drugs for gas chromatographic analysis, *Clin. Chim. Acta*, 126, 203. 1982.

104. **Bidlingmeyer, B. A., Korpi, J., and Little, J. N.,** Determination of tricyclic antidepressants using silica gel with a reversed-phase eluent, *Chromatographia*, 15, 83, 1982.

105. **Rao, S. N., Dhar, A. K., Kutt, H., and Okamoto, M.,** Determination of diazepam and its pharmacologically active metabolites in blood by Bond-Elut column chromatography and reversed-phase high-performance liquid chromatography, *J. Chromatogr.*, 231, 341, 1982.

106. **Good, T. J. and Andrews, J. S.,** Use of bonded-phase extraction columns for rapid sample preparation of benzodiazepines and metabolites from serum for h.p.l.c. analysis, *J. Chromatogr., Sci.*, 19, 562, 1981.

107. **Leferink, J. G., Dankers, J., and Maes, R. A. A.,** Time-saving method for the determination of the β_2 sympathomimetics terbutaline, salbutamol and fenoterol. Preliminary results, *J. Chromatogr.*, 229, 217, 1982.

108. **Lawrence, R. and Allwood, M. C.,** Development of an analytical method to measure cyclosporin A concentrations in body fluids using h.p.l.c., *J. Pharm. Pharmacol.*, 32 (Suppl.), 100, 1980.

109. **Jarc, H.,** Rapid extraction of oestrogen-active substances from animal tissues, *Fleishwirtschaft*, 60, 676, 1980.

110. **Hsieh, J. Y. K., Yang, R. K., and Davis, K. L.,** Improved sample preparation before liquid chromatographic determination of probenecid in cerebrospinal fluid, *Clin. Chem.*, 29, 213, 1983.

111. **Andersson, S. H. G., Axelson, M., Sahlberg, B. L., and Sjovall, J.,** Simplified method for isolation and analysis of ethynylsteroids in urine, *Analyt. Lett.*, 14, 783, 1981.

112. **Pallante, S. L., Stogniew, M., Colvin, M., and Liberato, D. J.,** Elution of disposable octadecylsilane cartridges with hydrophobic organic solvents (application in determination of chloramphenicol in biological materials, *Analyt. Chem.*, 54, 2612, 1982.

113. **Robert, J.,** Extraction of anthracyclines from biological fluids for h.p.l.c. evaluation, *J. Liq. Chromatogr.,* 3, 1561, 1980.

114. **Allan, R. J., Goodman, H. T., and Watson, T. R.,** Two high-performance liquid-chromatographic determinations for mebendazole and its metabolites in human plasma using rapid Sep-Pak C$_{18}$ extraction, *J. Chromatogr.,* 183, 311, 1980.

115. **George, R. C.,** Improved sample treatment before liquid chromatographic determination of anticonvulsants in serum *Clin. Chem.,* 17, 198, 1981.

116. **Elahi, N.,** Magnetically propelled, encapsulated XAD-2 and microphase extraction technique for rapid quantitation of basic drugs in serum, *J. Chromatogr. Sci.,* 20, 483, 1982.

117. **Oates, N. S., Shah, R. R., Idle, J. R., and Smith, R. L.,** The urinary disposition of phenformin and 4-hydroxyphenformin and their rapid simultaneous measurement, *J. Pharm. Pharmacol.,* 32, 731, 1980.

118. **Caplan, Y. H., Backer, R. C., Stajic, M., and Thompson, B. C.,** Detection of drugs using XAD-2 resin. III. Routine screening procedure for bile, *J. Forensic Sci.,* 24, 745, 1979.

119. **Elahi, N.,** Encapsulated XAD-2 extraction technique for rapid screening of drugs of abuse in urine, *J. Analyt. Toxicol.* 4, 26, 1980.

120. **Balkon, J., Donnelly, B., and Prendes, D.,** Rapid isolation technique for drugs from tissues and fluids: use of the Du Pont Prep. 1 system, *J. Forensic Sci.,* 27, 23, 1982.

121. **Ishida, T., Oguri, K., and Yoshimura, H.,** Determination of oxycodone metabolites in urine and faeces of several mammalian species, *J. Pharmacobio-Dyn.,* 5, 521, 1982.

122. **Beckett, A. H.,** The Times, January 7, 1994.

123. **Lafolie, P.,** Measurement of creatinine in urine screening for drugs of abuse *Clin. Chem.,* 39, 699, 1993.

124. **Simpson, D., Jarvie, D. R., and Moore, F. M. L.,** Measurement of creatinine in urine screening for drugs of abuse, *Clin. Chem.,* 39, 698, 1993.

125. **Beckett, A. H.,** Distribution and metabolism in man of some narcotic analgesics and some amphetamines, in *Scientific Basis of Drug Dependence,* Steinberg, H., Ed., Churchill, London, 1969, 129.

126. **Cone, E. J., Huestis, M. A., and Mitchell, J. M.,** Do consecutive urine catches differ in marijuana metabolite concentration? *J. Analyt. Toxicol.,* 17, 186, 1993.

127. **Galeano Diaz, T., Guiberteau Cabanillas, A., Lopez Martinez, L., and Salinas, F.,** Polagraphic behaviour and determination of furaltadone in its formulations, milk and urine by differential pulse polarography, *Anal. Chim. Acta,* 273, 351, 1993.

128. **Ohkubo, T., Shimoyama, R., and Sugawara, K.,** High-performance liquid-chromatographic determination of levomepromazine in human breast milk and serum using solid-phase extraction, *Biomed. Chromatogr.,* 7, 227, 1993.

129. **Ohkubo, T., Shimoyama, R., and Sugawara, K.,** Determination of chlorpromazine in human breast milk and serum by high-performance liquid chromatography, *J. Chromatogr.,* 614, 328, 1993.

130. **Tcyzkowska, K. L., Voyksner, R. D., Anderson, K. L., and Aronson, A. L.,** Determination of ceftiofur and its metabolite desfuroylceftiofur in bovine serum and milk by ion-paired liquid chromatography, *J. Chromatogr.,* 614, 123, 1993.

131. **Lambert, W. E., Vanneste, L., and Leenheer, A. P.,** Enzymic sample hydrolysis and HPLC in a study of phylloquinone concentration in human milk, *Clin. Chem.,* 38, 1743, 1993.

132. **Heintz, R. C., Stebler, T., Lunell, N. O., Mueller, S., and Guentert, T. W.,** Excretion of tenoxicam and 5'-hydroxy-tenoxicam into human milk, *J. Pharmacol. Med.,* 3, 57, 1993.

133. **Venn, R. F., Capper, S. J., Morley, J. S., and Miles, J. B.,** Methods and problems in the assay of CSF for β-endorphin and other endogenous peptides, in *Bioactive Analytes, Including CNS Drugs Peptides and Enantiomers,* Reid, E., Scales, B., and Wilson, I. D., Eds., Plenum Press, New York, 1986, 55.

134. **Desiderio, D. M., Kai, M., Tanzer, F. S., Trimble, J., and Wakelyn, C.,** Measurement of enkephalin peptides in canine brain regions, teeth and cerebrospinal fluid with high-performance liquid chromatography and mass spectrometry, *J. Chromatogr.,* 297, 245, 1984.

135. **Wahba, W. W., Winek, C. L., and Rozin, L.,** Distribution of morphine in body fluids of heroin users, *J. Analyt. Toxicol.,* 17, 123, 1993.

136. **Krol, G. J., Noe, A. J., and Yeh, S. C.,** Gas and liquid chromatographic analysis of nimodipine calcium antagonist in blood plasma and cerebrospinal fluid, *J. Chromatogr.,* 305, 105, 1984.
137. **Schramm, W., Smith, R. H., Craig, P. A., Paek, S. H., and Kuo, H. H.,** Determination of free progesterone in an ultrafiltrate of saliva collected in situ, *Clin. Chem.,* 36, 1488, 1990.
138. **Schramm, W., Annesley, T. M., Siegel, G. J., Sackellares, J. C., and Smith, R. H.,** Measurement of phenytoin and carbamazepine in an ultrafiltrate of saliva, *Ther. Drug Monitor.,* 13, 452, 1991.
139. **Mandel, I. W.,** Relation of saliva and plaque to caries, *J. Dental Res.,* 53, 246, 1974.
140. **Feller, K., LePetit, G., and Marks, U.,** Zur Verteilung von Pharmaka zwischen Speicheln und Blutplasma, *Pharmazie,* 31, 745, 1976.
141. **Anavekar, S. N., Saunders, R. H. Wardell, W. M., Shoulson, I., Emmings, F. G., Cook, C. E., and Gringeri, A. J.,** Parotid and whole saliva in the prediction of serum total and free phenytoin concentrations, *Clin. Pharmacol. Ther.,* 24, 629, 1978.
142. **Westenberg, H. G. M., de Zeeuw, R. A., van der Kleijn, E., and Oei, T. T.,** Relationship between carbamazepine concentration in plasma and saliva in man as determined by liquid chromatography, *Clin. Chim. Acta,* 79, 155, 1977.
143. **DiGregorio, G. J., Piraino, A. J., and Ruch, E.,** Diazepam concentration in parotid saliva, mixed saliva and plasma, *Clin. Pharmacol. Ther.,* 24, 720, 1978.
144. **Horning, M. G., Brown, L., Nowlin, J., Lertranangkoon, K., Kellaway, P., and Zion, T. E.,** Use of saliva in therapeutic drug monitoring, *Clin. Chem.,* 23, 157, 1977.
145. **Mucklow, J. C., Bending, M. R., Kahn, G. C., and Dollery, C. T.,** Drug concentration in saliva, *Clin. Pharmacol. Ther.,* 24, 563, 1978.
146. **Troupin, A. S. and Friel, P.,** Anticonvulsant level in saliva, serum and cerebrospinal fluid, *Epilepsia,* 16, 223, 1975.
147. **Schmidt, D. and Kupferberg, H. J.,** Diphenylhydantoin, phenobarbital and primidone in saliva, plasma, and cerebrospinal fluid, *Epilepsia,* 16, 735, 1975.
148. **Cook, C. E., Amerson, E., Poole, W. K., Lesser, P., and O'Tuama, L.,** Phenytoin and phenobarbital concentrations in saliva and plasma measured by radioimmunoassay, *Clin. Pharmacol. Ther.,* 18, 742, 1976.
149. **Matin, S. B., Wan, S. H., and Karam, J. H.,** Pharmacokinetics of tolbutamide: prediction by concentration in saliva, *Clin. Pharmacol. Ther.,* 16, 1052, 1974.
150. **Vesell, E. S., Passananti, G. T., Glenwright, A., and Dvorchik, B. H.,** Studies on the disposition of antipyrine, aminopyrine, and phenacetin using plasma, saliva and urine, *Clin. Pharmacol. Ther.,* 18, 259, 1975.
151. **Koup, J. R., Jusko, W. J., and Goldfarb, A. L.,** pH-Dependent secretion of procainamide into saliva, *J. Pharm. Sci.,* 64, 2008, 1975.
152. **Vermeer, B. J., Reman, F. C., and Van Gent, C. M.,** The determination of lipids and proteins in suction blister fluid, *J. Invest. Dermatol.,* 73, 303, 1979.
153. **Volden, G., Thorsrud, A. K., Bjørnson, I., and Jellum, E.,** Biochemical composition of suction blister fluid determined by high resolution multicomponent analysis (capillary gas chromatography-mass spectrometry and two-dimensional electrophoresis, *J. Invest. Dermatol.,* 75, 421, 1980.
154. **Walker, J. S., Knihinicki, R. D., Seideman, P., and Day, R. O.,** Pharmacokinetics of ibuprofen enantiomers in plasma and suction blister fluid in healthy volunteers, *J. Pharm. Sci.,* 82, 787, 1993.
155. **Blagbrough, I. S., Daykin, M. M., Doherty, M., Pattrick, M., and Shaw, P. N.,** High-performance liquid-chromatographic determination of naproxen, ibuprofen and diclofenac in plasma and synovial fluid in man, *J. Chromatogr.,* 578, 251, 1992.
156. **Young, M. A., Aarons, L., and Toon, S.,** The pharmacokinetics of the enantiomers of flurbiprofen in patients with rheumatoid arthritis, *Br. J. Clin. Pharmacol.,* 31, 102, 1991.
157. **Radermacher, J., Jentsch, D., Scholl, M. A., Lustinetz, T., Frölich, J. C.,** Diclofenac concentrations in synovial fluid and plasma after cutaneous application in inflammatory joint disease, *Br. J. Clin. Pharmacol.,* 31, 537, 1991.
158. **Carlucci, G., Biordi, L., and Bologna, M.,** Human plasma and aqueous humour determination of imipenem by liquid chromatography with ultra-violet detection, *J. Liq. Chromatogr.,* 16, 2347, 1993.

159. **Dutton, G.,** *Glucuronidation of Drugs and Other Compounds,* CRC Press, Boca Raton, Florida, 1980, 5.
160. **Williams, R. T.,** Detoxication mechanisms and the design of drugs, in *Symposium on Biological Approaches to Cancer Chemotherapy,* Academic Press, London, 1960, 21.
161. **Caldwell, J. and Hutt, A. J.,** Methodology for the isolation and characterization of conjugates of xenobiotic carboxylic acids, *Progress in Drug Metabolism,* Vol. 9, Bridges, A. W. and Chasseaud, L., Eds., Taylor and Francis, London, 1986.
162. **Mohler, W., Bletz, I., and Reiser, M.,** Die Struktur von Aussheidungsprodukten des 1-Hexyl-3,7-dimethylxanthins, *Archiv. der Pharmazie,* 299, 448, 1966.
163. **Hinze, H.-J., Bedessem, G., and Söder, A.,** Struktur der Aussheidungsprodukte des 3,7-Dimethyl-1(5-oxo-hexyl)-xanthins (BL191) beim Menshen, *Arzneim. Forsch.,* 22, 1144, 1972.
164. **Brodie, B. B., Gillette, J. R., and Ladu, B. N.,** Enzymatic metabolism of drugs and other foreign compounds, *Annu. Rev. Biochem.,* 27, 427, 1958.
165. **Mazel, P., Henderson, J. F., and Axelrod, J.,** S-Demethylation by microsomal enzymes, *J. Pharmacol.,* 143, 1, 1964.
166. **Thauer, R. K., Stoffler, G., and Uehleker, H.,** N-Hydroxylierung von sulfanilamid zu *p*-hydroxylaminobenzolsulfonamid durch lebermikrosomen, *Arch. Exp. Pathol. Pharmakol.,* 252, 32, 1965.
167. **Robinson, J. D., Wilson, I. D., Bevan, C. D., and Chamberlain, J.,** A novel use of RIA and HPLC in the identification of plasma metabolites of HR 158, in *Drug Metabolite Isolation and Determination,* Reid, E. and Leppard, J. P., Eds., Plenum Press, New York, 1983, 111.
168. **Uzan, A., Gueremy, C., and Le Fur, G.,** Biotransformation de la (quinuclidinyl-3-méthyl)-10 phénothiazine (LM 209) un nouvel anti-allergique, et distribution et excretion des metabolites, *Xenobiotica,* 6, 649, 1976.
169. **Uzan, A., Gueremy, C., and Le Fur, G.,** Absorption, distribution et excretion de la (Quinuclidinyl-3-méthyl)-10 phénothiazine (LM 209) un nouvel anti-allergique, *Xenobiotica,* 6, 663, 1976.
170. **Spector, E. and Shideman, F. E.,** Metabolism of of thiopyrimidine derivatives; thiamylal, thiopental and thiouracil, *Biochem. Pharmacol.,* 2, 182, 1959.
171. **Kou, B.-S., Poole, J. C., Huang, K. K., and Cheng, H.,** Pharmacokinetics and metabolic interconversion of 4-amino-5-chloro-2-[(methylsulfinyl)ethoxy]-N[2-(diethylamino)ethyl]benzamide and its sulfide and sulfone metabolites in rats, *J. Pharm. Sci.,* 82, 694, 1993.
172. **Charalampous, K. D., Orengo, A., Walker, K. E., and Kinross-Wright, J.,** Metabolic fate of β-(3,4,5-trimethoxyphenyl)-ethylamine (mescaline) in humans: isolation and identification of 3,4,5-trimethoxyphenylacetic acid, *J. Pharmacol. Exp. Ther.,* 145, 242, 1964.
173. **Culp, H. W. and McMahon, R. E.,** Reductase for aromatic aldehydes and ketones. The partial purification and properties of a reduced triphosphopyridine nucleotide-dependent reductase from rabbit cortex kidney, *J. Biol. Chem.,* 243, 848, 1968.
174. **Fouts, J. R., Kamm, J. J., and Brodie, B. B.,** Enzymatic reduction of prontosil and other azo dyes, *J. Pharmacol.,* 120, 291, 1957.
175. **Tompsett, S. L.,** Nitrazepam (Mogadon) in blood serum and urine and Librium in urine, *J. Clin. Pathol.,* 21, 366, 1968.
176. **Holmes, R. S. and Masters, C. J.,** The developmental multiplicity and isoenzyme status of cavian esterases, *Biochim. Biophys. Acta,* 132, 379, 1967.
177. **Holmes, R. S. and Masters, C. J.,** A comparative study of the multiplicity of mammalian esterases, *Biochim Biophys. Acta,* 151, 147, 1968.
178. **Mark, L. C., Kayden, H. J., Steele, J. M., Cooper, J. R., Berlin, I., Rovenstine, E. A., and Brodie, B. B.,** Physiological disposition and cardiac effects of procainamide, *J. Pharmacol. Exp. Ther.,* 102, 5, 1951.
179. **Wilson, I. D., Bhatti, A., Illing, H. P. A., Bryce, T. A., and Chamberlain, J.,** in *Drug Metabolite Isolation and Determination,* Reid, E. and Leppard, J. P., Eds., Plenum Press, New York, 1983, 181.
180. **Kato, K., Ide, H., Hirohata, I., and Fishman, W. H.,** Biosynthetic preparation of the NO-glucosiduronic acid of N-acetyl-N-phenylhydroxylamine, *Biochem. J.,* 103, 647, 1967.
181. **Wakabayashi, M., Wotiz, H. H., and Fishman, W. H.,** Action of β-glucuronidase on a steroid enol-β-glucosiduronic acid, *Biochim. Biophys. Acta,* 48, 198, 1961.

182. **Colucci, D. F. and Buyske, D. A.,** The biotransformation of a sulfonamide to a mercaptan and to mercapturic acid and glucuronide conjugates, *Biochem. Pharmacol.,* 14, 457, 1965.

183. **Hornke, I., Fehlhaber, H.-W., Girg, M., and Gantz, H.,** Metabolism of nomifensine: isolation and identification of the conjugates of nomifensine-$_{14}$C from human urine, *Arzneim. Forsch.,* 28, 58, 1978.

184. **Chamberlain, J. and Hill, H. M.,** A simple gas chromatographic method for the determination of nomifensine in plasma and a comparison of the method with other available techniques, *Br. J. Clin. Pharmacol.,* 4, 117S, 1977.

185. **Tang, B. K., Kalow, W., and Grey, A. A.,** Metabolic fate of phenobarbital in man. N-Glucoside formation, *Drug Metab. Dispos.,* 7, 315, 1979.

186. **Tang, B. K., Yilmaz, B., and Kalow, W.,** Determination of phenobarbital, p-hydroxyphenobarbital and phenobarbital-N-glucoside in urine by gas chromatography chemical ionisation mass spectrometry, *Biomed. Mass Spectromet.,* 11, 462, 1984.

187. **Soine, W. H., Bhargava, V. O., and Garretson, A. K.,** Direct detection of 1-(β-glucopyranosyl)phenobarbital in human urine, *Drug Metab. Dispos.,* 12, 792, 1984.

188. **Bhargava, V. O. and Garrettson, L. K.,** Development of phenobarbital glucosidation in the human neonate, *Dev. Pharmacol. Ther.,* 11, 8, 1988.

189. **Tang, B. K., Kalow, W., Endrenyi, L., and Chan, F.-Y.,** An assessment of short-cut procedures for studying drug metabolism in vivo using amobarbital as a model drug, *Eur. J. Clin. Pharmacol.,* 22, 229, 1982.

190. **Bannerjee, R. K. and Roy, A. B.,** Sulphotransferases of guinea pig liver, *Mol. Pharmacol.,* 2, 56, 1966.

191. **Lipmann, F.,** Development of the acetylation problem, a personal account, *Science,* 120, 855, 1954.

192. **Lindsay, B.,** A layman's guide to health in a highly technological society; you can't get there from here, *Drug Metab. Rev.,* 13, 5, 1982.

193. **Brodie, B. B. and Axelrod, J.,** Fate of acetanilide in man, *J. Pharmacol. Exp. Ther.,* 94, 29, 1948.

194. **Brodie, B. B. and Axelrod, J.,** Fate of acetophenetidin (phenacetin) in man and methods for estimation of acetophenetidin and its metabolites in biological material, *J. Pharmacol. Exp. Ther.,* 14, 1013, 1949.

195. **Field, J. B., Ohta, M., Boyle, C., and Remer, A.,** Potentiation of acetohexamide hypoglycemia by phenylbutazone, *N. Engl. J. Med.,* 277, 889, 1967.

196. **Elion, G. B., Yu, T.-F., Gutman, A. B., and Hitchings, G. H.,** Renal clearance of oxipurinol, the chief metabolite of allopurinol, *Am. J. Med.,* 45, 69, 1968.

197. **Stella, V. J.,** Prodrugs and site-specific drug delivery, in *Xenobiotic Metabolism and Disposition. Proceedings of the Second International ISSX Meeting,* New York, Taylor and Francis, 1989, 109.

198. **Eadie, M. J. and Tyrer, J. H.,** in *Anticonvulsant Therapy,* 2nd ed., Churchill Livingdstone, London, 1980, 142.

199. **Brogden, R. N., et al.,** *Drugs,* 20, 161, 1980.

200. **Baker, E. M.,** The metabolic fate of codeine in man, *J. Pharmacol. Exp. Ther.,* 114, 251, 1955.

201. **Dasberg, H. H., Kleijn, E., Guelen, P. J. R., and Praag, H. M.,** Plasma concentrations of diazepam and of its metabolite *N*-desmethyldiazepam in relation to anxiolytic effect, *Clin. Pharmacol. Ther.,* 15, 473, 1974.

202. **Rubens, R., Verhaegen, H., Brugmans, J., and Schuermans, V.,** Difenoxine, the active metabolite of diphenoxylate. Part 5: clinical comparison of difenoxine and diphenoxylate, *Arzneim. Forsch.,* 22, 526, 1972.

203. **Ambre, J. J. and Fischer, L. J.,** Glutethimide intoxication: plasma levels of glutethimide and a metabolite in humans, dogs and rats, *Res. Commun. Chem. Pathol. Pharmacol.,* 4, 307, 1972.

204. **Ambre, J. J. and Fischer, L. J.,** Identification and activity of the hydroxy metabolite that accumulates in the plasma of humans intoxicated with glutethimide, *Drug Metab. Dispos.,* 2, 151, 1974.

205. **Mitchell, J. R., Thorgeirsson, U. P., Black, M., Timbrell, J. A., Snodgrass, W. R., Potter, W. Z., Jollow, D. J., and Keiser, H. R.,** Increased incidence of isoniazid hepatitis in rapid acetylators: possible relation to hydrazine metabolites, *Clin. Pharmacol. Ther.,* 18, 70, 1975.

206. **Strong, J. M., Mayfield, D. E., Atkinson, A. J., Burris, B. C., Raymon, F., and Webster, L. T.,** Pharmacologic activity, metabolism, and pharmacokinetics of glycinexylidide, *Clin. Pharmacol. Ther.,* 17, 184, 1975.

207. **Gallagher, B. B., Baumel, I. P., and Mattson, R. H.,** Metabolic disposition of primidone and its metabolites in epileptic subjects after single and repeated administration, *Neurology,* 22, 1186, 1972.

208. **Forrest, I. S. and Forrest, F. M.,** Metabolism and action mechanism of the phenothiazine drugs, *Exp. Med. Surg.,* 21, 231, 1963.

209. **Curry, S. H.,** Assay of chlorpromazine and some of its metabolites in biological fluids, in *Assays of Drugs and Other Trace Compounds in Biological Fluids,* Reid, E., Ed., North-Holland, Amsterdam, 1976, 185.

210. **Illing, H. P. A. and Fromson, J. M.,** Species difference in the disposition and metabolism of 6,11-dihydro-11-oxodibenz(be)oxepin-2-acetic acid (isoxepac) in rat, rabbit, dog, rhesus monkey and man, *Drug Metab. Dispos.,* 6, 510, 1978.

211. **Dayton, P. G., Pruitt, A. W., McNay, J. L., and Steinworst, J.,** Studies with triamterene, a substituted pteridine. Unusual brain to plasma ratio in mammals, *Neuropharmacology,* 11, 435, 1972.

212. **Pruitt, A. W., Winkel, J. S., and Dayton, P. G.,** Variations in the fate of triamterene, *Clin. Pharmacol. Ther.,* 21, 610, 1977.

213. **Vereczkey, L., Bianchetti, G., Rovei, V., and Frigerio, A.,** Gas chromatographic method for the determination of nomifensine in human plasma, *J. Chromatogr.,* 116, 451, 1976.

214. **Bailey, E., Fenoughty, M., and Richardson, L.,** Automated high-resolution gas chromatographic analysis of psychtropic drugs in biological fluids using open-tubular glass capillary columns. I. Determination of nomifensine in human plasma, *J. Chromatogr.,* 131, 347, 1977.

215. **Bryce, T. A. and Burrows, J. L.,** Determination of oxpentifylline and a metabolite 1-(5′-hydroxyhexyl)-3,7-dimethylxanthine, by gas liquid chromatography using a nitrogen selective detector, *J. Chromatogr.,* 181, 355, 1980.

216. **Rupp, W., Badian, M., Christ, O., Hajdu, P., Kulkarni, R. D., Taueber, K., Uihlein, M., Bender, R., and Vanderbeke, O.,** Pharmacokinetics of single and multiple doses of clobazam in humans, *Br. J. Clin. Pharmacol.,* 7, 51S, 1979.

217. **Jusko, W. J.,** Role of tobacco smoking in pharmacokinetics, *J. Pharmacokinet. Biopharm.,* 6, 7, 1978.

218. **Miller, L. G.,** Recent developments in the study of the effects of cigarette smoking on clinical pharmacokinetics and clinical pharmacodynamics, *Clin. Pharmacokinet.,* 17, 90, 1989.

219. **Miller, L. G.,** Cigarettes and drug therapy: pharmacokinetic and pharmacodynamic considerations, *Clin. Pharmacol.,* 9, 125, 1990.

220. **Hill, R.,** Salicylate interference with measurement of acetaminophen, *Clin. Chem.,* 29, 590, 1983.

221. **Stearns, F. M.,** Determination of procainamide and *N*-acetyl-procainamide by high performance liquid chromatography, *Clin. Chem.,* 27, 2064, 1981.

222. **Najolia, M.,** Determination of chloramphenicol in serum using liquid chromatography, *Chromatogr. Newsletter,* 7, 7, 1979.

223. **Foreman, J. M., Griffiths, W. C., Dextraze, P. G., and Diamond, I.,** Simultaneous assay of diazepam, chlordiazepoxide, *N*-desmethyl-diazepam, *N*-demethylchlordiazepoxide and demoxepam in serum by high-performance liquid chromatography, *Clin. Biochem.,* 13, 122, 1980.

224. **Wirtz, P. G. and Street, T. L.,** Thioridazine interference in chromatographic and enzyme immunoassays for imipramine in serum, *Clin. Chem.,* 29, 724, 1983.

225. **Kuss, H. J. and Feistenauer, E.,** Quantitative high-performance liquid chromatographic assay for determination of maprotiline and its 2-hydroxy-3-(methylamino)propyl analogue oxaprotiline in human plasma, *J. Chromatogr.,* 204, 349, 1981.

226. **Thielemann, H.,** Dünnschichtchromatographischer Nachweis von Meprobamat (2-Methyl-2-propyl-propandiol-(1,3)-dicarbamat) im Harn, *Scientia Pharmaceutica,* 48, 387, 1980.

227. **Breithaupt, H., Eenzlen, E., and Goebel, G.,** Rapid high-pressure liquid-chromatographic determination of methotrexate and its metabolites, 7-hydroxymethotrexate and 2,4-diamino-N^{10}-methylpteroic acid in biological fluids, *Analyt. Biochem.,* 121, 103, 1982.

228. **Kelly, R., Christmore, D., Smith, R., Doshier, L., and Jacobs, S. L.,** Mexilitine in plasma by high-pressure liquid chromatography, *Ther. Drug Monitor.,* 3, 279, 1981.

229. **Svenneby, G., Wedege, E., and Karlsen, R. L.,** Pholcodine interference in immunoassay for opiaates in urine, *Forensic Sci. Int.,* 21, 223, 1983.

230. **Jahnchen, E. and Levy, G.,** The determination of phenylbutazone in plasma, *Clin. Chem.,* 18, 984, 1972.

231. **Hahn, E.,** Dimenhydrinate interferes with radioimmunoassay of theophylline, *Clin. Chem.,* 26, 1759, 1980.

232. **Broussard, L. A.,** Theophylline determination by high-pressure liquid chromatography, *Clin. Chem.,* 27, 1931, 1981.

233. **Soine, W. H., Soine, P. J., England, T. J., Ferkany, J. W., and Agriesti, B. E.,** Identification of phenobarbital N-glucosides as urinary metabolites of phenobarbital in mice, *J. Pharm. Sci.,* 80, 99, 1990.

234. **Christensen, J. M. and Stalker, D.,** Ibuprofen piconol hydrolysis in vitro in plasma, whole blood, and serum using different anticoagulants, *J. Pharm. Sci.,* 80, 29, 1990.

235. **Howanitz, P. J., McBride, J. H., Kilewer, K. E., and Rogerson, D. O.,** Prevalence of antibodies to HTLV-III in quality assurance sera, *Clin. Chem.,* 32, 773, 1986.

236. **Bove, J. R., DePalma, L., and Weirich, F.,** Anti-HTLV-III/LAV in pooled sera, *Clin. Chem.,* 33, 308, 1987.

237. **Tersmette, M., de Goede, R. E. Y., Over, J., de Jonge, E., Radema, H., Lucas, C. J., Huisman, H. G., and Miedema, F.,** Thermal inactivation of human immunodeficiency virus in lyophilised blood products evaluated by ID50 titration, *Vox Sang.* 51, 239, 1986.

238. **Horowitz, B., Wiebe, M. E., Lippin, A., and Stryker, M. H.,** Inactivation of viruses in labile blood derivatives. II. Physical methods, *Transfusion,* 25, 516, 1985.

239. **Good, S. S., Reynolds, D. J., and de Miranda, P.,** The analysis of zidovudine and its glucuronide metabolite by HPLC, *Analysis for Drugs and Metabolites, Including Anti-Infective Agents,* Reid, E. and Wilson, I. D., Eds., Royal Society of Chemistry, Cambridge, 1990, 173.

240. **de Bree, H. and van Berkel, M. P.,** *Analysis for Drugs and Metabolites, Including Anti-Infective Agents,* Reid, E. and Wilson, I. D., Eds., Royal Society of Chemistry, Cambridge, 1990, 221.

241. **Jennison, T. A., Wozniak, E., Nelson, G., and Urry, F. M.,** Quantitayive conversion of morphine 3-β-D-glucuronide to morphine using β-glucuronidase obtained from *Patella vulgata* as compared to acid hydrolysis, *J. Analyt. Toxicol.,* 17, 208, 1993.

242. **Sharifi, S., Michaelis, H. C., Lotterer, E., and Bircher, J.,** Determination of coumarin, 7-hydroxycoumarin, 7-hydroxycoumarin-glucuronide and 3-hydroxycoumarin by high-performance liquid chromatography, *J. Liq. Chromatogr.,* 16, 1263, 1993.

243. **Su, S. Y., Shiu, G. K., Simmons, J., Viswanathan, C. T., and Skelly, J. P.,** High-performance liquid-chromatographic analyis of six conjugated and unconjugated oestrogens in serum, *Biomed. Chromatogr.,* 6, 265, 1992.

244. **Castillo, M. and Smith, P. C.,** Direct determination of ibuprofen and ibuprofen acylglucuronide in plasma by high-performance liquid chromatography using solid-phase extraction, *J. Chromatogr.,* 125, 109, 1993.

245. **Vree, T. B., Van den Biggelaar-Martea, M., and Verwey-van Wissen, C. P. W. G. M.,** Determination of indomethacin, its metabolites and their glucuronides in human plasma and urine by means of direct gradient high-performance liquid-chromatographic analysis. Preliminary pharmacokinetics and effect of probenecid, *J. Chromatogr.,* 616, 271, 1993.

246. **Hartley, R., Green, M., and Levene, M. I.,** Analysis of morphine and its 3- and 6-glucuronides by high-performance liquid chromatography with fluorimetric detection following solid-phase extraction from neonatal plasma, *Biomed. Chromatogr.,* 7, 34, 1993.

247. **Nishikawa, M., Nakajima, K., Igarashi, K., Kasuya, F., Fukui, M., and Tsuchihashi, H.,** Determination of morphine 3-glucuronide in human urine by LC-AP CIMS, *Jpn. J. Toxicol. Environ. Health,* 38, 121, 1992.

248. **Andersen, J. V. and Hansen, S. H.,** Simultaneous quantitative determination of naproxen, its metabolite 6-*O*-desmethylnaproxen and their five conjugates in plasma and urine samples by high-performance liquid chromatography on dynamically modified silica, *J. Chromatogr.,* 577, 325, 1992.

249. **Vree, T. b., Van den Biggelaar-Martea, M., and Verwey-Van Wissen, C. P. W. G. M.,** Determination of naproxen and its metabolite *O*-desmethylnaproxen with their acy glucuronides in human plasma and urine by means of direct gradient high-performance liquid-chromatography, *J. Chromatogr.,* 578, 239, 1992.

250. **Andersen, J. V. and Hansen, S. H.,** Simultaneous quantitative determination of naproxen, its metabolite 6-*O*-desmethylnaproxen and their five conjugates in plasma and urine samples by high-performance liquid chromatography on dynamically modified silica, *J. Chromatogr.,* 577, 325, 1992.

251. **Tatsuno, M., Nishikawa, M., Katagi, M., Tsuchihashi, H., Igarashi, K., Kasuya, F., and Fukui, M.,** Determination of oxazepam glucuronide in human urine by liquid chromatography — AP CI — mass spectrometry, *Jpn. J. Toxicol. Environ. Health,* 38, 162, 1992.

252. **Ockzenfels, H., Köhler, F., and Meiser, W.,** Teratogene Wirkung und Stereospezifität eines Thalidomid-Metaboliten, *Pharmazie,* 31, 492, 1976.

253. **Blaschke, G. and Graute, W. F.,** Enantiomere des Kongigurationsstabilen C-Methylthalidomids, *Liebigs Ann. Chem.,* 647, 1987.

254. **Schmahl, H. J., Heger, W., and Nau, H.,** The enantiomers of the teratogenic thalidomide-analogue EM 12; chemical stability, stereoselective metabolism and renal excretion in the marmoset monkey, *Toxicol. Lett.,* 45, 23, 1989.

255. **Barrett, A. M. and Cullum, V. A.,** The biological properties of the optical isomers of propranolol and their effect on cardiac arrhythmias, *Br. J. Pharmacol.,* 34, 43, 1968.

256. **Adams, S. S., Bresloff, P., and Mason, G. G.,** Pharmacological difference between the optical isomers og ibuprofen: evidence for the metabolic inversion of the (–) isomer, *J. Pharm. Pharmacol.,* 28, 256, 1976.

257. **Hermansson, J. and von Bahr, C.,** Determination of (R)- and (S)-alprenolol and (R)- and (S)-metoprolol as their diastereomeric derivatives in human plasma by reversed-phase liquid chromatography, *J. Chromatogr.,* 227, 113, 1982.

258. **Dayer, P., Leeman, T., Kupfer, A., Kronbach, T., and Meyer, U. A.,** Stereo- and regioselectivity of hepatic oxidation in man — effect of the debrisoquine/sparteine phenotype on buuralol hydroxylation, *Eur. J. Clin. Pharmacol.,* 31, 313, 1986.

259. **Hermansson, J. and Eriksson, M.,** Determination of (R)- and (S)-disopyramide in human plasma using a chiral alpha-1-acid glycoprotein column, *J. Chromatogr.,* 336, 321, 1984.

260. **Bonde, J., Pedersen, L. E., Nygaard, E., Ramsing, T., Angelo, H. R., and Kampmann, J. P.,** Stereoselective pharmacokinetics of disopyramide and interaction with cimetidine, *Br. J. Clin. Pharmacol.,* 31, 708, 1991.

261. **Richards R. P., Caccia, S., Jori, A., Ballabio, M., De Ponte, P., and Garratini, S.,** in *Bioactive Analytes, Including CNS Drugs Peptides and Enantiomers,* Reid, E., Scayles, B., and Wilson, I. D., Eds., Plenum Press, New York, 1986, 273.

262. **Kroemer, H. K., Turgeon, J., Parker, R. A., and Roden, D. M.,** Flecainide enantiomers: disposition in human subjects and electrophysiological action in vitro, *Clin. Pharmacol. Ther.,* 46, 584, 1989.

263. **Knadler, M. P., Braten, D. C., and Hall, S. D.,** Stereoselective disposition of flurbiprofen in normal volunteers, *Br. J. Clin. Pharmacol.,* 33, 369, 1992.

264. **Kennedy, K. A. and Fischer, L. J.,** Quantitative and stereochemical aspects of glutethimide metabolism in humans, *Drug Metab. Dispos.,* 7, 319, 1979.

265. **Gimenez, F., Aubry, A.-F., Farinotti, R., Kirkland, K., and Wainer, I. W.,** The determination of the enantiomers of halofantrine and monodesbutylhalofantrine in plasma and whole blood using sequential achiral/chiral high-performance liquid chromatography, *J. Pharm. Biomed. Anal.,* 10, 245, 1992.

266. **Lee, E. J. D., Williams, K., Day, R. O., Graham, G., and Champion, D.,** Stereoselective disposition of ibuprofen enantiomers in man, *Br. J. Clin. Pharmacol.,* 19, 669, 1985.

267. **Lennard, M. S., Tucker, G. T., Silas, J. H., Freestone, S., Ramsay, L. E. Woods, H. F.,** Differential stereoselective metabolism of metoprolol in extensive and poor debrisoquin metabolizers, *Clin. Pharmacol. Ther.,* 34, 732, 1983.

268. **Grech-Belanger, O., Turgeon, J., and Gilbert, M.,** Stereoselective disposition of mexiletine in man, *Br. J. Clin. Pharmacol.,* 21, 481, 1986.

269. **Williams, K. M.,** Kinetics of misonidazole enantiomers, *Clin. Pharmacol. Ther.*, 36, 817, 1984.

270. **Godbillon, J., Richard, A., Gerardin, A., Meinertz, T., Kasper, W., and Jahnchen, E.,** Pharmacokinetics of the enantiomers of acenocoumarol, *Br. J. Clin. Pharmacol.*, 12, 621, 1981.

271. **Thijssen, H. H. W., Janssen, G. M. J., and Baars, L. G. M.,** Lack of the effect of cimetidine on pharmacodynamics and kinetics of single oral doses of R- and S-acenocoumarol, *Eur. J. Clin. Pharmacol.*, 30, 619, 1986.

272. **Soons, P. A. and Breimer, D. D.,** Stereoselective pharmacokinetics of oral and intravenous nitrendipine in healthy male subjects, *Br. J. Clin. Pharmacol.*, 32, 11, 1991.

273. **Soons, P. A., De Boer, A. G., Van Brummelen, P., and Breimer, D. D.,** Oral absorption profile of nitrendipine in healthy subjects: a kinetic and dynamic study, *Br. J. Clin. Pharmacol.*, 27, 179, 1989.

274. **Drayer, D. E., Cook, C. E., Seltzman, T. P., and Lorenzo, B.,** Stereoselective elimination of phenobarbital enantiomers in normal human subjects, *Clin. Res.*, 33, 528A, 1985.

275. **Jahnchen, E., Meinertz, T., Gilfrich, H., Groth, U., and Martini, A.,** The enantiomers of phenprocoumon: pharmacodynamic and pharmacokinetic studies, *Clin. Pharmacol. Ther.*, 20, 342, 1976.

276. **Straka, R. J., Lalonde, R. L., and Wainer, I. W.,** Measurement of underivatised propranolol enantiomers in serum using a cellulose-tris(3,5-dimethylphenylcarbamate) high-performance liquid chromatographic (HPLC) chiral stationary phase, *Pharm. Res.*, 5, 187, 1988.

277. **Kawashima, K., Levy, A., and Spector, S.,** Stereospecific radioimmunoassay for propranolol isomers, *J. Pharmacol. Exp. Ther.*, 196, 517, 1976.

278. **Jackman, G. P., McLean, A. J., Jeninings, G. L., and Bobik, A.,** No stereoselective first-pass hepatic extraction of propranolol, *Clin. Pharmacol. Ther.*, 30, 291, 1981.

279. **Silber, B., Holford, N. H., and Riegelman, S.,** Stereoselective disposition and glucuronidation of propranolol in humans, *J. Pharm. Sci.*, 71, 699, 1982.

280. **von Bahr, C., Hermansson, J., and Tawara, K.,** Plasma levels of (+) and (−)-propranolol and 4-hydroxypropranolol after administration of racemic (±)-propranolol in man, *Br. J. Clin. Pharmacol.*, 14, 79, 1982.

281. **Olanoff, L., Walle, T., Walle, K., Cowart, T. D., and Gaffney, T. E.,** Stereoselective clearance and distribution of intravenous propranolol, *Clin. Pharmacol. Ther.*, 35, 755, 1984.

282. **Notterman, D. A., Drayer, D. E., Metakis, L., and Reidenberg, M. M.,** Stereoselective renal tubular secretion of quinidine and quinine, *Clin. Pharmacol. Ther.*, 40, 511, 1986.

283. **Sedman, A. J., Gal, J., Mastropaolo, W., Johnson, P., Maloney, J. D., and Moyer, T. P.,** Serum tocainide enantiomer concentration in human subjects, *Br. J. Clin. Pharmacol.*, 17, 113, 1984.

284. **Edgar, B., Heggelund, A., Johansson, L., Nyberg, G., and Regardh, C. G.,** The pharmacokinetics of R- and S-tocainide in healthy subjects, *Br. J. Clin. Pharmacol.*, 17, 216P, 1984.

285. **Vogelsgang, B., Echizen, H., Schmidt, E., and Eichelbaum, M.,** Stereoselective first-pass metabolism of highly cleared drugs: studies of the bioavailability of L-a and D-verapamil examined with a stable isotope technique, *Br. J. Clin. Pharmacol.*, 18, 733, 1984.

286. **Eichelbaum, M., Mikus, G., and Vogelgesang, B.,** Pharmacokinetics of (+), (−), and (±) verapamil after intravenous administration, *Br. J. Clin. Pharmacol.*, 17, 453, 1984.

287. **Haegele, K. D., Schoun, J., Alken, R. G., and Huebert, N. D.,** Determination of the R(−) and S(+) enantiomers of gamma vinyl-gamma-aminobutyric acid in human body fluids by gas chromatography-mass spectrometry, *J. Chromatogr.*, 274, 103, 1983.

288. **O'Reilly, R. A.,** Studies on the optical enantiomorphs of warfarin in man, *Clin. Pharmacol. Ther.*, 16, 348, 1974.

289. **Hewick, D. S. and Mcewen, J.,** Plasma half-lives, plasma metabolites and anticoagulant efficacies of the enantiomers of warfarin in man, *J. Pharm. Pharmcol.*, 25, 458, 1973.

290. **Wingard, L., O'Reilly, R., and Levy G.,** Pharmacokinetics of warfarin enantiomers: a search for intrasubject correlation, *Clin. Pharmacol. Ther.*, 23, 212, 1978.

291. **Ariëns, E. J.,** Stereochemistry, a basis for sophisticated nonsense in pharmacokinetics and clinical pharmacology, *Eur. J. Clin. Pharmacol.*, 26, 663, 1984.

292. **Gross, M.,** in *Annual Reports in Medicinal Chemistry,* Vol. 25, Bristol, J. A., Ed., Academic Press, San Diego, 1991, 299.

293. Guidelines for Submitting Supporting Documentation in Drug Applications for the manufacture of Drug Substances, Food and Drug Administration, 1987.
294. **Cope, M. J.,** Aspects of chiral high-performance liquid chromatography in pharmaceutical analysis: scientific and regulatory considerations, *Analyt. Proc.,* 29, 180, 1992.
295. **Felsted, R. L. and Bachur, N. R.,** Mammalian carbonyl reductases, *Drug Metab. Rev.,* 11, 1, 1980.
296. **Moreland, T. A. and Hewick, D. S.,** Studies on a ketone reductase in human and rat liver and kidney soluble fraction using warfarin as a substrate, *Biochem. Pharmacol.* 24, 1953, 1975.
297. **Kerr, B. M., Boddy, A. V., Moreland, T. A., and Levy, R. H.,** Pharmacokinetic consequences of product stereoselectivity in the metabolism of nafimidone: estimation of fraction metabolized, *J. Pharm. Sci.,* 80, 812, 1990.
298. **Dickinson, R. G., Hooper, W. D., Dunstan, P. R., and Eadie, M. J.,** First dose and steady-state pharmacokinetics of oxcarbazepine and its 10-hydroxymetabolite, *Eur. J. Clin. Pharmacol.* 37, 69, 1989.
299. **Jenner, P. and Testa, B.,** The influence of stereochemical factors on drug disposition, *Drug Metab. Rev.,* 2, 117, 1974.
300. **Testa, B.,** Novel drug metabolites produced by functionalization reactions: chemistry and toxicology, *Trends Pharmacol. Sci.,* 7, 60. 1986.
301. **Jamali, F., Mehvar, R., and Pasutto, F. M.,** Enantioselective aspects of drug action and disposition: therapeutic pitfalls, *J. Pharm. Sci.,* 78, 695, 1989.
302. **Whelpton, R.,** Isotope-labelled materials as internal standards, in *Drug Determination in Therapeutic and Forensic Contexts,* Reid, E. and Wilson, I. D., Eds., Plenum Press, New York, 1984, 39.
303. **Blakey, D. C. and Thorpe, P. E.,** *Antibody, Immunoconjugates, Radiopharmaceuticals,* 1, 1, 1988.
304. **Patience, R. L. and Rees, L. H.,** Comparison by gel filtration chromatography and reversed-phase high-performance liquid chromatography of the immunoreactive growth hormone composition of a human pituitary extract, *J. Chromatogr.,* 324, 385, 1985.
305. **Chaiken, I. M., Kanmer, T., Sequeira, R. P., and Swaisgood, H. E.,** High-performance liquid chromatography and studies of neurophysio-neurohypophysial hormone pathways, *J. Chromatogr.,* 336, 63, 1984.
306. **Hermann, K., Lang, R. E., Unger, T., Bayer, C., and Ganten, D.,** Combined high-performance liquid chromatography-radioimmunoassay for the characterisation and quantitative measurement of neuropeptides, *J. Chromatogr.,* 312, 273, 1984.
307. **Stanton, P. G., Simpson, R. J., Lambrou, F., and Hearn, M. T. W.,** High-performance liquid chromatography of amino acids, peptides and proteins. XLVII. Analytical and semi-preparative separation of several pituitary proteins by high-performance ion-exchange chromatography, *J. Chromatogr.,* 266, 273, 1983.
308. **Grego, B. and Hearn, M. T. W.,** High-performance liquid chromatography of amino acids, peptides and proteins. LXIII. Reversed-phase high-performance liquid chromatographic characterisation of several polypeptide and protein hormones, *J. Chromatogr.,* 336, 25, 1984.
309. **Patience, R. L.,** Protein separations by reversed-phase HPLC, *Analyt. Proc.,* 22, 296, 1985.
310. **Stenman, U.-H., Laatikainen, T., Salminen, K., Huhtala, M.-L., and Leppäluoto, J.,** Rapid extraction and separation of plasma β-endorphin by cation-exchange high-performance liquid chromatography, *J. Chromatogr.,* 297, 399, 1984.
311. **Murphy, J. B., Furness, J. B., and Costa, M.,** Measurement and chromatographic characterization of vasoactive intestinal peptides from guinea-pig enteric nerves, *J. Chromatogr.,* 336, 41, 1984
312. **Bonsner, A. M. and Garcia-Webb, P.,** C-Peptide measurement and its clinical usefulness, *Annu. Clin. Biochem.,* 18, 200, 1981.
313. **Conlon, J. M. and Deacon, C. F.,** Determination of substance P and neurokinins by a combined HPLC/RIA procedure, in *Bioactive Analytes, Including CNS Drugs, Peptides and Enantiomers,* Reid, E., Scales, B., and Wilson, I. D. Eds. Plenum Press, New York, 1986, 45
314. **Pilosof, D., Kim, H. Y., Dyckes, F. E., and Vestal, M. L.,** Determination of non-derivatised peptides by thermospray liquid chromatography/mass spectrometry, *Analyt. Chem.,* 56, 1236, 1984.

315. **Glazko, A. J., Edgerton, W. H., Dill, W. A., and Lenz, W. R.,** Chloromycetin palmitate, *Antibiot. Chemother.,* 2, 234, 1952.

316. **Edmondson, H. T.,** Parenteral and oral clindamycin therapy in surgical infections: a preliminary report, *Ann. Surg.,* 178, 637, 1973.

317. **Dittert, L. W., Caldwell, H. C., Ellison, T., Irwin, G. M., Rivard, D. E., and Swintosky, J. V.,** Carbonate ester prodrugs of salicylic acid. synthesis, solubility characteristicsi in vitro enzymatic hydrolysis rates, and blood levels of total salicylate following oral administration to dogs, *J. Pharm. Sci.,* 57, 828, 1968.

318. **Daehne, W. V., Frederiksen, E., Gundersen, E., Lund, F., Morch, P., Peterson, H. J., Roholt, K., Tybring, L., and Godtfredsen, W. O.,** Acyl oxymethyl esters of ampicillin, *J. Med. Chem.,* 13, 607, 1970.

319. **Garceau, Y., Davis, I., and Hasegawa, J.,** Plasma propranolol levels in beagle dogs after administration of propranolol hemisuccinate ester, *J. Pharm. Sci.,* 67, 1360, 1978.

320. **Boshes, B.,** Sinemet and the treatment of Parkinsonism, *Ann. Intern. Med.,* 94, 364, 1981.

321. **Goddard, P., O'Mullane, J., Ambler, L., Daw, A., Brookman, L., Lee, A., and Petrak, K.,** R-(N-Acetyl)eglin c:poly(oxyethylene) conjugates: preparation, plasma persistence, and urinary excretion, *J. Pharm. Sci.,* 80, 1171, 1990.

322. **Muller, E. W. and Albers, E.,** Effect of hydrotropic substances on the complexation of sparingly soluble drugs with cyclodextrin derivatives and the influence of cyclodextrin complexation on the pharmacokinetics of the drugs, *J. Pharm. Sci.,* 80, 599, 1990.

323. **Embree, L., Gelmon, K. A., Lohr, A., Mayer, L. D., Coldman, A. J., Cullis, P. R., Palaitis, W., Pilkiewicz, F., Hudon, N. J, Heggie, J. R., and Goldie, J. H.,** Chromatographic analysis and pharmacokinetics of liposome-encapsulated doxorubicin in non-small-cell lung cancer patients, *J. Pharm. Sci.,* 82, 627, 1982.

324. **Gifford, L.,** Spectrophotometry and fluorimetry, in *Therapeutic Drug Monitoring,* Richards, A. and Marks, V., Eds., Churchill Livingstone, Edinburgh, 1981, 61.

325. **McConnell, J. B., Smith, H., Davis, M., and Williams, R.,** Plasma rifampicin assay by an improved solvent-extraction technique, *Br. J. Clin. Pharmacol.,* 8, 506, 1979.

326. **Broughton, P. M. G.,** A rapid ultra-violet spectrophotometric method for the detection estimation and identification of barbiturates inbiological material, *Biochem. J.,* 63, 207, 1956.

327. **Stevens, H. M.,** Colour tests, in *Isolation and Identification of Drugs,* Clarke, E. G. C., Ed., The Pharmaceutical Press, London, 1969, 123.

328. **Schroeder, S., Noeschel, H., and Bonow, A.,** Photometric determination of cefalothin in biological fluids, *Pharmazie,* 35, 544, 1980.

329. **Sohr, R., Schmeck, G., Preiss, R., and Mathias, M.,** Methodology of colorimetric determination of alkylating cytostatic agents in plasma with 4-(4-nitrobenzyl)pyridine, *Pharmazie,* 37, 779, 1982.

330. **Itinose, A. M. and Sznelwar, R. B.,** Determination of paracetamol in serum by visible spectrometry, *Rev. Pharmacol. Bioquim. Univ. Sao Paulo,* 18, 164, 1982.

331. **Fukumoto, M.,** Spectrophotometric determination of bromazepam, *Chem. Pharmacol. Bull.,* 28, 3678, 1980.

332. **Miller, I. R. and Bajaj, K. L.,** A sensitive colorimetric determination of biphenyl-2,2'-diol, *Mikrochim. Acta,* 1, 65, 1980.

333. **Schmid, J., Fedorcak, A., and Koss, F. W.,** Sensitive, specific colorimetric method for determination of paracetamol in human plasma, *Arzneim. Forsch.,* 30, 996, 1980.

334. **Kabanga, K. and Binda, B.,** Colorimetric determination of ketamine, *Ann. Anaesthesiol. Fr.,* 20, 337, 1979.

335. **Roberts, E. C. and Rossano, A. J.,** Sensitive colorimetric determination of 3,3'-dichlorobenzidine in urine, *Am. Ind. Hyg. Assoc.,* 43, 80, 1982.

336. **Buffoni, F., Coppi, C., and Santoni, G.,** Determination of thioethers in urine, *Clin. Chem.,* 28, 248, 1982.

337. **Chavetz, L., Daly, R. E., Schriftman, H., and Lomner, J. J.,** Selective colorimetric determination of acetaminophen, *J. Pharm. Sci.,* 60, 463, 1971.

338. **Glynn, J. P. and Kendall, S. E.,** Paracetamol measurement, *Lancet,* 1, 1147, 1975.

339. **Walberg, C. B.,** Determination of acetaminophen in serum, *J. Analyt. Toxicol.,* 1, 79, 1977.

340. **Bailey, D. N.,** Colorimetry of serum acetaminophen (paracetamol) in uremia, *Clin. Chem.,* 28, 187, 1982.
341. **Swanson, M. B. and Walters, M. I.,** Rapid colorimetric assay for acetaminophen without salicylate or phenylephrine interference, *Clin. Chem.,* 28, 1171, 1982.
342. **Stewart, J. T. and Settle, D. A.,** Colorimetric determination of isoniazide with 9-chloroacridine, *J. Pharm. Sci.,* 64, 1403, 1975.
343. **Bratton, A. C. and Marshall, E. K.,** A new coupling component for sulfanilamide determination, *J. Biol. Chem.,* 128, 537, 1939.
344. **Sitar, D. S., Graham, D. N., Rangno, R. E., Dusfresne, L. R., and Ogilvie, R. I.,** Modified colorimetric method for procainamide in plasma, *Clin. Chem.,* 22, 379, 1976.
345. **Boxenbaum, H. G. and Riegelman, S.,** Determination of isoniazid and metabolites in biological fluids, *J. Pharm. Sci.,* 63, 1191, 1974.
346. **Fellenberg, A. J. and Pollard, A. C.,** A rapid and sensitive spectrophotometric procedure for the micro determination of carbamazepine in blood, *Clin. Chim. Acta,* 69, 423, 1976.
347. **Burns, J. J., Rose, R. K., Chenkin, T., Goldman, A., Schulbert, A., and Brodie, B. B.,** The physiological disposition of phenylbutazone in man and a method for its estimation in biological material, *J. Pharmacol. Exp. Ther.,* 109, 346, 1953.
348. **Doyle, T. D. and Fazzari, F. R.,** Determination of drugs in dosage forms by difference spectrophotometry, *J. Pharm. Sci.,* 63, 1921, 1974.
349. **Gupta, R. C. and Lundberg, G. D.,** Quantitative determination of theophylline in blood by differential spectrophotometry, *Analyt. Chem.,* 45, 2403, 1973.
350. **Hammond, V. J. and Price, W. C.,** Dereivative spectroscopy: theoretical aspects, *J. Opt. Soc. Am.,* 43, 924, 1953.
351. **Giese, A. T. and French, G. S.,** The analysis of overlapping spectral absorption bands by derivative spectroscopy, *Applied Spectosc.,* 9, 78, 1955.
352. **Martinez, D. and Paz Gimenez, M.,** Determination of benzodiazepines by derivative spectroscopy, *J. Analyt. Toxicol.,* 5, 10, 1981.
353. **Jarvie, D. R., Fell, A. F., and Stewart, M. J.,** Rapid method for the emergency analysis of paraquat in plasma using second derivative spectroscopy, *Clin. Chim. Acta* 117, 153, 1981.
354. **Cruz, A., Bermejo, A., Lopez-Rivadulla, M., and Fernandez, P.,** Simulataneous determination of methaqualone and diazepam in plasma by derivative spectroscopy, *Analyt. Lett.,* 25, 253, 1992.
355. **Soriano, J., Jimenez, F., Jimenez, A., and Arias, J. J.,** Simultaneous determination of barbitone and 2-thiobarbituric acid by derivative spectrophotometry, *Spectrosc. Lett.,* 25, 257, 1992.
356. **Randez-Gil, F., Salvador, A., and De La Guardia, M.,** Influence of the differentiation system on the analytical parameters for the spectrophotometric determination of clonazapam in urine, *Microchem. J.,* 44, 249, 1991.
357. **Abdel-Khalek, M. M., Malrous, M. S., Daabees, H. G., and Beltagy, Y. A.,** Use of derivative spectrophotometry for the *in vitro* determination of phenylbutazone and oxphenbutazone in human plasma, *Analyt. Lett.,* 25, 1851, 1992.
358. **Teale, F. W. J. and Weber, G.,** Ultraviolet fluorescence of the aromatic amino acids, *Biochem. J.,* 65, 467, 1957.
359. **Williams, R. T.,** The fluorescence of some aromatic compounds in aqueous solution, *J. Royal Instit. Chem.,* 83, 611, 1959.
360. **Williams, R. T. and Bridges, J. W.,** Fluorescence of solutions: a review, *J. Clin. Pathol.,* 17, 371, 1964.
361. **Häussler, A. and Hajdu, P.,** 4-Chloro-*N*-(2-furylmethyl)-5-sulfamoylanthranilic acid diuretic, *Arzneim. Forsch.,* 14, 1710, 1964.
362. **Brodie, B. B. and Udenfriend, S.,** The estimation of quinine in human plasma with a note on the estimation of quinidine, *J. Pharmacol. Exp. Ther.,* 78, 154, 1943.
363. **McBay, A. J. and Algeri, E. J.,** Ataraxics and non-barbiturate sedatives, in *Progress in Chemical Toxicology,* Stolman, S. A., Ed., Academic Press, New York, 1963, 158.
364. **Capomacchia, A. C. and Vallner, J. J.,** Human plasma levels of propranolol: fluorimetric measurement in a hydrosolvatic system, *J. Pharm. Sci.,* 69, 1463, 1980.
365. **Dekirmenjian, H. Javaid, J. I., Liskevych, U., and Davis, J. M.,** Determination of butaperazine in plasma and red blood cells by fluorimetry, *Analyt. Biochem.,* 105, 6, 1980.

366. **Chang, S. F., Miller, A. M., Jernberg, M. J., Ober, R. E., and Conard, G. J.,** Measurement of flecainide acetate in human plasma by an extraction-spectrophotometric method, *Arzneim. Forsch.,* 33, 251, 1983.
367. **Axelrod, J., Brady, R. O., Witkop, B., and Evarts, E. V.,** The distribution and metabolism of lysergic acid diethylamide, *Ann N. Y. Acad. Sci.,* 66, 435, 1957.
368. **Farina, A., Aballe, F., Doldo, A., and Quaglia, M. G.,** Spectrofluorimetric assay of temafloxacin in pharmaceutical formulations and in serum, *Spectrosc. Lett.,* 24, 1219, 1991.
369. **Jusko, W. J.,** Fluorimetric analysis of ampicillin in biological fluids, *J. Pharm. Sci.,* 60, 728, 1971.
370. **Ragland, J. W., Kinross-Wright, V. J., and Ragland, R. S.,** Determination of phenothiazines in biological samples, *Analyt. Biochem.,* 12, 60, 1965.
371. **Dill, W. A., Leung, A., Kinkel, A. W., and Glazko, A. J.,** Simplified fluorimetric assay for diphenylhydantoin in plasma, *Clin. Chem.,* 22, 908, 1976.
372. **Kim, B. K. and Koda, R. T.,** Fluorimetric determination of methyldopa in biological fluids, *J. Pharm. Sci.,* 66, 1632, 1977.
373. **Koechlin, B. A. and D'Arconte, L.,** Determination of chlordiazepoxide and a metabolite of lactam character in plasma of humans, dogs, and rats by a specific spectrofluorimetric micro method, *Analyt. Biochem.,* 5, 195, 1963.
374. **Brodie, B. B., Udenfriend, S., Dill, W., and Chenkin, T.,** The estimation of basic organic compounds in biological material. III. Estimation of vconversion to fluorescent compounds, *J. Biol. Chem.,* 168, 319, 1947.
375. **Hajdu, P. and Damm, D.,** Eine neue fluorimetrische Methode zur Bestimmung von Pharmaka in Biologischem Material, *Arzneim. Forsch.,* 26, 2141, 1976.
376. **Hill, H. M., Chamberlain, J., Hajdu, P., and Damm, D.,** Rapid fluorimetric procedure for the analysis of fendosal in plasma and data following oral dosing, *Biopharm. Drug Dispos.,* 1, 97, 1980.
377. **Kohn, K. W.,** determination of tetracyclines by extraction of fluorescent complexes, *Analyt. Chem.,* 33, 862, 1961.
378. **Chang, W.-B., Zhao, Y.-B., Ci, Y.-X., and Hu, L.-Y.,** Spectrofluorimetric determination of tetracycline and anhydrotetracycline in serum and urine, *Analyst,* 117, 1377, 1992.
379. **Ibsen, K. H., Saunders, R. L., and Urist, M. R.,** Fluorometric determination of oxytetracycline in biological material, *Analyt. Biochem.,* 5, 505, 1963.
380. **Djozan, D. and Farajzadeh, M. A.,** Use of fluorescamine (Fluram) in fluorimetric trace analysis of primary amines of pharmaceutical and biological interest, *J. Pharm. Biomed. Anal.,* 10, 1063, 1992.
381. **Pütter, J.,** A fluorimetric method for the determination of praziquantel in blood-plasma and urine, *Eur. J. Drug Metab. Pharmacokinet.,* 3, 143, 1979.
382. **Frei, R. W. and Lawrence, J. F.,** Fluorigenic labelling in high-speed liquid chromatography, *J. Chromatogr.,* 83, 321, 1973.
383. **Scouten, W. H., Lubcher, R., and Baughman, W.,** N-Dansylaziridine: a new fluorescent modification for cysteine thiols, *Biochem. Biophys. Acta* 336, 421, 1974.
384. **Abdel-Hay, M. H., Galal, S. M., Bedair, M. M., Gazy, A. A., and Wahbi, A. A. M.,** Spectrofluorimetric determinatin of guanethidine sulphate, guanoxan sulphate and amiloride hydrochloride in tablets and in biological fluids using 9,10-phenanthraquinone, *Talanta,* 39, 1369, 1992.
385. **Vanhoof, F. and Heyndrickx, A.,** Thin layer chromatographic spectrophotofluorimetric analysis of amphetamine and amphetamine analogs after reaction with 4-chloro-7-nitrobenzo-1,2,5-oxadiazole, *Analyt. Chem.,* 46, 286, 1974.
386. **Wahbi, A.-A., Bedai, M. M., Galal, S. M., Abdel-Hay, M. H., and Gazy, A. A.,** Spectrofluorimetric determination of guanethidine sulphate, guanfacine hydrochloride, guanoclor sulphate and guanoxan sulphate in tablets and biological fluids, using benzoin, *Mikrochim. Acta,* 111, 83, 1993.
387. **Ettre, L. S.,** Evolution of liquid chromatography, in *High Performance Liquid Chromatography. Advances and Perspectives,* Horvath, C., Ed., Academic Press, New York, 1980.
388. **Neil, M. W. and Payton, J. E.,** Application of petroleum ether-methanol solvent system to the analysis of common barbiturates by paper chromatography, *Medicine Sci. Law,* 2, 4, 1961.

389. **Ahmed, Z. F., El-Darawy, Z. I., Aboul-Emein, M. N., Abu El-Naga, M. A., and El-Leithy, S. A.,** Identification of some barbiturates by paper and thin-layer chromatography, *J. Pharm. Sci.,* 55, 433, 1966.
390. **Curry, A. S. and Powell, H.,** Paper chromatographic examination of the alkaloid extract in toxicology, *Nature,* 173, 1143, 1954.
391. **Moss, M. S. and Jackson, J. V.,** A furfural reagent of high specificity for the detection of carbamateson paper chromatograms, *J. Pharm. Pharmacol.,* 13, 361, 1961.
392. **Christensen, F.,** Paper chromatographyof dicoumarol and some related substances with a method for the quantitaive determination of dicoumarol on paper chromatograms, *Acta Pharmacol. Toxicol.,* 21, 23, 1964.
393. **Algeri, E. J. and Walker, J. T.,** Paper chromatography for identification of the common barbiturates, *Am. J. Clin. Pathol.,* 22, 37, 1952.
394. **Johnson, C. A. and Fowler, S.,** Detection and identification of 17,21-dihydroxy-20-oxosteroids in corticosteroids, *J. Pharm. Pharmacol.,* 16, 17T, 1964.
395. **Pilsbury, V. B. and Jackson, J. V.,** Identification of thiazide diuretic drugs, *J. Pharm. Pharmacol.,* 18, 713, 1966.
396. **Street, H. V.,** Rapid separation of drugs and poisons by high-temperature reversed-phase paper chromatography. II. Phenothiazine tranquillizers and imipramine, *Acta Pharmacol. Toxicol.,* 19, 312, 1962.
397. **Street, H. V.,** Rapid separation of drugs and poisons by high-temperature reversed-phase paper chromatography. III. Alkaloids, *Acta Pharmacol. Toxicol.,* 19, 325, 1962.
398. **Street, H. V.,** Separation of barbiturates by reversed-phase paper chromatography at elevated temperatures, *Forensic Sci. Soc. J.,* 2, 118, 1962.
399. **Street, H. V.,** The paper chromatographic separation of codeine, morphine and nalorphine, *J. Pharm. Pharmacol.,* 14, 56, 1962.
400. **Bush, I. E.,** *The Chromatography of Steroids,* Pergamon Press, Oxford, 1961, 245.
401. **Chandrasekaran, B., Vijayendran, R., Purushothaman, K. K., and Nagarajan, B.,** New methods for urinary estimation of antitumor compounds echitamine and plumbagin, *Ind. J. Biochem. Biophys.,* 19, 148, 1982.
402. **Shraiber, M. S.,** Beginnings of thin-layer chromatography, *J. Chromatogr.,* 73, 367, 1972.
403. **Roveri, P., Cavrini, V., Andrisano, V., and Gatti, R.,** Thin-layer chromatography of aliphatic thiols after fluorescent labelling with methyl 4-(6-methoxynaphthalen-2-yl)-4-oxobut-2-enoate, *J. Liq. Chromatogr.,* 16, 1859, 1993.
404. **Stahl, E.,** Must find correct reference, *Chem. Ztg.,* 82, 232, 1958.
405. **Chavan, J. D. and Khatri, J. M.,** Simultaneous determination of retinol and _-tocopherol in human plasma by HPTLC, *J. Planar Chromatogr. Mod. TLC,* 5, 280, 1992.
406. **Moffat, A. C.,** Thin-layer chromatography, in *Clarke's Isolation and Identification of Drugs,* Moffat, A. C., Ed., 2nd ed., The Pharmaceutical Press, London, 1986, 160–177.
407. **Tivert, A.-M. and Backman, A.,** Separation of the enantiomers of β-blocking drugs by TLC with a chiral mobile phase additive, *J. Planar Chromatogr.,* 6, 216, 1993.
408. **Bushan, R. and Ali, I.,** Resolution of racemic mixtures of hyoscamine and colchicine on impregnated silica gel layers, *Chromatographia,* 35, 679, 1993.
409. **Tyihak, E., Mincsovics, E., and Kalasz, H.,** New planar liquid chromatography technique; overpressured thin-layer chromatography, *J. Chromatogr.,* 174, 75, 1979.
410. **Mincsovics, E., Tyihak, E., and Kalasz, H.,** Resolution and retention behaviour of some dyes in over-pressured thin-layer chromatography, *J. Chromatogr.,* 191, 293, 1980.
411. **Tyihak, E., Mincsovics, E., and Kalasz, H.,** Optimization of operating parameters in overpressured thin-layer chromatography, *J. Chromatogr.,* 211, 45, 1981.
412. **Nyiredy, S. and Szepesi, G.,** Planar chromatography: current status and future perspectives in pharmaceutical analysis (short review) — II. Special techniques and future perspectives in planar chromatography, *J. Pharm. Biomed. Anal.,* 10, 1017, 1992.
413. **Kovacs-Hadady, K.,** Overpressured layer chromatographic study of retention behavior of various benzodiazepine derivatives on layers impregnated with tricaprylmethylammonium chloride, *J. Pharm. Biomed. Anal.,* 10, 1025, 1992.

414. **Vegh, Z.,** Optimization and validation of the determination of potassium canrenoate by reversed-phase ion-pair OPLC, *J. Planar Chromatogr. Mod. TLC,* 6, 228, 1993.

415. **Pothier, J., Galand, N., and Viel, C.,** Determination of some narcotic and toxic alkaloidal compounds by overpressured thin-layer chromatography with ethyl acetate as eluent, *J. Chromatogr.,* 634, 356, 1993.

416. **Chamberlain, J.,** *The Analysis of Drugs in Biological Fluids,* CRC Press, Boca Raton, 1985, 184.

417. **Nyiredy, S, Meszaros, S. Y., Dallenbach-Tolke, K. Nyiredy-Mikita, K., and Sticher, O.,** Ultra-microchamber rotation planar chromatography (U-RPC): a new analytical and preparative forced-flow method, *J. Planar Chromatogr.,* 1, 54, 1989.

418. **Nyiredy, S., Botz, L., and Sticher, O.,** ROTACHROM: new instrument for rotation planar chromatography (RPC), *J. Planar Chromatogr.,* 2, 53, 1989.

419. **Ojanpera, I.,** Toxicological drug screening by thin-layer chromatography, *Trends Analyt. Chem.,* 11, 222, 1992.

420. **Cochin, J. and Daly, J. W.,** Use of thin-layer chromatography for the analysis of drugs. Identification and isolation of phenothiazine tranquillizers and antihistamines in body fluids and tissues, *J. Pharmacol. Exp. Ther.,* 139, 160, 1963.

421. **Bogan, J., Rentoul, E., and Smith, H.,** The detection of barbiturates and related drugs by thin-layer chromatography, *Forens. Sci. J.,* 4, 147, 1964.

422. **Jack, D. R., Dean, S., Kendall, M. A., and Laugher, S.,** Detection of some antihypertensive drugs and their metabolites in urine by thin-layer chromatography. II. A further five beta-blockers and dihydralazine, *J. Chromatogr.,* 196, 189, 1980.

423. **Kanter, S. L., Hollister, L. E., and Zamora, J. U.,** Marijuana metabolites in urine of man. XI. Detection of unconjugated and conjugated ⁹-tetrahydrocannabinol-11-oic acid by thin-layer chromatography, *J. Chromatogr.,* 235, 507, 1982.

424. **Johnston, E. J. and Jacobs, A. L.,** Thin layer chromatography of cardiac glycosides, *J. Pharm. Sci.,* 55, 531, 1979.

425. **Aderjan, R., Doster, S., Petri, H., and Schmidt, G.,** Cardiac glycosides and metabolites — problems of recovery from tissue extracts. Separation of visible zones in the nanogram range by thin-layer chromatography, *Z. Rechstmed.,* 83, 201, 1979.

426. **Hall, A.,** Thin-layer chromatography of corticosteroids, *J. Pharm. Pharmacol.,* 16, 9T, 1964.

427. **Medina, M. B. and Nagdy, N.,** Improved thin-layer chromatographic detection of diethylstilbestrol and zeranol in plasma and tissues isolated with alumina and ion-exchange membrane columns in tandem, *J. Chromatogr.,* 614, 315, 1993.

428. **French, W. N. and Wehrli, A.,** Thin-layer chromatography of ergot alkaloids in pharmaceutical preparations, *J. Pharmacol. Sci.,* 54, 1515, 1965.

429. **Mulé, S. J.,** Determination of narcotic analgesics in human biological materials. Application of ultraviolet spectrophotometry, thin layer, and gas chromatography, *Analyt. Chem.,* 36, 1907, 1964.

430. **Korczak-Fabierkiewicz, C., Kofoed, J., and Lucas, G. H. W.,** Sulfoxides as an additional rapid test of phenothiazine tranquilizers, *J. Forens. Sci.,* 10, 308, 1965.

431. **Budd, R. D., Mathis, D. F., and Leung, W. J.,** Screening and confirmation of opiates by thin-layer chromatography, *Clin. Toxicol.,* 16, 61, 1980.

432. **Hsu, L. S. F., Sharrard, J. I., Love, C., and Marrs, T. C.,** Rapid method for screening urine samples in suspected abuse of cocaine, *Annu. Clin. Biochem.,* 18, 368, 1981.

433. **Stevens, H. M., and Moffat, A. C.,** A rapid screening procedure for quaternary ammonium compounds in fluids and tissues with special reference to suxamethonium (succinylcholine), *J. Forens. Sci. Soc.,* 14, 141, 1974.

434. **Cummings, A. J. and King, M. L.,** Urinary excretion of acetylsalicylic acid in man, *Nature,* 209, 620, 1966.

435. **Shaw, I. C.,** Micro-scale thin-layer chromatographic method for screening of urine samples for aspirin metabolites, *Analyst,* 107, 1090, 1982.

436. **De Clercq, H., Massart, D. L., and Dryon, L.,** Evaluation and optimal combination of TLC systems for quantitative identification I: sulfonamides, *J. Pharm. Sci.,* 66, 1269, 1977.

437. **Klein, S. and Kho, B. T.,** Thin-layer chromatography in drug analysis. I. Identification procedure for various sulfonamides in pharmaceutical combinations, *J. Pharm. Sci.,* 51, 966, 1962.

438. **Getz, M. E. and Wheeler, H. G.,** Thin-layer chromatography of organophosphorus insecticides with several adsorbents and ternary solvent systems, *J.Assoc. Off. Analyt. Chem.,* 51, 1101, 1968.

439. **Genest, K. and Farmilio, C. G.,** The identification and determination of lysergic acid diethylamide in narcotic seizures, *J. Pharm. Pharmacol.,* 16, 250, 1964.

440. **Quintens, I., Eykens, J., Roets, E., and Hoogmartens, J.,** Identification of cephalosporins by thin-layer chromatography and colour reactions, *J. Planar. Chromatogr. Mod. TLC,* 6, 181, 1993.

441. **Steyn, J. M.,** A thin-layer chromatographic method for the determination of acebutolol and its major metabolite in serum, *J. Chromatogr.,* 120, 4665, 1976.

442. **Zhang, A. H., Bu, Z. Y., Zhang, Y., Wang, Q., Li, W. B., and Qi, L.,** Thin-layer chromatography scanning in measuring alprazolam concentration in plasma, *Zhongguo Yaoxue Zazhi,* 28, 233, 1993.

443. **Guo, Y., Wu, X., Yang, L., and Nan, G.,** Determination of amiodarone and de-ethylamiodorone in serum by thin-layer scanning method, *Zhongguo Yaoxue Zazhi,* 28, 290, 1993.

444. **Klimes, J. and Kastner, P.,** Thin-layer chromatography of benzodiazepines, *J. Planar Chromatogr.,* 6, 168, 1993.

445. **Hundt, H. K. L. and Clark, E. C.,** Thin-layer chromatographic method for determining carbamazepine and two of its metabolites in serum, *J. Chromatogr.,* 107, 149, 1975.

446. **Wad, N., Weidkuhn, E., and Rosenmund, H.,** A quantitative method for the simultaneous determination of carbamazepine, mephentoin, phenylethylmalonamide, phenobarbital, phenytoin and primidone in serum by thin-layer chromatography. Additional comments to the analysis of carbamazepine-10,11-epoxide, *J. Chromatogr.,* 183, 387, 1980.

447. **Davis, C. M. and Fenimore, D. C.,** Rapid micro-analysis of anticonvulsants by high-performance thin-layer chromatography, *J. Chromatogr.,* 222, 265, 1981.

448. **DeAngelis, R. L., Robinson, M. M., Brown, A. R., Johnson, T. E., and Welch, R. M.,** Quantitation of the anticonvulsant cinromide (3-bromo-N-ethylcinnamamide) and its major plasma metabolites by thin-layer chromatography, *J. Chromatogr.,* 221, 353, 1980.

449. **Steyn, J. M. and Hundt, H. K. L.,** Drugs in serum, in *Densitometry in Thin-Layer Chromatography. Practice and Applications,* Touchstone, J. C. and Sherma, J., Eds., Wiley Inter-Science, New York, 1979.

450. **Riedel, E., Kreutz, G., and Hermsdorf, D.,** Quantitative thin-layer chromatographic determination of dihydroergot alkaloids, *J. Chromatogr.,* 229, 417, 1982.

451. **Somona, M. G. and Grandjean, E. M.,** Simple high-performance thin-layer chromatography method for the determination of disopyramide and its mono-N-dealkylated metabolite in serum, *J. Chromatogr.,* 224, 532, 1981.

452. **Wesley-Hadzija, B. and Mattox, A. M.,** Thin-layer chromatographic determination of frusemide and 4-chloro-5-sulphamoylanthranilic acid in plasma and urine, *J. Chromatogr.,* 229, 425, 1982.

453. **Boddy, A. V. and Idle, J. R.,** Combined thin-layer chromatography-photography-densitometry for the quantification of ifosfamide and its principal metabolite in urine, cerebrospinal fluid and plasma, *J. Chromatogr.,* 575, 137, 1992.

454. **Taha, I. A. K., Ahmad, R. A. J., and Rogers, H. J.,** Melphalan estimation by quantitative thin-layer chromatography. Observations on melphalan hydrolysis in vitro and pharmacokinetics in rabbits, *Cancer Chemother. Pharmacol.,* 5, 181, 1981.

455. **Gattavecchia, E., Tonelli, D., and Breccia, A.,** Determination of metronidazole, misomidazole and its metabolite in seruma and urine on RP-18 high performance thin-layer chromatographic plates, *J. Chromatogr.,* 224, 465, 1981.

456. **Zhou, J. and Ruan, Y.,** Thin-layer chromatographic determination of morphine hydrochloride in serum, *Yaowu Fenxi Zazhi,* 11, 293, 1991.

457. **Wolff, K., Sanderson, M. J., Hay, A. W. M., and Barnes, I.,** Evaluation of the measurement of drugs of abuse by commercial and in-house horizontal thin-layer chromatography, *Ann. Clin. Biochem.,* 30, 163, 1993.

458. **Schaeffer-Korting, M. and Mutschler, E.,** Fluorodensitometric determination of nadolol in plasma and urine, *J. Chromatogr.,* 230, 461, 1982.

459. **Hundt, H. K. L. and Barlow, E. C.,** Thin-layer chromatographic method for the quantitative analysis of nalidixic acid in human plasma, *J. Chromatogr.,* 223, 165, 1981.

460. **Kuunz, F. R., Jork, H., and Keller, H. E.,** Determination of netilmicin in serum by thin-layer densitometry and fluorescence polarization immunoassay, *Fresenius' J. Analyt. Chem.,* 346, 847, 1993.

461. **Kresse, M., Schley, J., and Mueller-Oerlinghausen, B.,** Reliable routine method for determination of perazine in serum by thin-layer chromatography with an internal standard, *J. Chromatogr.,* 183, 475, 1980.

462. **Martin, P., Morden, W., Wall, P., and Wilson, I.,** TLC combined with tandem mass spectrometry: application to the analysis of antipyrine and its metabolites in extracts of human urine, *J. Planar Chromatogr.,* 5, 255–258, 1992.

463. **Steyn, J. M. and Hundt, H. K. L.,** A thin-layer chromatographic method for the quantitative determination of quinidine in human serum, *J. Chromatogr.,* 111, 463, 1975.

464. **Colthup, P. V., Dallas, F. A. A., Saynor, D. A., Carey, P. F., Skidmore, I. F., and Martin, L. E.,** Determination of salbutamol in human plasma and urine by high-performance thin-layer chromatography, *J. Chromatogr.,* 345, 111, 1985.

465. **Le Roux, A. M., Wium, C. A., Joubert, J. R., and Van Jaarsveld, P. P.,** Evaluation of high-performance thin-layer-chromatographic technique for the determination of salbutamol serum levels in clinical trials, *J. Chromatogr.,* 581, 306, 1992.

466. **Heizmann, P. and Haefelfinger, P.,** Determination of sulfadiazine and sulfamethoxazole in urine and plasma by thin-layer chromatography, *Fresenius' Z. Anal. Chem.,* 302, 410, 1980.

467. **Heilweil, E. and Touchstone, J. C.,** Theophylline analysis by direct application of of serum to thin-layer chromatograms, *J. Chromatogr. Sci.,* 19, 594, 1981.

468. **Selles, J. P., Fuseau, E., Bres, J., Godard., Montoya, F., and Brun, S.,** Micro-assay for pharmacokinetics and clinical pharmacy studies, *Trav. Soc. Pharmacol. Montpellier,* 40, 81, 1980.

469. **Vukusic, I. and Vodopivec, P.,** Quantitative thin-layer chromatographic determination of ticrynafen in plasma of the dog, *J. Chromatogr.,* 222, 324, 1981.

470. **Danielson, N. D., Holeman, J. A., Bristol, D. C., and Kirzner, D. H.,** Simple methods for the qualitative identification and quantitative determination of macrolide antibiotics, *J. Pharm. Biomed. Anal.,* 11, 121, 1993.

471. **Chamberlain, J.,** Gas chromatographic determination of levels of aldadiene in human plasma and urine following therapeutic doses of spironolactone, *J. Chromatogr.,* 55, 249, 1971.

472. **Horning, M. G., Lertratanangkoon, K., Nowlin, J., Stilwell, W. G., Stilwell, R., Zion, T., Kellaway, P., and Hill, R. M.,** Anticonvulsant drug monitoring by GC-MS-COM techniques, *J. Chromatogr. Sci.,* 12, 630, 1974.

473. **Budd, R. D.,** Studies of barbiturate degradation following methylation with dimethylsulphate, *J. Chromatogr.,* 237, 155, 1982.

474. **Kupferberg, H. J.,** Quantitative estimation of diphenylhydantoin, primidone and phenobarbital in plasma by gas-liquid chromatography, *Clin. Chim. Acta,* 29, 283, 1970.

475. **Solow, E. B., Metaxas, J. M., and Summers, T. R.,** Antiepileptic drugs: a current assessment of simultaneous determination of multiple drug therapy by gas-liquid chromatography on column methylation, *J. Chromatogr. Sci.,* 12, 256, 1974.

476. **Hishta, C., Mays, D. L., and Garofalo, M.,** Gas chromatographic determination of penicillins, *Analyt. Chem.,* 43, 1530, 1971.

477. **Brown, L. W. and Bowman, P. B.,** Gas chromatographic assay for the antibiotic spectinomycin, *J. Chromatogr. Sci.,* 12, 373, 1974.

478. **Harvey, D. J. and Paton, W. D. M.,** Use of trimethylsilyl and other homologous trialkylsilyl derivatives for the separation and characterisation of mono- and dihydroxycannabinoids by combined gas chromatography and mass spectrometry, *J. Chromatogr.,* 109, 73, 1975.

479. **Van Giessen, B. and Tsuji, K.,** Gas-liquid chromatographic assay method for neomycin in petrolatum-based ointment, *J. Pharm. Sci.,* 60, 1068, 1971.

480. **Smith, R. V. and Stocklinski, A. W.,** Gas chromatographic determination of apomorphine in urine and faeces, *Analyt. Chem.,* 47, 1321, 1975.

481. **Margosis, M.,** Analysis of antibiotics by gas chromatography. II. Chloramphenicol, *J. Chromatogr.,* 47, 341, 1970.

482. **Wilkinson, G. R. and Way, E. L.,** Submicrogram estimation of morphine in biological fluids by gas-liquid chromatography, *Biochem. Pharmacol.,* 18, 1435, 1969.

483. **Wu, H. L., Masada, M., and Uno, T.,** Gas chromatographic and gas chromatographic-mass spectrometric analysis of ampicillin, *J. Chromatogr.,* 137, 127, 1977.

484. **Pierce, A. E.,** *Silylation of Organic Compounds,* Pierce chemicals Company, Rockford, 1968.

485. **Knapp, D. R.,** *Handbook of Analytical Derivatization Reactions,* Wiley-Interscience, New York, 1979.

486. **Delargy, H. and Temple, D. J.,** An improved method for the analysis of flurazepam and its main metabolite in human plasma, *J. Pharm. Pharmacol.,* 32, 96P, 1980.

487. **De Silva, J. A. F., Puglisi, C. V., Brooks, M. A., and Hackman, M. R.,** Determination of flurazepam (Dalmane) and its major metabolites in blood by electron-capture gas-liquid chromatography and in urine by differential pulse polarography, *J. Chromatogr.,* 99, 461, 1974.

488. **Hajdu, P., Uihlein, M., and Damm, D.,** Quantitative determination of clobazam in serum and urine by gas chromatography, thin-layer chromatography and fluorometry, *J. Clin. Chem. Clin. Biochem.,* 18, 209, 1980.

489. **Quaglio, M. P. and Bellini, A. M.,** Simultaneous determination of some butyrophenones and benzodiazepines in human plasma by g.l.c., *Farmaco Ed. Prat.,* 36, 487, 1981.

490. **Sioufi, A. and Pommier, F.,** Gas-chromatographic determination of clioquinol (Vioform) in human plasma, *J. Chromatogr.,* 226, 219, 1981.

491. **Johnson, G. R.,** Fatal case involving hydroxyzine, *J. Analyt. Toxicol.,* 6, 69, 1982.

492. **Turcant, A., Premel-Cabic, A., Cailleux, A., and Allain, P.,** Micro-method for automated identification and quantitation of fifteen barbiturates in plasma by gas-liquid chromatography, *J. Chromatogr.,* 229, 222, 1982.

493. **Cooper, S. F., Dugal, R., and Bertrand, M. J.,** Determination of loxapine in human plasma and urine and identification of three urinary metabolites, *Xenobiotica,* 9, 405, 1979.

494. **Ranise, A., Benassi, E., and Besio, G.,** Rapid gas-chromatographic method for the determination of carbamazepine and unrearranged carbamazepine 10,11-epoxide in human plasma, *J. Chromatogr.,* 222, 120, 1981.

495. **Stavchansky, S. and Loper, A.,** Nitrogen-phosphorus detection of phencyclidine in blood serum, *J. Pharm. Sci.,* 71, 194, 1982.

496. **Bredersen, J. E., Ellingsen, O. F., and Karlsen, J.,** Rapid isothermal gas-liquid chromatographic determination of tricyclic antidepressants in serum with use of a nitrogen-selective detector, *J. Chromatogr.,* 204, 361, 1981.

497. **Kaa, E.,** Rapid gas-chromatographic method for emergency determination of paracetamol in human serum, *J. Chromatogr.,* 221, 414, 1980.

498. **Sioufi, A. and Pommier, F.,** Gas-chromatographic determination of isosorbide dintrate in human plasma and urine, *J. Chromatogr.,* 229, 347, 1982.

499. **Rajagopalan, T. G., Anjaneyulu, B., Shanbag, V. D., and Grewal, R. S.,** Electron-capture gas-chromatographic assay for primaquine in blood, *J. Chromatogr.,* 224, 265, 1981.

500. **Jones, A. B., Elsohly, M. A., Bedford, J. A., and Turner, C. E.,** Determination of cannabidiol in plasma by electron-capture gas chromatography, *J. Chromatogr.,* 226, 99, 1981.

501. **Kasai, S., Kobayashi, K., and Takayama, Y.,** Gas-chromatographic enzymic determination of amygdalin, *J. Chromatogr.,* 210, 342, 1981.

502. **Van den Bosch, N., Driessen, O., Edmunds, A., Van Oosterom, A. T., Timmermans, P. J. A., De Dos, V., and Slee, P. H. T. J.,** Determination of plasma concentrations of underivatised cyclophosphamide by capillary gas chromatography, *Methods Find. Exp. Clin. Pharmacol.,* 3, 377, 1981.

503. **Pateman, A. J.,** Sensitive gas-chromatographic method for determination of alphadolone in plasma, *J. Chromatogr.,* 226, 213, 1981.

504. **De Boer, A. G., Breimer, D. D., and Gubbens-Stibbe, J. M.,** Assay of propranolol in human and rat plasma by capillary gas chromatography with nitrogen-selective detection, *Pharmacol. Weekbl. Sci. Ed.,* 2, 1105, 1980.

505. **Knowles, J. A., White, G. R., Kick, C. J., Spangler, T. B., and Ruelins, H. W.,** Gas-chromatographic determination of guanabenz in biological fluids by electron-capture detection, *J. Pharm. Sci.,* 71, 710, 1982.

506. **McWilliam, I. G. and Dewar, R. A.,** Flame ionisation detector for gas chromatography, *Nature,* 181, 760, 1958.

507. **Geary, R. and Stavchansky, S.,** Gas-liquid-chromatographic determination of almitrine in rat intestinal fluid using flame-ionization detection, *Analyt. Lett.,* 25, 1843, 1992.

508. **Kinberger, B., Holmen, A., and Wahrgren, P.,** Determination of barbiturates and some neutral drugs in serum using quartz-glass capillary gas chromatography, *J. Chromatogr.,* 224, 449, 1981.

509. **Verheesen, P. E., Brombacher, P. J., Cremers, H. M. H. G., and De Boer, R.,** Determination of low levels of bupivacaine (Marcaine) in plasma during epidural analgesia, *J. Clin. Chem. Clin. Biochem.,* 18, 351, 1980.

510. **Riva, R., Albani, F., and Baruzzi, A.,** Rapid quantitative determination of underivatised carbamazepine, phenytoin, phenobarbitone and *p*-hydroxyphenobarbitone in biological fluids by packed-column gas chromatography, *J. Chromatogr.,* 221, 75, 1980.

511. **Davidson, W. J. and Wilson, A.,** Determination of nanogram quantities of disulfiram in human and rat plasma by gas-liquid chromatography, *J. Study Alcohol.,* 40, 1073, 1979.

512. **Kumps, A. and Mardens, Y.,** Simplified simulataneous determination of valproic acid and ethosuximide in serum by gas-liquid chromatography, *Clin. Chem.,* 26, 1759, 1980.

513. **Gillespie, T. J., Gandolfi, A. J., Maiorino, R. M., and Vaughn, R. W.,** Gas-chromatographic determination of fentanyl and its analogues in human plasma, *J. Analyt. Toxicol.,* 5, 133, 1981.

514. **Orzalesi, G. Mari, F., Bertol, E., Selleri, R., and Pisaturo, G.,** Anti-Inflammatory agents: determination of ibuproxam and its metabolites in humans. Correlation between bioavailability, tolerance and chemico-physical characteristics, *Arzneim.-Forsch.,* 30, 1607, 1980.

515. **Karch, F. E. and Chmielewski, K. F.,** A g.l.c. assay for lignocaine in human plasma, *J. Pharm. Sci.,* 70, 229, 1981.

516. **Gupta, R. N. and Eng, R. F.,** Gas-chromatographic and high-performance liquid-chromatographic determination of meprobamate in plasma, *J. High Res. Chromatogr. Chromatogr. Commun.,* 3, 419, 1980.

517. **Peat, M. A. and Finkle, B. S.,** Determination of methaqualone and its major metabolite in plasma and saliva after single oral doses, *J. Analyt. Toxicol.* 4, 114, 1980.

518. **Said, S. A.,** Gas-liquid chromatographic determination of methoxsalen in human serum and urine, *Pharmazie,* 37, 557, 1982.

519. **Chan, K., Latham, S. M., and Tse, J.,** Rapid and simple method for determination of mexiletine in human plasma and urine by g.l.c., *J. Pharm. Pharmacol.,* 32, 98P, 1980.

520. **Verebey, K., De Pace, A. Jukofsky, D., Volavka, J. V., and Mulé, S. J.,** Quantitative determination of 2-hydroxy-3-methoxy-6β-naltrexol, naltrexolone and 6β-hydroxynaltrexol in human plasma, red cells, saliva and urine by gas-liquid chromatography, *J. Analyt. Toxicol.,* 4, 33, 1980.

521. **Cooper, J. D. H. and Turnell, D. C.,** Rapid method for determination of perhexiline in serum using gas-liquid chromatography, *Ann. Clin. Biochem.,* 17, 155, 1980.

522. **Bailey, D. N. and Guba, J. J.,** Measurement of phencyclidine in saliva, *J. Analyt. Toxicol.,* 4, 311, 1980.

523. **Schroeter, A. and Zschiesche, M.,** Gas-chromatographic determination of acetylsalicylic acid and salicylic acid in serum, *Z. Med. Labor.,* 22, 215, 1981.

524. **Norman, T. R., Burrows, G. D., Davies, B. M., and Wurm, J. M. E.,** Determination of viloxazine in plasma by gas-liquid chromatography, *Br. J. Clin. Pharmacol.,* 8, 169, 1979.

525. **Lovelock, J. E. and Lipsky, S. R.,** Electron affinity spectroscopy — a new method for the identification of functional groups in chemical compounds separated by gas chromatography, *J. Am. Chem. Soc.,* 82, 431, 1960.

526. **Beharrel, G. P., Hailey, D. M., and McLaurin, M. K.,** Determination of nitrazepam (Mogadon) in plasma by electron-capture gas-liquid chromatography, *J. Chromatogr.,* 70, 45, 1972.

527. **Greenblatt, D. J.,** Electron-capture GLC determination of clobazam and desmethylclobazam in plasma, *J. Pharm. Sci.,* 69, 1351, 1980.

528. **Badcock, N. R. and Pollard, A. C.,** Micro-determination of clonazepam in plasma or serum by electron-capture gas-liquid chromatography, *J. Chromatogr.,* 230, 353, 1982.

529. **Loescher, W.,** Rapid gas-chromatographic measurement of diazepam and its metabolites, desmethyldiazepam, oxazepam and 3-hydroxydiazepam (temazepam) in small samples of plasma, *Ther. Drug Monitor.,* 4, 315, 1982.

530. **Beischlag, T. V. and Inaba, T.,** Determination of nonderivatized *para*-hydroxylated metabolites of diazepam in biological fluids with a GC megabore column system, *J. Analyt. Toxicol.,* 16, 236, 1992.
531. **Ikeda, M., Kawase, M., Hiramatsu, M., Hirota, K., and Ohmori, S.,** Improved gas-chromatographic method of determining diclofenac in plasma, *J. Chromatogr.,* 183, 41, 1980.
532. **Aderjan, R., Fritz, P., and Mattern, R.,** Determination and pharmacokinetics of flurazepam metabolites in human blood, *Arzneim.-Forsch.,* 30, 1944, 1980.
533. **Sumirtapura, Y. C., Aubert, C., and Cano, J. P.,** Highly specific and sensitive method for determination of flunitrazepam in plasma by electron-capture gas-liquid chromatography, *Arzneim.-Forsch.,* 32, 252, 1982.
534. **Gjerde, H., Dahlin, E., and Christophersen, A. S.,** Simultaneous determination of common benzodiazepines in blood using capillary gas chromatography, *J. Pharm. Biomed. Anal.,* 10, 317, 1992.
535. **Rosseel, M. T. and Bogaert, M. G.,** Simultaneous determination of isosorbide dinitrate and its mononitrates in human plasma by capillary column GLC, *J. Pharm. Sci.,* 68, 659, 1979.
536. **Luedecke, F., Hennig, B., and Vetter, B.,** Gas-chromatographic analysis of trace amounts of isosorbide dinitrate and glycerol trinitrate in human plasma after solid-phase extraction, *Pharmazie,* 47, 640, 1992.
537. **Heizmann, P. and Von Alten, R.,** Determination of midazolam and its _-hydroxy-metabolite in plasma by gas chromatography with electron-capture detection, *J. High Res. Chromatogr. Chromatogr. Commun.,* 4, 266, 1981.
538. **Le Guellec, C., Bun, H., Giocanti, M., and Durand, A.,** Determination of nifedipine in plasma by a rapid capillary gas-chromatographic method, *Biomed. Chromatogr.,* 6, 20, 1992.
539. **Jorgensen, M., Andersen, M. P., and Hansen, S. H.,** Simultaneous determination of nitroglycerin and its dinitrate metabolites by capillary gas chromatography with electron-capture detection, *J. Chromatogr.,* 577, 167, 1992.
540. **Han, C., Gumbleton, M., Lau, D. T. W., and Benet, L. Z.,** Improved gas chromatographic assay for the simultaneous determination of nitroglycerin and its mono- and dinitrate metabolites, *J. Chromatogr.,* 579, 237, 1992.
541. **Van Boven, M., Daenens, P., and Vandereycken, G.,** Determination of nitromethaqualone in blood by electro-capture gas chromatography, *J. Chromatogr.,* 182, 435, 1980.
542. **Timm, U., Zell, M., and Herzfeld, L.,** Sensitive gas-liquid chromatographic method for the determination of oxiconazole in plasma, *J. Chromatogr.,* 229, 111, 1982.
543. **Jones, C. R., Ryle, P. R., and Weatherley, B. C.,** Measurement of pyrimethamine in human plasma by gas-liquid chromatography, *J. Chromatogr.,* 224, 492, 1981.
544. **Divoll, M. and Greenblatt, D. J.,** Plasma concentrations of temazepam, a 3-hydroxybenzodiazepine, determined by electron-capture gas-liquid chromatography, *J. Chromatogr.,* 222, 125, 1981.
545. **Jochemsen, R. and Breimer, D. D.,** Assay of triazolam in plasma by capillary gas chromatography, *J. Chromatogr.,* 223, 438, 1981.
546. **Edecki, T., Robin, D. W., Prakash, C., Blair, I. A., and Wood, A. J. J.,** Sensitive assay for triazolam in plasma following low oral doses, *J. Chromatogr.,* 577, 190, 1992.
547. **Poole, C. F., Johansson, L., and Vessman, J.,** Determination of bifunctional compounds. VIII. Formation of electron-capturing derivatives of alprenolol by transboronation. Application to the determination of alprenolol in plasma, *J. Chromatogr.,* 194, 365, 1980.
548. **Vessman, J., Johansson, L., and Groningsson, K.,** Determination of moroxydine in biological fluids by electron-capture gas chromatography, *J. Chromatogr.,* 229, 227, 1982.
549. **Pillai, G. K., Axelson, J. E., and McErlane, K. M.,** Electron-capture gas-liquid chromatographic determination of tocainide in biological fluids using fused-silica capillary columns, *J. Chromatogr.,* 229, 103, 1982.
550. **Lantz, R. J., Farid, K. Z., Koons, J., Tenbarge, J. B., and Bopp, R. J.,** Determination of fluoxetine and norfluoxetine in human plasma by capillary gas chromatography with electron-capture detection, *J. Chromatogr.,* 614, 175, 1993.
551. **Sioufi, A., Pommier, F., Mangoni, P., Gauron, S., and Metayer, J. P.,** The gas-chromatographic determination of phentolamine (Regitine) in human plasma and urine, *J. Chromatogr.,* 222, 429, 1981.

552. **Guerret, M., Julien-Larose, C., Kiechel, J. R., and Lavene, D.,** Determination of 3-hydroxyguanfecine in biological fluids by electron-capture gas-liquid chromatography, *J. Chromatogr.,* 233, 181, 1982.

553. **Terada, M., Yamamoto, T., Yoshida, T., Kuroiwa, Y., and Yoshimura, S.,** Rapid and highly sensitive method for the determination of methylamphetamine and amphetamine in urine by electron-capture gas chromatography, *J. Chromatogr.,* 237, 285, 1982.

554. **Nash, J. F., Bupp, R. J., Carmichael, R. H., Favid, K. Z., and Lemberger, L.,** Determination of fluoxetine and norfluoxetine in plasma by gas chromatography with electron-capture detection, *Clin. Chem.,* 28, 2100, 1982.

555. **Cooper, S. F., and Lapierre, Y. D.,** Gas-liquid chromatographic determination of pipothiazine in plasma of psychiatric patients, *J. Chromatogr.,* 222, 191, 1981.

556. **Sioufi, A. and Pommier, F.,** Gas-chromatographic determination of amantadine hydrochloride (Symmetrel) in human plasma and urine, *J. Chromatogr.,* 183, 33, 1980.

557. **Lo, L. Y., Land, G., and Bye, A.,** Sensitive assay for pseudoephedrineand its metabolite, norpseudoephedrine, in plasma and urine using gas-liquid chromatography with electron-capture detection, *J. Chromatogr.,* 222, 297, 1981.

558. **Guerret, M., Lavene, D., and Kiechel, J. R.,** Determination of pindolol in biological fluids by electron-capture g.l.c., *J. Pharm. Sci.,* 69, 1191, 1980.

559. **Karmen, A. and Giuffrida, L.,** Enhancement of the response of the hydrogen flame ionization detector to compounds containing halogen and phosphorus, *Nature,* 201, 1204, 1964.

560. **Clatworthy, A. J.,** Notes and comments relating to instrumental techniques, in *Assays of Drugs and Other Trace Compounds in Biological Fluids,* Reid, E., Ed., North-Holland, Amsterdam, 1976, 10.

561. **Baune, A., Bromet, N., Courte, S., and Voisin, C.,** Trace determinations of almitrine in plasma by gas-liquid chromatography using a nitrogen-phosphorous detector, *J. Chromatogr.,* 241, 29, 1982.

562. **Polgar, M. and Vereczkey, L.,** Determination of apovincaminic acid in human plasma by gas-liquid chromatography, *J. Chromatogr.,* 241, 29, 1982.

563. **Coudore, F., Alazard, J.-M., Paire, M., Andraud, G., and Lavarenne, J.,** Rapid toxicological screening of barbiturates in plasma by wide-bore capillary gas chromatography and nitrogen-phosphorus detection, *J. Analyt. Toxicol.,* 17, 109, 1993.

564. **Lau, O. W., Wong, Y. C., and Chan, K.,** Use of a packed column for the determination of bupivacaine in human plasma by gas chromatography: application in a pharmacokinetic study of bupivacaine, *Forensic Sci. Int.,* 53, 125, 1992.

565. **Delbeke, F. T. and Debackere, M.,** Determination of butanilicaine in horse plasma and urine by extractive benzoylation and gas chromatography with a nitrogen-phosphorous detector, *J. Chromatogr.,* 237, 344, 1982.

566. **Gonzalez, R. and Kraml, K.,** Butriptyline: improved gas-liquid chromatographic method using a nitrogen-phosphorous detector for its determination in serum, *Clin. Biochem.,* 13, 141, 1980.

567. **Levandoski, P. and Flanagan, T.,** Use of nitrogen-specific detector for g.l.c. determination of caramiphen in whole blood, *J. Pharm. Sci.,* 69, 1353, 1980.

568. **Woestenborghs, R., Michielsen, L., Lorreyne, W., and Heykants, J.,** Sensitive gas-chromatographic method for the determination of cinnarizine and flunarizine in biological samples, *J. Chromatogr.,* 232, 85, 1982.

569. **Curvall, M., Kazemi-Vala, E., and Enzell, K. R.,** Simultaneous determination of nicotine and cotinine in plasma using capillary gas chromatography with nitrogen-sensitive detection, *J. Chromatogr.,* 232, 283, 1982.

570. **Antal, E., Mercik, S., and Kramer, P. A.,** Technical considerations in gas-chromatographic analysis of desipramine, *J. Chromatogr.,* 183, 149, 1980.

571. **Javaid, J. I., Dekirmenjian, H., Liskevych, U., Lin, R. L., and Davis, J. M.,** Fluphenazine determination in human plasma by a sensitive gas-chromatographic method using a nitrogen detector, *J. Chromatogr. Sci.,* 19, 439, 1981.

572. **Abernathy, D. R., Greenblatt, D. J., and Ochs, H. R.,** Lidocaine determination in human plasma with application to single low dose pharmacokinetic studies, *J. Chromatogr.,* 232, 180, 1982.

573. **Jacob, P., Rigod, J. F., Pond, S. M., and Benowitz, N. L.,** Determination of methadone and its primary metabolite in biological fluids using gas chromatography with nitrogen-phosphorous detection, *J. Analyt. Toxicol.,* 5, 292, 1981.
574. **Pilling, M., Tse, J., and Chan, K.,** Modified gas-liquid chromatographic assay to monitor plasma mexiletine in a tinnitus study, *Methods Find. Exp. Clin. Pharmacol.,* 4, 243, 1982.
575. **Lachatre, G. F., Nicot, G. S., Merle, L. J., and Valette, J. P.,** Determination of mianserin in human plasma by gas-liquid chromatography, *Ther. Drug Monitor.,* 4, 359, 1982.
576. **Chang, S. F., Hansen, C. S., Fox, J. M., and Ober, R. E.,** Quantitative determination of nefopam in human plasma, saliva and cerebrospinal fluid by gas-liquid chromatography using a nitrogen-selective detector, *J. Chromatogr.,* 226, 79, 1981.
577. **Davison, S. C., Hyman, N., Prentis, R. A., Deghan, A., and Chan, K.,** Simultaneous monitoring of plasma levels of neostigmine and pyridostigmine in man, *Methods Find. Exp. Clin. Pharmacol.,* 2, 77, 1980.
578. **Kapil, R. P., Padovani, P. K., King, S.-Y. P., and Lam, G. N.,** Nanogram level quantitation of oxycodone in human plasma by capillary gas chromatography using nitrogen-phosphorus selective detection, *J. Chromatogr.,* 577, 283, 1992.
579. **Tse, J., and Chan, K.,** Simultaneous determination of pethidine and norpethidine in bio-fluids by nitrogen-selective gas chromatography, *Methods Find. Exp. Clin. Pharmacol.,* 3, 99, 1981.
580. **Miceli, J. N., Bowman, D. B., and Aravind, M. K.,** Improved method for quantitation of phencyclidine in biological samples utilising nitrogen-detection gas chromatography, *J. Analyt. Toxicol.,* 5, 29, 1981.
581. **Chan, K., Murray, G. R., Rostron, C., Calvey, T. N., and Williams, N. E.,** Quantitative gas-liquid chromatographic method for determination of phenoperidine in human plasma, *J. Chromatogr.,* 223, 213, 1981.
582. **Thoma, M., Farinello, L., and Mueller, A.,** Gas-chromatographic determination of therapeutic levels of prajmalium bitartrate in human plasma, *Arzneim.-Forsch.,* 31, 1020, 1981.
583. **Dean, K., Land, B., and Bye, A.,** Analysis of procyclidine in human plasma and urine by gas-liquid chromatography, *J. Chromatogr.,* 221, 408, 1980.
584. **Jourdil, N., Pinteur, B., Vincent, F., Marka, C., and Bessard, G.,** Simultaneous determination of trimipramine and desmethyl- and hydroxy-trimipramine in plasma and red blood cells by capillary gas chromatography with nitrogen-selective detection, *J. Chromatogr.,* 613, 59, 1993.
585. **Bailey, E. and Barron, E. J.,** Determination of tranylcypromine in human plasma and urine using high-resolution gas-liquid chromatography with nitrogen-sensitive detection, *J. Chromatogr.,* 183, 25, 1980.
586. **Gillespie, T. J. and Sipes, I. G.,** Sensitive gas-chromatographic determination of trifluoperazine in human plasma, *J. Chromatogr.,* 223, 95, 1981.
587. **Rossi, S.-A., Johnson, J. V., and Yost, R. A.,** Optimization of short-column gas chromatography-electron ionization mass spectrometry conditions for the determination of underivatized anabolic steroids, *Biol. Mass Spectromet.,* 21, 420, 1992.
588. **Gjerde, H., Hasvold, I., Pettersen, G., and Christophersen, A. S.,** Determination of amphetamine and methamphetamine in blood by derivatization with perfluoro-octanoyl chloride and gas chromatography-mass spectrometry, *J. Analyt. Toxicol.,* 17, 65, 1993.
589. **Aderjan, R. E., Schmitt, G., Wu, M., and Meyer, C.,** Determination of cocaine and benzoylecgonine by derivatization with iodomethane-D$_3$ or PFPA-HFIP in human blood and urine using GC-MS (EI or PCI mode), *J. Analyt. Toxicol.* 17, 51, 1993.
590. **Dill, D. N. and Eilbacher, B.,** Gas-chromatographic-mass spectrometric method for the determination of bifemelane in human plasma at therapeutic doses, *J. Chromatogr.,* 613, 150, 1993.
591. **Kokatsu, J., Yomoda, R., and Suwa, T.,** Selected ion monitoring for the determination of bromovalerylurea in human plasma, *Chem. Pharm. Bull.,* 40, 1517, 1992.
592. **De Giovanni, N., and Fucci, N.,** Gas-chromatographic-mass-spectrometric analysis of buflomedil hydrochloride in biological samples after acute intoxication, *Forensic Sci. Int.,* 51, 125, 1991.
593. **Bronner, W. E., and Xu, A. S.,** Gas-chromatographic-mass-spectrometric methods of analysis for detection of 11-nor{delta}9-tetrahydrocannabinol-9-carboxylic acid in biological matrices. *J. Chromatogr.,* 580, 63, 1992.

594. Gjerde, H., Fongen, U., Gundersen, H., and Christophersen, A. S., Evaluation of a method for simultaneous quantification of codeine, ethylmorphine and morphine in blood, *Forensic Sci. Int.*, 51, 105, 1991.

595. Fuller, D. C. and Anderson, W. H., Simplified procedure for the determination of free codeine, free morphine, and 6-acetylmorphine in urine, *J. Analyt. Toxicol.*, 16, 315, 1992.

596. West, R. E. and Ritz, D. P., GC-MS analysis of five common benzodiazepine metabolites in urine as t-butyl-dimethylsilyl derivatives, *J. Analyt. Toxicol.*, 17, 114, 1993.

597. Duthel, J. M., Constant, H., Vallon, J. J., Rochet, T., and Miachon, S., Quantitation by gas chromatography with selected-ion-monitoring mass spectrometry of "natural" diazepam, *N*-demethyldiazepam and oxazepam in normal human serum, *J. Chromatogr.*, 579, 85, 1992.

598. Li, Y. W., Li, J., and Zhou, T. H., GC-MS study on diuretics in urine. II. detection method using trimethylsilylation, *Chin. Chem. Lett.*, 2, 19, 1991.

599. Shioya, H., Shimojo, M., and Kawahara, Y., Determination of enalapril and its active metabolite enalaprilate in plasma and urine by gas chromatography-mass spectrometry, *Biomed. Chromatogr.*, 6, 59, 1992.

600. Takamatsu, T., Yamazaki, K., Kayano, M., Takenaka, F., Hasui, M., and Ohkawa, T., Determination of eperisone in human plasma by gas chromatography-mass spectrometry, *J. Chromatogr.*, 584, 261, 1992.

601. Anderson, L. W., Parker, R. J., Collins, M. J., Ahlgren, J. D., Wilkinson, D., and Strong, J. M., Gas-chromatographic-mass-spectrometric method for routine monitoring of 5-fluouracil in plasma of patients receiving low-level protracted infusions, *J. Chromatogr.*, 581, 195, 1992.

602. Goldberger, B. A., Darwin, W. D., Grant, T. M., Allen, A. C., Caplan, Y. H., and Cone, E. J., Measurement of heroin and its metabolites by isotope-dilution electron-impact mass spectrometry, *Clin. Chem.*, 39, 670, 1993.

603. Ackermann, R., Kaiser, G., Schueller, F., and Dieterle, W., Determination of the antidepressant levoprotiline and its *N*-desmethyl metabolite in biological fluids by gas chromatography-mass spectrometry, *Biol. Mass Spectromet.*, 20, 709, 1991.

604. Hagedorn, H. W., Schultz, R., and Friedrich, A., Detection of methandienone (methandrostenolone) and metabolites in horse urine by gas chromatography-mass spectrometry, *J. Chromatogr.*, 577, 195, 1992.

605. Quaglio, M. P., Bellini, A. M., Minozzi, A M., Frisina, G., and Testoni, F., Simultaneous determination of propranolol or metoprolol in the presence of butyrophenones in human plasma by gas chromatography with mass spectrometry, *J. Pharm. Sci.*, 82, 87, 1993.

606. Goto, N., Kamata, T., and Ikegami, K., Trace analysis of quinapril and its active metabolite, quinaprilat, in human plasma and urine by gas chromatography-negative-ion chemical ionization mass spectrometry, *J. Chromatogr.*, 578, 195, 1992.

607. Susanto, F., Humfield, S., Niederau, C. M., and Reinauer, H., Method for the determination of theophylline in serum by isotope dilution mass spectrometry, *Fresenius J. Analyt. Chem.*, 344, 549, 1992.

608. Gaetani, E., Laureri, C. F., and Vitto, M., Ion-trap detector-capillary gas chromatography of valproic acid and its mono-unsaturated metabolites in serum using methyl ester derivatives, *J. Pharm. Biomed. Anal.*, 10, 193, 1992.

609. Fisher, E., Wittfoht, W., and Nau, H., Quantitative determination of valproic acid and 14 metabolites in serum and urine by gas chromatography-mass spectrometry, *Biomed. Chromatogr.*, 6, 24, 1992.

610. Catlin, D., Cowan, D., Donike, M., Fraisse, D., Oftebro, H., and Rendic, S., Testing urine for drugs, *J. Autom. Chem.*, 14, 85–92, 1992.

611. Siren, H., Saarinen, M., Hainari, S., Lukkari, P., and Riekkola, M.-L., Screening of β-blockers in human serum by ion-pair chromatography and their identification as methyl or acetyl derivatives by gas chromatography-mass spectrometry, *J. Chromatogr.*, 632, 215, 1993.

612. Kuhlman, J. J., Levine, B., Klette, K. L., Magluilo, J., Kalasinsky, S., and Smith, M. L., Measurement of azacyclonol in urine and serum of humans following terfenadine (Seldane) administration using gas chromatography-mass spectrometry, *J. Chromatogr.*, 116, 207, 1992.

613. **Goto, N., Sato, T., Shigetosi, M., and Ikegami, K.,** Determination of dioxopiperazine metabolites of quinapril in biological fluids by gas chromatography-mass spectrometry, *J. Chromatogr.*, 578, 203, 1992.

614. **Matsuoka, M., Horimoto, S., Mabuchi, M., and Banno, K.,** Determination of three metabolites of a new angiotensin-converting enzyme inhibitor, imidapril, in plasma and urine by gas chromatography-mass spectrometry using multiple ion detection, *J. Chromatogr.*, 581, 65, 1992.

615. **Davoli, E., Zucchetti, M., D'Incalci, M., Sessa, C., Vavali, F., and Fanelli, R.,** Mass-spectrometric identification and analysis of some aphidicolin metabolites in cancer patients, *Biol. Mass Spectromet.*, 22, 351, 1993.

616. **Millard, B. J.,** Mass-spectrometric quantitation in gas chromatography, and other gas chromatography advances, in *Assays of Drugs and Other Trace Compounds in Biological Fluids*, Reid, E., Ed., North-Holland, Amsterdam, 1976, 1.

617. **Deutsch, J. C. and Kolhouse, J. F.,** Ascorbate and dehydroascorbate measurements in aqueous solutions and plasma determined by gas chromatography-mass spectrometry, *Analyt. Chem.*, 65, 321, 1993.

618. **Kishida, K., Manabe, R., Bando, K., and Miwa, Y.,** Application of mass fragmentography to determination of acetazolamide in body fluids, *Anal. Lett.*, 14, 335, 1981.

619. **Prakash, C., Adedoyin, A., Wilkinson, G. R., and Blair, I. A.,** Enantiospecific quantification of hexobarbitone and its metabolites in biological fluids by gas-chromatography-electron-capture negative-ion chemical-ionization mass spectrometry, *Biol. Mass Spectromet.*, 20, 559, 1991.

620. **Midha, K. K., Roscoe, R. M. H., Hall, K., Hawes, E. M., Cooper, J. K., McKay, G., and Shetty, H. U.,** Gas chromatographic/mass spectrometric assay for plasma trifluoperazine concentrations following single doses, *Biomed. Mass Spectromet.*, 9, 186, 1982.

621. **Mitchum, R. K., Moler, G. F., and Korfmacher, W. A.,** Combined capillary gas chromatography-atmospheric pressure negative chemical-ionisation mass spectrometry for determination of 2,3,7,8-tetrachlorodibenzo-p-dioxin in tissue, *Analyt. Chem.*, 52, 2278, 1980.

622. **Marunaka, T. and Umeno, Y.,** Determination of fluorouracil and pyrimidine bases in plasma by gas chromatography-chemical ionisation mass fragmentography, *J. Chromatogr.*, 221, 382, 1980.

623. **Julien-Larose, C., Lange, C., Davene, D., Kiechel, J. R., and Basselier, J. J.,** Determination of guanfacine in human plasma by g.l.c.-m.s. with electron-capture negative chemical ionisation, *Int. J. Mass Spec. Ion Phys.*, 48, 221, 1983.

624. **Idzu, G., Ishibashi, M., Miyazaki, H., and Yamamoto, K.,** Determination of glyceryl trinitrate in human plasma by gas chromatography-negative-ion chemical-ionisation selected ion monitoring, *J. Chromatogr.*, 229, 327, 1982.

625. **Reimer, M. L. J., Mamer, O. A., Zavitsanos, A. P., Siddiqui, A. W., and Dadgar, D.,** Determination of amphetamine, methamphetamine and desmethyldeprenyl in human plasma by gas chromatography-negative-ion chemical ionization mass spectrometry, *Biol. Mass Spectromet.*, 22, 235, 1993.

626. **Morris, M. J., Gilbert, J. D., Hsieh, J. Y.-K., Matuszewski, B. K., Ramjit, H. G., and Bayne, W. F.,** Determination of the HMG-CoA reductase inhibitors simvastatin, lovastatin and pravastatin in plasma by gas chromatography-chemical-ionization mass spectrometry, *Biol. Mass Spectromet.*, 22, 1, 1993.

627. **Bennett, M. J., Sherwood, W. G., Bhaha, A., and Hale, D. E.,** Identification of urinary metabolites of (±)-2-(p-isobutylphenyl)propionic acid (Ibuprofen) by routine organic acid screening, *Clin. Chim. Acta*, 210, 55, 1992.

628. **Bastide, M., Chabard, J.-L., Lartigue, C., Bargnoux, H., Petit, J., Berger, J.-A., Mansour, H. A., Lesieur, D., and Busch, N.,** Identification of several human urinary metabolites of 6-benzoylbenzoxazolinone by gas chromatography-mass spectrometry, *Biol. Mass Spectromet.*, 20, 484, 1991.

629. **Kalasinsky, K. S., Levine, B., and Smith M. L.,** Feasibility of using GC-FTIR for drug analysis in the forensic toxicology laboratory, *J. Analyt. Toxicol.*, 16, 332, 1992.

630. **Platoff, G. E., Hill, D. W., Koch, T. R., and Caplan, Y. H.,** Serial capillary gas chromatography-Fourier-transform infra-red spectrometry-mass spectrometry (GC-IR-MS): qualitative and quantitative analysis of amphetamine, methamphetamine and related analogues in human urine, *J. Analyt. Toxicol.*, 16, 389, 1992.

631. **Minagawa, T., Suzuki, S., and Suzuki, T.,** Analysis of benzophenones by gas chromatography-Fourier-transform infra-red spectrometry, *Forensic Sci. Int.,* 51, 179, 1991.

632. **Shinohara, Y., Magara, H., and Baba, S.,** Stereoselective pharmacokinetics and inversion of suprofen enantiomers in humans, *J. Pharm. Sci.,* 80, 1075, 1990.

633. **Shinohara, Y., Kirii, N., Tamaoki, H. Magara, H., and Baba, S.,** Determination of the enantiomers of suprofen and [^{2}H$_3$]suprofen in plasma by capillary gas chromatography-mass spectrometry, *J. Chromatogr.,* 525, 93, 1990.

634. **Torok-Both, G. A., Baker, J. B., Coutts, R. T., McKenna, K. F., and Aspeslet, L. J.,** Simultaneous determination of fluoxetine and norfluoxetine enantiomers in biological samples by gas chromatography with electron-capture detection, *J. Chromatogr.,* 579, 99, 1992.

635. **Jack, D. S., Rumble, R. H., Davies, N. W., and Francis, H. W.,** Enantiospecific gas-chromatographic-mass-spectrometric procedure for the determination of ketoprofen and ibuprofen in synovial fluid and plasma: application to protein binding studies, *J. Chromatogr.,* 584, 189, 1992.

636. **Changchit, A., Gal, J., and Zirrolli, J. A.,** Stereospecific gas-chromatographic-mass-spectrometric assay of the chiral labetalol metabolite 3-amino-1-phenylbutane, *Biol. Mass Spectromet.,* 20, 751, 1991.

637. **Frank, H., Nicholson, G. J., and Bayer, E.,** Rapid gas chromatographic separation of amino acid enantiomers with a novel chiral stationary phase, *J. Chromatogr. Sci.,* 15, 174, 1977.

638. **König, W. A., Benecke, I., and Sievers, S.,** New results in the gas chromatographic separation of enantiomers of hydroxy acids and carbohydrates, *J. Chromatogr.,* 217, 71, 1981.

639. **Koskielski, T., Sybilska, D., Belniak, S., and Jurczak, J.,** Application of a gas-liquid chromatography system with _-cyclodextrin for monitoring the stereochemical course of β-pinene hydrogenation, *Chromatographia,* 21, 413, 1986.

640. **Koenig, W. A., Icheln, D., Runge, T., Pfaffenberger, B., Ludwig, P., and Huehnerfuss, H.,** Gas-chromatographic enantiomer separation of agrochemicals using modified cyclodextrins, *J. High Res. Chromatogr.,* 14, 530, 1991.

641. **König, W. A., Lutz, S., and Wenz, G.,** Modified cyclodextrins — novel, highly enantioselective stationary phases for gas chromatography, *Angew. Chem., (Int, Edn.)* 27, 979, 1988.

642. **Schurig, B. and Nowotny, H.-P.,** Separation of enantiomers on diluted permethylated β-cyclodextrin by high-resolution gas chromatography, *J. Chromatogr.,* 441, 155, 1988.

643. **Roboz, J., Nieves, E., and Holland, J. F.,** Separation and quantification by gas chromatography-mass spectrometry of arabinitol enantiomers to aid the differentiatial diagnosis of disseminated canidiasis, *J. Chromatogr.,* 500, 413, 1990.

644. **Koenig, W. A., Icheln, D., and Hardt, I.,** Unusual retention behaviour of methyl lactate and methyl 2-hydroxybutyrate enantiomers on a modified cyclodextrin, *J. High Res. Chromatogr.,* 14, 694, 1991.

645. **König, W. A., Benecke, I., and Ernst, K.,** Separation of chiral ketones by enantioselective gas chromatography, *J. Chromatogr.,* 253, 267, 1982.

646. **Schramm, T. M., McKinnon, G. E., and Eadie, M. J.,** Gas chromatographic assay of vigabatrin enantiomers in plasma, *J. Chromatogr.,* 616, 39, 1993.

647. **St. Louis, P., Rybczynski, J., Bissonnette, B., and Hartley, E. J.,** Determination of bupivacaine in plasma by high-performance liquid chromatography. Levels after scalp infiltration in children, *Clin. Biochem.,* 24, 463, 1991.

648. **Neilsen, K. K. and Broesin, K.,** High-performance liquid chromatography of clomipramine hydrochloride and metabolites in human plasma and urine, *Ther. Drug Monitor.,* 15, 122, 1993.

649. **Mohammed, S. S., Butschkau, M., and Derendorf, H.,** Reversed-phase liquid-chromatographic method for the determination of codeine in biological fluids with applications, *J. Liq. Chromatogr.,* 16, 2325, 1993.

650. **Nickmilder, M., Verbeeck, R., and Lhoest, G.,** Isolation and mass spectrometric identification of cyclosporin A isomeric dihydrodiol precursors from erythromycin-induced rabbit liver microsomes, *Pharmacol. Acta Helv.,* 67, 275, 1992.

651. **Hamilton, C. L. and Cornpropst, J. D.,** Determination of dapoxetine, an investigational agent with the potential for treating depression, and its mono- and di-desmethyl metabolites in human plasma using column-switching high-performance liquid chromatography, *J. Chromatogr.,* 612, 253, 1993.

652. **Jensen, B. H. and Larsen, C.,** Quantitation of diltiazem in human plasma by HPLC using an end-capped reversed-phase column, *Acta Pharm. Nord.,* 3, 179, 1991.

653. **Taylor, P. J., Charles, B. G., Norris, R., Salm, P., and Ravenscroft, P. J.,** Measurement of dothiepine and its major metabolites in plasma by high-performanceliquid chromatography, *J. Chromatogr.,* 581, 152, 1992.

654. **Vree, T. B., Van Ewijk-Beneken Lolmer, E. W. J., and Nouws, J. F. M.,** Direct-gradient high-performance liquid-chromatographic analysis and preliminary pharmacokinetics of flumequine and flumequine acylglucuronide in humans: effect of probenicid, *J. Chromatogr.,* 579, 131, 1992.

655. **Palmisano, F., Bernardi, F., De Lena, M., Guerrieri, A., Lorusso, V., and Zambonin, P. G.,** Simultaneous determination of 5'-deoxy-5-fluorouridine, 5-fluorouracil and its main main metabolites in body fluids by HPLC, *Chromatographia,* 33, 413, 1992.

656. **Thomare, P., Wang, K., Van der Meersch-Mougeot, V., and Diquet, B.,** Sensitive micro-method for column liquid-chromatographic determination of fluoxetine and norfluoxetine in human plasma, *J. Chromatogr.,* 583, 217, 1992.

657. **Haertter, S., Wetzel, H., and Hiemke, C.,** Automated determination of fluvoxamine in plasma by column-switching high-performance liquid chromatography, *Clin. Chem.,* 38, 2082, 1992.

658. **Ohkubo, T., Shimoyama, R., and Sugawara, K.,** Measurement of haloperidol in human breast milk by high-performance liquid chromatography, *J. Pharm. Sci.,* 81, 947, 1992.

659. **Kubo, H., Umiguchi, Y., and Kinoshita, T.,** Fluorimetric determination of indomethacin in serum by high-performance liquid chromatography using online oxidation with hydrogen peroxide, *J. Liq. Chromatogr.,* 16, 465, 1993.

660. **Kondo, T., Buss, D. C., and Routledge, P. A.,** Method for rapid determination of lorazepam by high-performance liquid chromatography, *Ther. Drug Monitor.,* 15, 35, 1993.

661. **Kato, Y., Kaneko, H., Matsushita, T., Inamori, K., Egi, S., Togawa, A., Yokohama, T., and Mohri, K.,** Direct injection analysis of melphalan in plasma using column-switching high-performance liquid chromatography, *Ther. Drug Monitor.,* 14, 66, 1992.

662. **Kubo, H., Umiguchi, Y., Fukumoto, M., and Kinoshita, T.,** Fluorimetric determination of methotrexate in serum by high-performance liquid chromatography using inline oxidation with hydrogen peroxide, *Analyt. Sci.,* 8, 789, 1992.

663. **Qian, M. X. and Gallo, J. M.,** High-performance liquid-chromatographic determination of the calcium channel blocker nimodipine in monkey plasma, *J. Chromatogr.,* 578, 316, 1992.

664. **Hassan, E. and Gallo, G. M.,** High-performance liquid-chromatographic analysis of the anticancer drug oxantrazole in rat whole blood and tissues, *J. Chromatogr.,* 582, 225, 1992.

665. **Nelis, H. J., Vandenbranden, J., De Kruif, A., Belpaire, F., and De Leenheer, A. P.,** Liquid-chromatographic determination of oxytetracycline in bovine plasma by double-phase extraction, *J. Pharm. Sci.,* 81, 1216, 1992.

666. **Luhmann, I., Szathmary, S. C., and Gruenert, I.,** Determination of pipamperone in human plasma by high-performance liquid chromatography, *Arzneim. Forsch.,* 42, 1069, 1992.

667. **Chmielowiec, D., Schuster, D., and Gengo, F.,** Determination of pindolol in human serum by HPLC, *J. Chromatogr. Sci.,* 29, 37, 1991.

668. **Milligan, P. A.,** Determination of piroxicam and its major metabolites in the plasma, urine, and bile of humans by high-performance liquid chromatography, *J. Chromatogr.,* 576, 121, 1992.

669. **Gonzalez-Esquivel, D. F., Morano Okuno, C., Sanchez Rodriguez, M., Sotelo Morales, J., and Jung Cook, H.,** Sensitive high-performance liquid-chromatographic assay for praziquantel in plasma, urine and liver homogenates, *J. Chromatogr.,* 613, 174, 1993.

670. **Yatscoff, R. W., Faraci, C., and Bolingbroke, P.,** Measurement of rapamycin in whole blood using reversed-phase high-performance liquid chromatography, *Ther. Drug Monitor.,* 14, 134, 1992.

671. **Carlucci, G., Mazzeo, P., Fanini, D., and Palumbo, G.,** Determination of rufloxacin, a new tricyclic fluoroquinolone, in biological fluids using high-performance liquid chromatography with ultra-violet detection, *Ther. Drug Monitor.,* 13, 448, 1991.

672. **Burger, D. M., Rosing, H., Van Gijn, R., Meenhorst, P.L., Van Tellingen, O., and Beijnen, J. H.,** Determination of stavudine, a new antiretroviral agent, in human plasma by reversed-phase high-performance liquid chromatography with ultra-violet detection, *J. Chromatogr.,* 584, 239, 1992.

673. **Lhoest, G., Maton, N., and Verbeeck, R.,** Isolation and mass-spectrometric identification of two metabolites of FK 506 from rat liver microsomal incubation media, *Pharmacol. Acta Helv.,* 67, 270, 1992.

674. **Reimer, G., Suarez, A., and Chui, Y. C.,** Liquid chromatographic procedure for the analysis of yohimbine in equine serum and urine, *J. Analyt. Toxicol.,* 17, 178, 1993.

675. **Molema, G., Jansen, R. W., Visser, J., and Meijer, D. K. F.,** Simultaneous analysis of azidothymidine and its mono-, di and tri-phosphate derivatives in biological fluids, tissue and cultured cells by a rapid high-performance liquid chromatographic method, *J. Chromatogr.,* 579, 107, 1992.

676. **Dell, D. and Chamberlain, J.,** Determination of molsidomine in plasma by high-pressure liquid chromatography, *J. Chromatogr.,* 146, 465, 1978.

677. **Brewster, J. D., Lightfield, A. R., and Barford, R. A.,** Evaluation of restricted access media for high performance liquid-chromatographic analysis of sulphonamide antibiotic residues in bovine serum, *J. Chromatogr.,* 598, 23, 1992.

678. **El-Yazigi, A. and Wahab, F. A.,** Expedient liquid-chromatographic analysis of Azathioprine in plasma by use of silica solid-phase extraction, *Ther. Drug Monitor.,* 14, 312, 1992.

679. **Shibl, A. M., Tawfik, A. F., El-Houfy, S., and Al-Shammary, F. J.,** Determination of lomefloxacin in biological fluids by high-performance liquid chromatography and a microbiological method, *J. Clin. Pharmacol. Ther.,* 16, 353, 1991.

680. **Funaki, T., Onadera, H., Ogawa, K., Ichihara, S., Fukazawa, H., and Kuruma, I.,** Simultaneous determination of a new anticancer drug galocitabine and its metabolites in blood by high-performance liquid chromatography, *J. Pharmacol. Biomed., Anal.,* 11, 379, 1993.

681. **Schill, G.,** Ion-pair HPLC of acidic and basic drugs metabolites and endogenous compounds, in *Blood Drugs and Other Analytical Challenges,* Reid, E., Ed., Ellis Horwood, Chichester, 1978, 195.

682. **Tomlinson, E., Jeffries, T. M., and Riley, C. M.,** Ion-pair high-performance liquid chromatography, *J. Chromatogr.,* 159, 315, 1978.

683. **Armstrong, D. W. and Henry, S. J.,** Use of an aqueous micellar mobile phase for separation of phenols and polynuclear aromatic hydrocarbons via h.p.l.c., *J. Liq. Chromatogr.,* 3, 657, 1980.

684. **Foley, J. P. and May, W. E.,** Optimization of secondary chemical equilibria in liquid chromatography: theory and verification, *Analyt. Chem.,* 59, 102, 1987.

685. **Foley, J. P. and May, W. E.,** Optimization of secondary chemical equilibria in liquid chromatography: variable influencing the self-selectivity, retention, and efficiency in acid-base systems, *Analyt. Chem.,* 59, 110, 1987.

686. **Landy, J. S. and Dorsey, J. G.,** Characterization of micellar mobile phases for reversed-phase chromatography, *Analyt. Chim. Acta,* 178, 179, 1985.

687. **Dorsey, J. G.,** Micellar liquid chromatography, in *Bioanalysis of Drugs and Metabolites, Especially Anti-Inflammatory and Cardiovascular,* Reid, E., Robinson, J. R., and Wilson, I. D., Eds., Plenum Press, New York, 1988, 235.

688. **Jacobs, M. H., Senior, P. M., and Kessler, G.,** Clinical experience with theophylline. Relationship between dosage, serum concentration, and toxicity, *J. Am. Med. Ass.,* 235, 1983, 1976.

689. **Habel, D., Guermouche, S., and Guermouche, M. H.,** Direct determination of theophylline in human serum by high-performance liquid chromatography using zwitterionic micellar mobile phase. comparison with an enzyme multiplied immunoassay technique, *Analyst,* 118, 1511–1513, 1993.

690. **Nicholls, G., Clark, B. J., and Brown, J. E.,** Optimization of the separation of anthrcyclines and their metabolites using reversed-phase liquid chromatography, *Analyt. Proc.,* 30, 103, 1993.

691. **Bogusz, M., Erkens, M., Franke, J. P, Wijsbeek, J., and De Zeeuw, R. A.,** Interlaboratory applicability of a retention index library of drugs for screening by reversed-phase HPLC in systematic toxicological analysis, *J. Liq. Chromatogr.,* 16, 1341–1354, 1993.

692. **Balikova, M.,** Selective system of identification and determination of antidepressants and neuroleptics in serum or plasma by solid-phase extraction followed by high-performance liquid chromatography with photodiode-array detection in analytical toxicology, *J. Chromatogr.,* 581, 75, 1992.

693. **Overzet, F., Ghijsen, R. T., Drenth, B. F. H., and De Zeeuw, R. A.,** Multi-channel diode array U.V.-visible spectrophotometer as detector in screening for unknown butoprozine metabolites in dog bile by high-performance liquid chromatography, *J. Chromatogr.,* 240, 190, 1982.

694. **de Zeeuw, R. A.,** Photodiode array HPLC detectors in metabolic profiling and other analytical screening techniques, in *Bioactive Analytes, Including CNS Drugs Peptides and Enantiomers,* Reid, E., Scales, B., and Wilson, I. D., Eds., Plenum Press, New York, 1986, 355.

695. **Tsai, T. and Chen, C.,** Ultra-violet spectral investigation of ferulic acid in rabbit plasma by HPLC and its pharmacokinetic application, *Int. J. Pharmacol.,* 80, 75, 1992.

696. **Helmlin, H.-J. and Brenneisen, R.,** Determination of psychtropic phenylalkylamine derivatives in biological matrices by high-performance liquid chromatography with photodiode-array detection, *J. Chromatogr.,* 593, 87, 1992.

697. **Liu, H., Forman, L. J., Montoya, J., Eggers, C., Barham, C., and Delgado, M.,** Determination of valproic acid by high-performance liquid chromatography with photodiode-array and fluorescence detection, *J. Chromatogr.,* 576, 163, 1992.

698. **Kintz, P., Traqui, A., and Mangin, P.,** Determination of nalbuphine using high-performance liquid chromatography coupled to photodiode-array detection and gas chromatography coupled to mass spectrometry, *J. Chromatogr.,* 579, 172, 1992.

699. **Lurie, I. S., Cooper, D. A., and Klein, R. F. X.,** High-performance liquid-chromatographic analysis of benzodiazepines using diode-array, electrochemical and thermospray mass-spectrometric detection, *J. Chromatogr.,* 598, 59, 1992.

700. **Bogusz, M.,** High-performance liquid-chromatographic determination of morphine, morphine 3-glucuronide, morphine 6-glucuronide and codeine in biological samples using multiwavelength forward optical detection: a reply, *J. Chromatogr.,* 579, 189, 1992.

701. **Liu, H., Delgado, M., Forman, L. J., Eggersw, C. M., and Montoya, J. L.,** Simultaneous determination of carbamazepine, phenytoin, phenobarbital, primidone and their principal metabolites by high-performance liquid chromatography with photodiode-array detection, *J. Chromatogr.,* 616, 105, 1993.

702. **Tracqui, A., Kintz, P., Kreissig, P., and Mangin, P.,** Simple and rapid screening procedure for twenty-seven neuroleptics using HPLC-DAD, *J. Liq. Chromatogr.,* 15, 1381, 1992.

703. **Tracqui, A., Kintz, P., and Mangin, P.,** High-performance liquid-chromatographic assay with diode-array detection for toxicological screening of zopiclone, zolpidem, suriclone and alpidem in human plasma, *J. Chromatogr.,* 616, 95, 1993.

704. **Tracqui, A., Mikail, I., Kintz, P., and Mangin, P.,** Nonfatal prolonged overdosage of pyrimethamine in an infant: measurement of plasma and urine levels using HPLC with diode-array detection, *J. Analyt. Toxicol.,* 17, 248, 1993.

705. **White, P. C.,** A rotating filter disc alternative to photodiode array detection systems, in *Bioactive Analytes, Including CNS Drugs Peptides and Enantiomers,* Reid, E., Scales, B., and Wilson, I. D., Eds., Plenum Press, New York, 1986, 373.

706. **Kabra, P. M, Bhatnagar, P. K., Nelson, M. A., Wall, J. H., and Marton, L. J.,** Liquid-chromatographic determination of tobramycin in serum with spectrophotometric detection, *Clin. Chem.,* 29, 672, 1983.

707. **Caturla, M. C., Cusido, E., and Westerlund, D.,** High-performance liquid chromatograpic method for the determination of aminoglycosides based on automated pre-column derivatization with phthalaldehyde, *J. Chromatogr.,* 593, 69, 1992.

708. **Fischer, D. H. and Bourque, A. J.,** Quantitation of amphetamine in urine: solid phase extraction, polymeric reagent derivatization and reversed-phase high-performance liquid chromatography, *J. Chromatogr.,* 614, 142, 1993.

709. **Boison, J. O., Korsrud, G. O., MacNeil, J. D., Keng, L., and Papich, M.,** Determination of penicillin G in bovine plasma by high-performance liquid chromatography after pre-column derivatization, *J. Chromatogr.,* 576, 315, 1992.

710. **Lai, F. and Sheehan, T.,** Enhancement of detection sensitivity and cleanup selectivity for tobramycin through pre-column derivatization, *J. Chromatogr.,* 173, 1992.

711. **Virtanen, V. and Lajunen, L. H. J.,** Determinations of clodronate in aqueous solutions by HPLC using post-column derivatization, *Talanta,* 40, 661, 1993.

712. **Reddinguis, R. J., De Jong, G. J., Brinkmann, U. A. T., and Frei, R. W.,** Simple extraction detector for liquid-chromatographic determination of secoverine in biological samples, *J. Chromatogr.,* 205, 77, 1981.

713. **Baba, W. I., Tudhope, G. R., and Wilson, G. M.,** Triamterene, a new diuretic drug. Studies in normal men and in adrenolectomised rats, *Br. Med. J.,* ii, 756, 1962.

714. **Brodie, R. R., Chasseaud, L. F., Taylor, T., and Walmsley, L. M.,** Determination of the diuretic triamterene in the plasma and urine of humans by high-performance liquid chromatography, *J. Chromatogr.,* 164, 527, 1979.

715. **Dave, K. J., Stobaugh, J. F., Rossi, T. M., and Riley, C. M.,** Reversed-phase liquid chromatography of the opioid peptides. 3. Development of a micro-analytical system for opioid peptides involving microbore liquid chromatography post-column derivatization and laser-induced fluorescence detection, *J. Pharm. Biomed. Anal.,* 10, 965, 1992.

716. **Rabel, S. R., Stobaugh, J. F., Heinig, R., and Bostick, J. M.,** Improvements in detector sensitivity for the determination of ivermectin in plasma using chromatographic techniques and laser-induced fluorescence detection with automated derivatization, *J. Chromatogr.,* 617, 79, 1993.

717. **Mascher, H., Kikuta, C., Metz, R., and Vergin, H.,** New, high-sensitivity high-performance liquid-chromatographic method for the determination of acyclovir in human plasma, using fluorimetric detection, *J. Chromatogr.,* 583, 122, 1992.

718. **Reeuwijk, H. J. E. M., Tjaden, U. R., and Van der Greef, J.,** Simultaneous determination of furosemide and amiloride in plasma using high-performance liquid chromatography with fluorescence detecion, *J. Chromatogr.,* 575, 269, 1992.

719. **Brooks, M. A., Strojny, N., Hackman, M. R., and De Silva, J. A. F.,** Determination of bromolasalocid in plasma by high-performance liquid chromatography with fluorimetric detection, *J. Chromatogr.,* 229, 167, 1982.

720. **Hippmann, D. and Takacs, F.,** Determination of celiprolol in biological material by h.p.l.c. with fluorescence detection, *Arzneim.-Forsch.,* 33, 8, 1983.

721. **Moncrieff, J.,** Extractionless determination of diclofenac sodium in serum using reversed-phase high-performance liquid chromatography with fluorimetric detection, *J. Chromatogr.,* 577, 185, 1992.

722. **Bozkurt, A., Basci, N. E., Isimer, A., and Kayaalp, S. O.,** Determination of debrisoquine and 4-hydroxydebrisoquine in urine by high-performance liquid chromatography with fluorescence detection after solid-phase extraction, *J. Pharm. Biomed. Anal.,* 11, 745, 1993.

723. **Bykadi, G., Flora, K. P., Cradock, J. C., and Poochikian, G. K.,** Determination of ellipticine in biological samples by high performance liquid chromatography, *J. Chromatogr.,* 231, 137, 1982.

724. **Stiff, D. D., Schwinghammer, T. L., and Corey, S. E.,** High-performance liquid-chromatographic analysis of etoposide in plasma using fluorescence detection, *J. Liq. Chromatogr.,* 15, 863, 1992.

725. **Brooks, M. A. and Dixon, R.,** Determination of extramustine and its 17-keto-metabolite in plasma by high-performance liquid chromatography, *J. Chromatogr.,* 182, 387, 1980.

726. **De Jong, J. W., Hegge, J. A. J., Harmsen, E., and De Tombe, P. P.,** Fluorimetric liquid-chromatographic assay of the antiarrhythmic agent flecainide in blood plasma, *J. Chromatogr.,* 229, 498, 1982.

727. **Rapaka, S., Roth, J., Vishwanathan, C., Goehl, T. J., Prasad, V. K., and Cabana, B. E.,** Improved method for analysis of furosemide in plasma by high-performance liquid chromatography, *J. Chromatogr.,* 227, 463, 1982.

728. **Adams, W. J., Skinner, G. S., Bombardt, P. A., Courtney, M., and Brewer, J. E.,** Determination of glyburide in human serum by liquid chromatography with fluorescence detection, *Analyt. Chem.,* 54, 1287, 1982.

729. **Oosterhuis, B., Van den Berg, M., and Van Boxtel, C. J.,** Sensitive high-performance liquid-chromatographic method for the determination of labetolol in human plasma using fluorimetric detection, *J. Chromatogr.,* 226, 259, 1981.

730. **De Jong, J., Vermorken, J. B., and Van der Vijgh, W. J. F.,** Analysis and pharmacokinetics of a new prodrug PN-1-leucyldoxorubicin and its metabolites in plasma using HPLC with fluorescence detection, *J. Pharm. Biomed. Anal.,* 10, 309, 1992.

731. **Egan, C. M., Jones, C. R., and McCluskey, M.,** Method for the determination of melphalan in biological samples by high-performance liquid chromatography with fluorescence detection, *J. Chromatogr.,* 224, 338, 1981.

732. **Frost, T.,** Determination of meptazinolin plasma by high-performance liquid chromatography with fluorescence detection, *Analyst,* 106, 999, 1981.

733. **Costa, P., Bressolle, F., Sarrazin, B., Mosser, J., and Galtier, M.,** Pharmacokinetics of moxisylyte in healthy volunteers after intravenous and intracavernous administration, *J. Pharm. Sci.,* 82, 729, 1993.

734. **Duchene, P., Le Dily, J., Bromet-Petit, M., Mosser, J., and Feniou, C.,** High-performance liquid chromatographic assay of the metabolites of thymoxamine, *J. Chromatogr.,* 424, 205, 1988.

735. **Woestenborghs, R., Embrechts, L., and Heykants, J.,** HPLC-fluorescence method for the determination of the new β_2-adrenoceptor blocking agent nebivolol in human plasma, in *Bioanalysis of Drugs and Metabolites, Especially Anti-Inflammatory and Cardiovascular,* Reid, E., Robinson, J. R., and Wilson, I. D., Eds., Plenum Press, New York, 1988, 215.

736. **Timm, U. and Wiedekamm, E.,** Determination of pyrimethamine in human plasma after administration of Fansidar or Fansidar-mefloquine by means of high-performance liquid chromatography with fluorescence detection, *J. Chromatogr.,* 230, 107, 1982.

737. **Oddie, C. J., Jackman, J. P., and Bobik, A.,** Determination of prenalterol in plasma by high-performance liquid chromatography with fluorescence detection, *J. Chromatogr.,* 231, 473, 1982.

738. **Lombardi, F., Ardemagni, R., Cozani, V., and Visconti, M.,** High-performance liquid chromatographic determination of rufloxacin and its main active metabolite in biological fluids, *J. Chromatogr.,* 576, 129, 1992.

739. **McCarthy, P.T., Atwal, S., Sykes, A. P., and Ayres, J. G.,** Measurement of terbutaline and salbutamol in plasma by high-performance liquid chromatography with fluorescence detection, *Biomed. Chromatogr.,* 7, 25, 1993.

740. **Colthup, P. V., Young, G. C., and Felgate, C. C.,** Determination of salmeterol in rat and dog plasma by high-performance liquid chromatography with fluorescence detection, *J. Pharm. Sci.,* 82, 323, 1993.

741. **Watson, E. and Kapur, P. A.,** High-performance liquid-chromatographic determination of verapamil in plasma by fluorescence detection, *J. Pharm. Sci.,* 70, 800, 1981.

742. **Rosseel, M. T. and Lefebvre, R. A.,** Sensitive determination of cinnarizine in human plasma by high-performance liquid chromatography and fluorescence detection, *Chromatographia,* 36, 356, 1993.

743. **Fois, R. A. and Ashley, J. J.,** Synthesis of a fluorescent derivative of cyclosporin A for high-performance liquid chromatography analysis, *J. Pharm. Sci.,* 80, 363, 1990.

744. **Niwa, T., Fujita, K., Goto, J., and Nambara, T.,** Separation of bile acid *N*-acetylglucosaminidines by high-performance liquid chromatography with pre-column fluorescence labelling, *Analyt. Sci.,* 8, 659, 1992.

745. **Redalieu, E., Coleman, J. M., Chan, K., Seaman, J., Degen, P. H., Flesch, G., Brox, A., and Batiste, G.,** Urinary excretion of aminohydroxypropylidene biphosphonate in cancer patients after single intravenous infusions, *J. Pharm. Sci.,* 82, 665, 1993.

746. **Nakashima, K., Suetsugu, K., Yoshida, K., Akiyama, S., Uzu, S., and Imai, K.,** High-performance liquid-chromatography with chemiluminescence detection of methamphetamine and its related compounds using 4-(*NN*-dimethylaminosulfonyl)-7-fluoro-2,1,3-benzoxadiazole, *Biomed. Chromatogr.,* 6, 149, 1992.

747. **Rhys Williams, A. T., Winfield, S. A., and Belloli, R. C.,** Dansyl derivatisation of anabolic agents with high-performance liquid chromatographic separation and fluorescence detection, *J. Chromatogr.,* 240, 224, 1982.

748. **Nagaoka, H., Nohta, H., Saito, M., and Ohkura, Y.,** Determination of 2′,3′-dideoxyinosine and 2′,3′-dideoxyadenosinr in rat plasma by high-performance liquid chromatography with pre-column fluorescence derivatization, *Chem. Pharmacol. Bull.,* 40, 2202, 1992.

749. **Ohta, M., Iwasaki, M., Kai, M., and Ohkura, Y.,** Determination of a biguanide, metformin, by high-performance liquid chromatography with pre-column fluorescence derivatization, *Analyt. Sci.,* 9, 217, 1993.

750. **Suckow, R. F., Zhang, M. F., and Cooper, T. B.,** Sensitive and selective liquid-chromatographic assay of fluoxetine and norfluoxetine in plasma with fluorescence detection after pre-column derivatization, *Clin. Chem.,* 38, 1756, 1992.

751. **Walker, S. E. and Coates, P. E.,** High-performance liquid-chromatographic method for determination of gentamicin in biological fluids, *J. Chromatogr.,* 223, 131, 1981.

752. **Uzu, S., Imai, K., Nakashima, K., and Akiyama, S.,** Determination of medroxyprogesterone acetate in serum by HPLC with peroxyoxalate chemiluminescence detection using a fluorogenic reagent 4-(*N,N*-dimethylaminosulfonyl)-7-hydrazino-2,1,3-benzoxadiazole, *J. Pharm. Biomed. Anal.,* 10, 979, 1992.

753. **Warren, D. J. and Sloerdal, L.,** Sensitive high-performance liquid-chromatographic method for the determination of mercaptopurine in plasma using pre-column derivatization and fluorescence detection, *Ther. Drug Monitor.,* 15, 25–30, 1993.

754. **Nishitani, A., Kanda, S., and Mai, K.,** Sensitive determination method for mexiletine derivatized with dansyl chloride in rat plasma utilizing a HPLC peroxyoxalate chemiluminescence detection system, *Biomed. Chromatogr.,* 6, 124, 1992.

755. **Mason, W. D. and Amick, E. N.,** High-pressure liquid-chromatographic analysis of phenylpropanolamine in human plasma following derivatisation with phthalaldehyde, *J. Pharm. Sci.,* 70, 707, 1981.

756. **Morris, R. G., Sallustio, B. C., Saccoia, N. C., Mangas, S., Fergusson, L. K., and Kassapidis, C.,** Application of an improved HPLC perhexiline assay to human plasma specimens, *J. Liq. Chromatogr.,* 15, 3219, 1992.

757. **Tod, M., Biarez, O., Nicolas, P., and Petijean, O.,** Sensitive determination of josamycin and rokitamycin in plasma by high-performance liquid chromatography with fluorescence detection, *J. Chromatogr.,* 575, 171, 1992.

758. **Sedman, A. J. and Gal, J.,** Pre-column derivatisation with fluorescamine and high-performance liquid-chromatographic analysis of drugs. Application to tocainide, *J. Chromatogr.,* 232, 315, 1982.

759. **Roy, I. M., Jefferies, T. M., Threadgill, M., and Dewar, G. H.,** Analysis of cocaine, benzoylecgonine, ecgonine methyl ester, ethylcocaine and norcocaine in human urine using HPLC with post-column ion-pair extraction and fluorescence detction, *J. Pharm. Biomed. Anal.,* 10, 943, 1992.

760. **Sar, F., Leroy, P., and Nicolas, A.,** Determination of amikacin in dog plasma by reversed-phase ion-pairing liquid chromatography with post-column derivatization, *Analyt. Lett.,* 25, 1235, 1992.

761. **Halvax, J. J., Wiese, G., Van Bennekom, W. P., and Bult, A.,** Online phase-transfer-catalysed dansylation of phenolic compounds followed by normal-phase liquid chromatography with fluorescence detection, *J. Pharm. Biomed. Anal.,* 10, 335, 1992.

762. **Zhou, F.-X., Krull, I. S., and Feibush, B.,** 9-Fluoreneacetyl-tagged solid-phase reagent for derivatization in direct plasma injection, *J. Chromatogr.,* 609, 103, 1992.

763. **Mascher, H. and Kikuta, C.,** High-performance liquid chromatographic determination of total thiamine in human plasma for oral bioavailability studies, *J. Pharm. Sci.,* 82, 56, 1993.

764. **Ishida, J., Sonezaki, S., Yamaguchi, M., and Yoshitake, T.,** Determination of dexamethasone in plasma by high-performance liquid chromatography with chemiluminescence detection, *Analyt. Sci.,* 9, 319, 1993.

765. **Supko, J. G. and Malspeis, L.,** Liquid-chromatographic analysis of 9-aminocamptothecin in plasma monitored by fluorescence induced upon post-column acidification, *J. Liq. Chromatogr.,* 15, 3261, 1992.

766. **Kubo, H., Umiguchi, Y., and Kinoshita, T.,** Fluorimetric determination of indomethacin in serum by high-performance liquid chromatography with inline alkaline hydrolysis, *Chromatographia,* 33, 321, 1992.

767. **Berrueta, L. A., Gallo, B., and Vicente, F.,** Analysis of oxazepam in urine using soild-phase extraction and high-performance liquid chromatography with fluorescence detection by post-column derivatization, *J. Chromatogr.,* 616, 344, 1993.

768. **Uihlein, M. and Schwab, E.,** A novel reactor for photochemical post-column derivatisation in HPLC, *Chromatographia,* 15, 140, 1982.

769. **Beck, O., Seideman, P., Wennberg, M., and Peterson, C.,** Trace analysis of methotrexate and 7-hydroxymethotrexate in human plasma and urine by a novel high-performance liquid-chromatographic method, *Ther. Drug Monitor.,* 13, 528, 1991.

770. **Urmos, I., Benko, S. M., and Klebovich, I.,** Simple and rapid determination of clomiphene cis and trans isomers in human plasma by high-performance liquid chromatography using online post-column photochemical derivatization and fluorescence detection, *J. Chromatogr.,* 617, 168, 1993.

771. **Kissinger, P. T.,** Trace-organic analysis by reverse-phase HPLC with amperometric detection, in *Blood, Drugs, and Other Analytical Challenges,* Reid, E., Ed., Ellis Horwood, Chichester, 1978, 213.

772. **Hamilton, M. and Kissinger, P. T.,** Determination of acetaminophen metabolites in urine by liquid chromatography-electrochemistry, *Analyt. Biochem.,* 125, 143, 1982.

773. **Melendez, V., Peggins, J. O., Brewer, T. G., and Theoharides, A. D.,** Determination of the antimalarial arteether and its deethylated metabolite dihydroartemisinin in plasma by high-performance liquid chromatography with reductive electrochemical detection, *J. Pharm. Sci.,* 80, 132, 1990.

774. **Karp, S., Helt, C. S., and Soujari, N. H.,** Solid-state post-column reactor for the electrochemical detection of ascorbic and dehydroascorbic acids in high-performance liquid chromatography, *Microchem. J.,* 47, 157, 1993.

775. **Betto, P., Meneguz, A., Ricciarello, G., and Pichini, S.,** Simultaneous high-performance liquid-chromatographic analysis of buspirone and its metabolite 1-(2-pyrimidinyl)piperazine in plasma using electrochemical detection, *J. Chromatogr.,* 575, 117, 1992.

776. **Perrett, D. and Drury, P. L.,** Determination of captopril in physiological fluids by use of high-performance liquid chromatography with electrochemical detection, *J. Liq. Chromatogr.,* 5, 97, 1982.

777. **Nissinen, E. and Taskinen, J.,** Simultaneous determination of carbidopa. levodopa and 3,4-dihydroxyphenylacetic acid using high-performance liquid chromatography with electrochemical detection, J. *Chromatogr.,* 231, 459, 1982.

778. **McKay, G., Hall, K., Cooper, J. K. Hawes, E. M., and Midha, K. K.,** Gas-chromatographic-mass spectrometric procedure for quantitation of chlorpromazine in plasma and its comparison with a new high-performance liquid-chromatographic assay with electrochemical detection, *J. Chromatogr.,* 232, 275, 1982.

779. **Digua, K., Kauffmann, J.-M., Ghanem, G., and Patriarche, G. J.,** Determination of cisplatin in human plasma by HPLC with a glassy carbon-based wall-jet amperometric detector, *J. Liq. Chromatogr.,* 15, 3295, 1992.

780. **Bouquet, S., Guyon, S., Chapelle, G., Perault, M. C., and Barthes, D.,** Reductive electochemical detection for artetherSensitive determination in plasma of imipramine and desipramine by high-performance liquid chromatography using electrochemical determination, *J. Liq. Chromatogr.,* 15, 1993, 1992.

781. **Suckow, R. F., and Cooper, T. B.,** Simultaneous determination of imipramine, desimipramine and their 2-hydroxy metabolites in plasma by ion-pair reversed-phase high-performance liquid chromatography with amperometric detection, *J. Pharm. Sci.,* 70, 257, 1981.

782. **Ruo, T. L., Wang, Z., Dordal, M. S., and Atkinson, A. J.,** Assay of inulin in biological fluids by high-performance liquid chromatography with pulsed amperometric detection, *Clin. Chim. Acta,* 204, 217, 1992.

783. **Jagota, N. K. and Stewart, J. T.,** Determination of mefenamic acid in human serum using solid-phase extraction and high-performance liquid chromatography-electrochemical detection, *LC-GC* 10, 688, 1992.

784. **Brown, L. W., Hundt, H. K., and Swart, K. J.,** Automated high-performance liquid-chromatographic method for the determination of mianserin in plasma using electrochemical detection, *J. Chromatogr.,* 582, 268, 1992.

785. **Tjaden, U. R., Langenberg, J. P., Ensing, K., Van Bennekom, W. P., De Bruijn, E. A., and Van Oosterom, A. T.,** Determination of mitomycin in plasma, serum and urine by high-performance liquid chromatography with ultra-violet and electrochemical detection, *J. Chromatogr.,* 232, 355, 1982.

786. **Hanisch, W. and Meyer, L. V.,** Determination of the heroin metabolite 6-monoacetylmorphine in urine by high-performance liquid chromatography with electrochemical detection, *J. Analyt. Toxicol.,* 17, 48, 1993.

787. **Vandenberghe, H., Maclead, S. M., Chinyanga, H., and Soldin, S. J.,** Analysis of morphine in serum by high-performance liquid chromatography with amperometric detection, *Ther. Drug Monitor.,* 4, 307, 1982.

788. **Reid, R. W., Deakin, A., and Leehey, D. J.,** Measurement of naloxone in plasma using high-performance liquid chromatography with electrochemical detection, *J. Chromatogr.,* 614, 117, 1993.

789. **Whelpton, R., Fernandes, K., Wilkinson, K. A., and Goldhill, D. R.,** Determination of paracetamol in blood and plasma using high-performance liquid chromatography with dual electrode coulometric quantification in the redox mode, *Biomed. Chromatogr.,* 7, 90, 1993.

790. **Curry, S. H., Brown, E. A., Hu, O. Y. P., and Perrin, J. H.,** Liquid-chromatographic assay of phenothiazine, thioxanthene and butyrophenone neuroleptics and antihistamines in blood and plasma with conventional and radial-compression columns and ultra-violet and electrochemical detection, *J. Chromatogr.,* 231, 361, 1982.

791. **Fox, A. R. and McLoughlin, D. A.,** Rapid, sensitive high-performance liquid-chromatographic method for the quantification of promethazine in human serum with electrochemical detection, *J. Chromatogr.,* 631, 255, 1993.

792. **Endoh, Y. S., Yoshimura, H., Sasaki, N., Ishihara, Y., Sasaki, H., Nakamura, S., Inoue, Y., and Nishikawa, M.,** High-performance liquid-chromatographic determination of pamaquine, primaquine and carboxyprimaquine in calf plasma using electrochemical detection, *J. Chromatogr.,* 579, 123, 1992.

793. **Aravagiri, M., Marder, S. R., Van Putten, T., and Midha, K. K.,** Determination of risperidone in plasma by high-performance liquid chromatography with electrochemical detection: application to therapeutic drug monitoring in schizophrenic patients, *J. Pharm. Sci.,* 82, 447, 1993.

794. **Tamisier-Karolak, L., Delhotal-Landes, B., Jolliet-Riant, P., Milliez, J., Jannet, D., Barre, J., and Flouvat, B.,** Plasma assay of salbutamol by means of high-performance liquid chromatography with amperometric determination using a loop column for injection of plasma extracts. Application to the evaluation of subcutaneous administration of salbutamol, *Ther. Drug Monitor.,* 14, 243, 1992.

795. **Sagar, K. A., Kelly, M. T., and Smyth, M. R.,** Analysis of salbutamol in human plasma by high-performance liquid chromatography with electrochemical detection using a micro-electrochemical flow cell, *Electroanalysis,* 4, 481, 1992.

796. **Sagar, K. A., Kelly, M. T., and Smyth, M. R.,** Simultaneous determination of salbutamol and terbutaline at overdose levels in human plasma by high-performance liquid chromatography with electrochemical detection, *Biomed. Chromatogr.,* 7, 29, 1993.

797. **Ramos, F., Conceiaco Castilho, M., Noronha da Silveira, M. I., Prates, J. A. M., and Dias Correia, J. H. R.,** Determination of salbutamol in rats at low concentrations using liquid chromatography with electrochemical detection, *Analyt. Chim. Acta,* 275, 279, 1993.

798. **Oosterhuis, B. and Van Boxtel, C. J.,** Determination of salbutamol in human plasma with bimodal high-performance liquid chromatography and a rotated amperometric detector, *J. Chromatogr.,* 232, 327, 1982.

799. **Andrew, P. D., Birch, H. L., and Phillpot, D. A.,** Determination of sumatriptan succinate in plasma and urine by high-performance liquid chromatography with electrochemical detection, *J. Pharm. Sci.,* 82, 73, 1993.

800. **Sagar, K. A., Kelly, M. T., and Smyth, M. R.,** Analysis of terbutaline in human plasma by high-performance liquid chromatography with electrochemical detection using a micro-electrochemical flow cell, *J. Chromatogr.,* 577, 109, 1992.

801. **Nordholm, L. and Dalgaard, L.,** Assay of trimethoprim in plasma and urine by high-performance liquid chromatography using electrochemical detection, *J. Chromatogr.,* 233, 426, 1982.

802. **Augustijns, P. and Verbeke, N.,** Microassay method for the determination of theophylline in biological samples using HPLC with electrochemical detection, *J. Liq. Chromatogr.,* 15, 1303, 1992.

803. **Murata, H., Okabe, K., Harada, K., Suzuki, M., Inagaki, K., Nagano, H., Akita, T., Yoshida, S., Matsuda, M., and Ishigure, H.,** Quantification of pentazocine in human plasma by HPLC with electrochemical detection, *J. Liq. Chromatogr.,* 15, 3247, 1992.

804. **Jones, P. R. and Yank, S. K.,** A liquid chromatograph/mass spectrometer interface, *Analyt. Chem.,* 47, 1000, 1975.

805. **Carroll, D. I., Dzidic, I., Stillwell, R. N., Haegele, K. D., and Horning, E. V.,** Atmospheric pressure ionisation mass spectrometry: corona discharge ion source for use in liquid chromatograph-mass spectrometer-computer analytical system, *Analyt. Chem.,* 47, 2369, 1975.

806. **Blakley, C. R. McAdams, M. J., and Vestal, M. L.,** Crossed-beam liquid chromatograph-mass spectrometer combination, *J. Chromatogr.,* 158, 261, 1978.

807. **Arpino, P., Baldwin, M. A., and McLafferty, F. W.,** Liquid chromatography-mass spectrometry. II. Continuous monitoring, *Biomed. Mass Spectomet.,* 1, 80, 1974.

808. **Scott, R. P. W., Scott, C. G., Munroe, M., and Hess, J.,** Interface for on-line chromatography-mass spectroscopy analysis, *J. Chromatogr.,* 99, 395, 1974.

809. **McFadden, W. H., Schwartz, H. L., and Evans, S.,** Direct analysis of liquid chromatographic effluents, *J. Chromatogr.,* 122, 389, 1976.

810. **Eckers, C., Skrabalak, D. S., and Henion, J.,** On-line direct liquid-introduction interface for micro-liquid chromatography-mass spectrometry: application to drug analysis, *Clin. Chem.,* 28, 1882, 1982.

811. **Martin, L. E., Oxford, J., and Tanner, R. J. N.,** Use of on-line high-performance liquid chromatography-mass spectrometry for identification of ranitidine and its metabolites in urine, *Xenobiotica,* 11, 831, 1981.

812. **Henion, J. D., Thomson, B. A., and Dawson, P. H.,** Determination of sulpha drugs in biological fluids by liquid chromatography-mass spectrometry mass spectrometry, *Analyt. Chem.,* 54, 451, 1982.

813. **Henion, J. D. and Haylin, G. A.,** Applications of a new commercially available LC/MS interface to equine drug analyzes, *Spectra 2000,* 8, 53, 1980.

814. **Pan, H.-T., Kumari, P., de Silva, J. A. F., and Lin, C.-C.,** Determination of ceftibuten in sputum by column-switching high-performance liquid chromatography on-line with thermospray mass spectrometry, *J. Pharm. Sci.,* 82, 52–55, 1993.

815. **Kaye, B., Clark, M. H. W., Cussans, N. J., Macrae, P. V., and Stopher, D. A.,** Sensitive determination of albanoquil in blood by high-performance liquid chromatography-atmospheric-pressure-ionization mass spectrometry, *Biol. Mass Spectromet.,* 21, 585, 1992.

816. **Verweij, A. M. A., Lipman, P. J. L., and Zweipfenning, P. G. M.,** Quantitative liquid chromatography, thermospray-tandem mass spectrometry (LC-TSP-MS-MS) analysis of some thermolabile benzodiazepines in whole-blood, *Forensic Sci. Int.,* 54, 67, 1992.

817. **Lindberg, C., Blomqvist, A., and Paulson, J.,** Determination of (22R,S)-budesonide in human plasma by automated liquid chromatography-thermospray mass spectrometry, *Biol. Mass Spectromet.,* 21, 525, 1992.

818. **Pichini, S., Altieri, I., Bacosi, A., Di Carlo, S., Zuccaro, P., Iannetti, P., and Pacifici, R.,** High-performance liquid-chromatographic-mass-spectrometric assay of busulphan in serum and cerebrospinal fluid, *J. Chromatogr.,* 581, 143, 1992.

819. **Sato, K., Kobayashi, K., Moore, C. M., Mizuno, Y., and Katsumata, Y.,** Semi-quantitative analysis of cefaclor in human serum by capillary high-performance liquid chromatography-fast-atom bombardment mass spectrometry, *Forensic Sci. Int.,* 59, 71, 1993.

820. **Abian, J., Stone, A., Morrow, M. G., Creer, M. H., Fink, L. M., and Lay, J. O.,** Thermospray high-performance liquid chromatography-mass-spectrometric determination of cyclosporins, *Rapid Commun. Mass Spectromet.,* 6, 684, 1992.

821. **Jajoo, H. K., Bennett, S. M., and Kornhauser, D. M.,** Thermospray liquid-chromatographic-mass-spectrometric analysis of anti-AIDS nucleosides: quantification of 2′,3′-dideoxycytidine in plasma samples, *J. Chromatogr.,* 577, 299, 1992.

822. **Gilbert, J. D., Olah, T. V., Barrish, A., and Greber, T. F.,** Determination of L 654066, a new 5α-reductase inhibitor in plasma by liquid chromatography-atmospheric pressure chemical-ionization mass spectrometry, *Biol. Mass Spectromet.,* 21, 341, 1992.

823. **Christensen, R. G. and Malone, W.,** Determination of oltipraz in serum by high-performance liquid chromatography with optical absorbance and mass-spectrometric detection, *J. Chromatogr.,* 584, 207, 1992.

824. **Van Bakergem, E., Van der Hoeven, R. A. M., Niessen, W. M. A., Tjaden, U. R., and Van der Greef, J.,** Online continuous-flow dialysis thermospray tandem mass spectrometry for quantitative screening of drugs in plasma: rogletimide, *J. Chromatogr.,* 598, 189, 1992.

825. **Auriola, S. O. K., Lepisto, A.-M., Naaranlahti, T., and Lapinjoki, S. P.,** Determination of taxol by high-performance liquid chromatography-thermospray mass spectrometry, *J. Chromatogr.,* 594, 153, 1992.

826. **Avery, M. J., Mitchell, D. Y., Falkner, F. C., and Fouda, H. G.,** Simultaneous determination of tenidap and its stable isotope analog in serum by high-performance liquid chromatography-atmospheric pressure chemical-ionization tandem mass spectrometry, *Biol. Mass Spectromet.,* 21, 353, 1992.

827. **Ohla, T. V., Gilbert, J. D., and Barrish, A.,** Determination of the μ-adrenergic blocker timolol in plasma by liquid chromatography-atmospheric pressure chemical-ionization mass spectrometry, *J. Pharm. Biomed. Anal.,* 11, 157, 1993.

828. **Hewitt, S. A., Blanchflower, W. J., McCaughey, W. J., Elliott, C. T., and Kennedy, D. G.,** Liquid chromatography-thermospray mass-spectrometric assay for trenbolone in bovine bile and faeces, *J. Chromatogr.,* 639, 185, 1993.

829. **Nachilobe, P., Boison, J. O., Cassidy, R. M., and Fesser, A. C. E.,** Determination of trimethoprim in bovine serum by high-performance liquid chromatography with confirmation by thermospray liquid chromatography-mass spectrometry, *J. Chromatogr.,* 616, 243, 1993.

830. **Spraul, M., Hofmann, M., Dvortsak, P., Nicholson, J. K., and Wilson, I. D.,** Liquid chromatography coupled with high-field proton NMR for profiling human urine for endogenous compounds and drug metabolites, *J. Pharm. Biomed. Anal.,* 10, 601, 1992.

831. **Bales, J. R., Higham, D. P., Howe, I., Nicholson, J. K., and Sadler, P. J.,** Use of high-resolution proton nuclear magnetic resonance spectroscopy for rapid multi-component analysis of urine, *Clin. Chem.,* 30, 426, 1984.

832. **Spurway, T. D., Gartland, K. P. R., Warrander, A., Pickford, R., Nicholson, J. K., and Wilson, I. D.,** Proton nuclear magnetic resonance of urine and bile from paracetamol dosed rats, *J. Pharm. Biomed. Anal.,* 8, 969, 1990.

833. **Spraul, M., Hofmann, M., Dvortsak, F., Nicholson, J. K., and Wilson, I. D.,** High-performance liquid chromatography coupled to high-field proton nuclear magnetic resonance spectroscopy: application to the urinary metabolites of ibuprofen, *Analyt. Chem.,* 65, 327, 1993.

834. **Spraul, M., Hofmann, M., Lindon, J. C., Nicholson, J. K., and Wilson, I. D.,** Liquid chromatography coupled with high-field proton nuclear magnetic resonance spectroscopy: current status and future prospects, *Analyt. Proc.,* 30, 390, 1993.

835. **Wilson, I. D., Nicholson, J. K., Hofmann, M., Spraul, M., and Lindon, J. C.,** Investigation of the human metabolism of antipyrine using coupled liquid chromatography and nuclear magnetic resonance spectroscopy of urine, *J. Chromatogr.,* 617, 324–328, 1993.

836. **Terabe, S., Yashima, T., Tanaka, N., and Araki, M.,** Separation of oxygen isotope benzoic acids by capillary zone electrophoresis based on isotope effects on the dissociation of the carboxyl group, *Analyt. Chem.,* 60, 1673, 1988.

837. **Lee, K.-J., Lee, J. J., and Moon, D. C.,** Determination of tricyclic antidepressants in human plasma by micellar electrokinetic capillary chromatography, *J. Chromatogr.,* 616, 135, 1993.

838. **Lukkari, P., Ennelin, A., Siren, H., and Riekkola, M.-L.,** Effect of temperature, effective capillary length and applied voltage on the migration of none β-blockers in micellar electrokinetic capillary chromatography, *J. Liq. Chromatogr.,* 16, 2069, 1993.

839. **Wätzig, H. and Dette, C.,** Capillary electrophoresis (CE) — a review. Strategies for method development and applications related to pharmaceutical and biological sciences, *Pharmazie,* 49, 83, 1994.

840. **Thormann, W., Lienhard, S., and Wernly, P.,** Strategies for the monitoring of drugs in body fluids by micellar electrokinetic capillary chromatography, *J. Chromatogr.,* 636, 137–148, 1993.

841. **Macka, M., Borak, J., Semenkova, L., Popl, M., and Mikes, V.,** Determination of acyclovir in blood serum and plasma by micellar liquid chromatography with fluorimetric detection, *J. Liq. Chromatogr.,* 16, 2359, 1993.

842. **Okafo, G. N. and Camilleri, P.,** Micellar electrokinetic capillary chromatography of amoxycillin and related molecules, *Analyst,* 117, 1421, 1992.

843. **Lee, K.-J., Heo, G. S., Kim, N. J., and Moon, D. C.,** Analysis of antiepileptic drugs in human plasma using micellar electrokinetic capillary chromatography, *J. Chromatogr.,* 608, 243, 1992.

844. **Bonet Domingo, E., Medina Hernandez, M. J., Ramis Ramos, G., and Garcia Alvarez-Coque, M. C.,** High-performance liquid-chromatographic determination of diuretics in urine by micellar liquid chromatography, *J. Chromatogr.,* 582, 189, 1992.

845. **Arrowood, S. and Hoyt, A. M.,** Determination of cimetidine in urine by capillary zone electrophoresis, *Microchem. J.,* 47, 90, 1993.

846. **Li, S., Fried, K., Wainer, I. W., and Lloyd, D. K.,** Determination of dextromethorphan and dextrorphan in urine by capillary zone electrophoresis: application to the determination of debrisoquin-oxidation metabolic phenotype, *Chromatographia,* 35, 216, 1993.

847. **Liu, Y.-M. and Sheu, S.-J.,** Determination of ephedrine alkaloids by capillary electrophoresis, *J. Chromatogr.,* 600, 370, 1992.

848. **Baillet, A., Pianetti, G. A., Taverna, M., Mahuzier, G., and Baylocq-Ferrier, D.,** Fosfomycin determination in serum by capillary zone electrophoresis with indirect ultra-violet detection, *J. Chromatogr.,* 616, 311, 1993.

849. **Shihabi, Z. K.,** Serum pentobarbital assay by capillary electrophoresis, *J. Liq. Chromatogr.,* 16, 2059, 1993.

850. **Charman, W. N., Humberstone, A. J., and Charman, S. A.,** Analysis of pilocarpine and its degradation products by micellar electrokinetic capillary chromatography, *Pharm. Res.,* 9, 1219, 1992.

851. **Baeyens, W., Weiss, G., Van Der Weken, G., Van Den Bossche, W., and Dewaele, C.,** Analysis of pilocarpine and its trans epimer, isopilocarpine, by capillary electrophoresis, *J. Chromatogr.,* 638, 319, 1993.

852. **Guzman, N. A., Moschera, J., Bailey, C. A., Iqbal, K., and Malick, A. W.,** Assay of protein drug substances present in solution mixtures by fluorescamine derivatization and capillary electrophoresis, *J. Chromatogr.,* 598, 123, 1992.

853. **Garcia, L. L. and Shihabi, Z. K.,** Suramin determination by capillary electrophoresis, *J. Liq. Chromatogr.,* 16, 2049, 1993.

854. **Lukkari, P., Siren, H., Pantsar, M., and Riekkola, M.-L.,** Determination of ten μ-blockers in urine by micellar electrokinetic capillary chromatography, *J. Chromatogr.,* 632, 143, 1993.

855. **Lee, K.-J., Heo, G. S., Kim, N. J., and Moon, D.-C.,** Separation of theophylline and its analogues by micellar electrokinetic chromatography: application to the determination of theophylline in human plasma, *J. Chromatogr.,* 577, 135, 1992.

856. **Tagliaro, F., Dorizzi, R., Ghielmi, S., Poiesi, C., Moretto, S., Archetti, S., and Marigo, M.,** Free-solution capillary electrophoresis of theophylline in serum, *Fresenius' J. Analyt. Chem.,* 343, 168, 1992.

857. **Schmutz, A. and Thormann, W.,** Determination of phenobarbital, ethosuximide and primidone in human serum by micellar electrokinetic capillary chromatography with direct sample injection, *Ther. Drug Monitor.,* 15, 310, 1993.

858. **Parker, C. E., Perkins, J. R., Tomer, K. B., Shida, Y., O'Hara, K., and Kono, M.,** Application of nanoscale packed capillary liquid chromatography (75 μm i.d.) and capillary zone electrophoresis-electrospray ionization mass spectrometry to the analysis of macrolide antibiotics, *J. Am. Soc. Mass Spectromet.,* 3, 563, 1992.

859. **Liu, H. and Wehmeyer, K. R.,** Solid-phase extraction with supercritical-fluid elution as a sample preparation technique for the ultra-trace analysis of flavone in blood plasma, *J. Chromatogr.,* 577, 61, 1992.

860. **Klesper, E., Corwin, A., and Turner, D.,** High pressure gas chromatography above critical temperatures, *J. Org. Chem.,* 27, 700, 1962.

861. **Novotny, M., Springston, S. R., Peaden, P. A., Fjelsted, J. C., and Lee, M. L.,** Capillary supercritical fluid chromatography, *Analyt. Chem.,* 53, 407A, 1981.

862. **Roberts, D. W. and Wilson, I. D.,** Bioanalytical supercritical fluid chromatography, *Analysis for Drugs and Metabolites, Including Anti-Infective Agents,* Reid, E. and Wilson, I. D., Eds., Royal Society of Chemistry, Cambridge, 1990, 257.

863. **Jagota, N. K. and Stewart, J. T.,** Separation of non-steroidal anti-inflammatory agents using supercritical-fluid chromatography, *J. Chromatogr.,* 604, 255, 1992.

864. **Jagota, N. K. and Stewart, J. T.,** Separation of chlordiazepoxide and selected chlordiazepoxide mixtures using capillary SFC, *J. Liq. Chromatogr.,* 16, 291, 1993.

865. **Jagota, N. K. and Stewart, J. T.,** Analysis of diazepam and chlordiazepoxide and their related compounds using supercritical-fluid chromatography, *J. Liq. Chromatogr.,* 15, 2429, 1992.

866. **Jagota, N. K. and Stewart, J. T.,** Supercritical-fluid chromatography of selected oestrogens, *J. Pharm. Biomed. Anal.,* 10, 667, 1992.

867. **Roston, D. A.,** Supercritical-fluid extraction — supercritical-fluid chromatography for analysis of a prostaglandin: HPMC dispersion, *Drug. Dev. Indust. Pharmacol.,* 18, 245, 1992.

868. **Gyllenhaal, O. and Vessman, J.,** Packed-column supercritical fluid chromatography of omeprazole and related compounds. Selection of column support with triethylamine- and methanol-modified carbon dioxide as the mobile phase, *J. Chromatogr.,* 628, 275, 1993.

869. **Biermanns, P., Miller, C., Lyon, V., and Wilson, W.,** Chiral resolution of β-blockers by packed-column supercritical fluid chromatography, *LC-GC,* 11, 744, 1993.

870. **Bargmann-Leyder, N., Siret, L., Tambute, A., and Caude, M.,** Direct and rapid enantiomeric separation of β-blockers by coupling carbon dioxide and supercritical-fluid chromatographic phase ChyRoSine-A, *Spectra 2000,* 171, 27, 1993.

871. **Lee, E. D., Hsu, S.-H., and Henion, J. D.,** Electron-ionization-like mass spectra by capillary supercritical fluid chromatography/charge exchange mass spectrometry, *Analyt. Chem.,* 60, 1990, 1988.

872. **Dalgleish, C. E.,** Optical resolution of aromatic amino acids on paper chromatograms, *J. Chem. Soc.,* 137, 3940, 1952.

873. **Wainer, I. W., Alembic, M. C., and Fisher, L. J.,** The determination of (*R*)- and (*S*)-glutethimide and the corresponding 4-hydroxyglutethimide metabolites in human serum and urine using a Pirkle-type chiral stationary phase, *J. Pharm. Biomed. Anal.,* 5, 735, 1987.

874. **Meese, C. O., Thalheimer, P., and Eichelbaum, M.,** High-performance liquid chromatographic method for the analysis of debrisoquine and its *S*-(+)- and *R*-(−)-hydroxy metabolites in urine, *J. Chromatogr.,* 423, 344, 1987.

875. **Delatour, P., Benoit, E., Caude, M., and Tambute, A.,** Species differences in the generation of the chiral sulfoxide metabolite of albendazole in sheep and rats, *Chirality,* 2, 156, 1990.

876. **Gimenez, E., Farinotti, R., Thuillier, G., Hazebroucq, G., and Wainer, I. W.,** The determination of the enantiomers of mefloquine in plasma and whole blood using a coupled achiral-chiral HPLC system, *J. Chromatogr.,* 529, 339, 1990.

877. **Oliveros, L., Minguillon, C., and Billaud, C.,** Resolution of several racemic 3-hydroxy-1,4-benzodiazepin-2-ones by high-performance liquid chromatography on a chiral silica-bonded stationary phase, *J. Pharm. Biomed. Anal.,* 10, 925, 1992.

878. **Haupt, D., Pettersson, C., and Westlund, D.,** Separation of (*R*)- and (*S*)-naproxen using micellar chromatography and an α₁-acid glycoprotein column: application for chiral monitoring in human liver microsomes by coupled-column chromatography, *J. Biochem. Biophys. Methods,* 25, 273, 1992.

879. **Pirkle, W. H. and Welch, C. J.,** An improved chiral stationary phase for the chromatographic separation of underivatized naproxen enantiomers, *J. Liq. Chromatogr.,* 15, 1947, 1992.

880. **Domenici, E., Bertucci, C., Salvadori, P., Felix, G., Cahagne, I., Motellier, S., and Wainer, I. W.,** Synthesis and chromatographic properties of an HPLC chiral stationary phase based upon human serum albumin, *Chromatographia,* 29, 170, 1990.

881. **Domenici, E., Bertucci, C., Salvadori, P., and Wainer. I. W.,** Use of a human serum albumin-based high-performance liquid chromatography chiral stationary phase for the investigation of protein binding: detection of the allosteric interaction between warfarin and benzodiazepine binding sites, *J. Pharm. Sci.,* 80, 164, 1990.

882. **Debowski, J., Jurczak, J., and Sybilska, D.,** Eine vollständige Racemattrennung durch Elutions-Chromatographie an Cellulose-tri-acetat, *J. Chromatogr.,* 282, 83, 1983.

883. **Armstrong, D. W. and Jin, H. L.,** Liquid chromatographic separation of anomeric forms of saccharides with cyclodextrin bonded phases, *Chirality,* 1, 27, 1989.

884. **He, J., Shibukawa, A., Nakagawa, T., Wada, H., Fujima, H., Imai, E., and Go-oh, Y.,** Direct injection analysis of atenolol enantiomers in plasma using an achiral-chiral coupled column HPLC system, *Chem. Pharmacol. Bull.,* 41, 544, 1993.

885. **Sztruhar, I., Ladanyi, L., Zalavari-Dosa, L., and Beck, I.,** Liquid-chromatographic resolution of the enantiomers of psychotropic drug levomepromazine with β-cyclodextrin-bonded chiral stationary phase, *J. Pharm. Biomed. Anal.,* 11, 263, 1993.

886. **Haginaka, J. and Wakai, J.,** β-Cyclodextrin-diol phase silica materials for direct-injection analysis of drug enantiomers in serum by liquid chromatography, *Analyt. Sci.,* 8, 137, 1992.

887. **Franzelius, C. and Besserer, K.,** Identification and quantitation of intact diastereoisomeric benzodiazepine glucuronides in biological samples by high-performance liquid chromatography, *J. Chromatogr.,* 613, 162, 1993.

888. **Armstrong, D. W., Chen, S., Chang, C., and Chang, S. C.,** New approach for the direct resolution of racemic β-adrenergic blocking agents by HPLC, *J. Liq. Chromatogr.,* 15, 545, 1992.

889. **Naidong, W. and Lee, J. W.,** Development and validation of a high performance liquid-chromatographic method for the quantitation of warfarin enantiomers in humna plasma, *J. Pharm. Biomed. Anal.,* 11, 785, 1993.

890. **Hesse, G. and Hagel, R.,** Complete resolution of a racemic mixture by elution chromatography on cellulose triacetate, *Chromatographia,* 6, 277, 1973.

891. **Wainer, I. W.,** HPLC chiral stationary phases for the stereochemical resolution of enantiomeric compounds, in *Drug Stereochemistry: Analytical Methods and Pharmacology,* 2nd ed., Wainer, I. W., Ed., Marcel Dekker, New York, 1993, 139.

892. **Aboul-Enein, H. J.,** Applications of cellulose-based chiral stationary phases in the resolution of some beta-adrenoceptor antagonists, *Analyt. Lett.,* 26, 271, 1993.

893. **Aboul-Enein, H. Y. and Bakr, S. A.,** Direct chromatographic resolution of racemic cyclohexylaminoglutethimide and its acetylated metabolite using cellulose-based chiral stationary phase, *Chirality,* 3, 204, 1991.

894. **Ohkubo, T., Uno, T., and Suguwara, K.,** Enantiomer separation of dihydropyridine derivatives by liquid chromatography with chiral stationary phase, *Chromatographia,* 33, 287, 1992.

895. **Sakamoto, T., Ohtake, Y., Itoh, M., Tabata, S., Kuriki, T., and Uno, K.,** Determination of felodipine enantiomers using chiral stationary phase liquid chromatography and gas chromatography-mass spectrometry, and the study of their pharmacokinetic profiles in human and dog, *Biomed. Chromatogr.,* 7, 99, 1993.

896. **Aboul-Enein, H. Y. and Bakr, S. A.,** Simple chiral liquid-chromatographic separation of flurbiprofen enantiomers in biological fluids, *J. Liq. Chromatogr.,* 15, 1983, 1992.

897. **Flesch, G., Francotte, E., Hell, F., and Degen, P. H.,** Determination of the R-(−) and S-(+) enantiomers of the monohydroxylated metabolite of oxcarbazepine in human plasma by enantioselective high-performance liquid chromatography, *J. Chromatogr.,* 119, 147, 1992.

898. **Aboul-Enein, H. Y. and Islam, M. R.,** Enantiomeric separation of ketamine hydrochloride in pharmaceutical formulation and human serum by chiral liquid chromatography, *J. Liq. Chromatogr.,* 15, 3285, 1992.

899. **Yamaguchi, M., Yamashita, K., Aoki, I., Tabata, T., Hirai, S., and Yashiki, T.,** Determination of manidipine enantiomers in human serum using chiral chromatography and column-switching liquid chromatography, *J. Chromatogr.,* 575, 123, 1992.

900. **Leloux, M. S.,** Rapid chiral separation of metoprolol in plasma — application to the pharmacokinetics/pharmacodynamics of metoprolol enantiomers in the conscious goat, *Biomed. Chromatogr.,* 6, 99, 1992.

901. **Heinig, R. and Blaschke, G.,** In vivo and in vitro stereoselective metabolism of mianserin in mice, *Arzneim. Forsch.,* 43, 5, 1993.

902. **Fischer, C. J., Schönberger, F., Mück, W., Heuk, K., and Eichelbaum, M.,** Simultaneous assessment of the intravenous and oral disposition of the enantiomers of racemic nimodipine by chiral stationary-phase high-performance liquid chromatography/mass spectroscopy combined with a stable isotope technique, *J. Pharm. Sci.,* 82, 244, 1993.

903. **Herring, V. L. and Johnson, J. A.,** Direct high-performance liquid-chromatographic determination in urine of the enantiomers of propranolol and its major basic metabolite 4-hydroxypropranolol, *J. Chromatogr.,* 612, 215, 1993.

904. **Elsing, B. and Blaschke, G.,** Achiral and chiral high-performance liquid-chromatographic determination of tramadol and its major metabolites in urine after oral administration of racemic tramadol, *J. Chromatogr.,* 612, 223, 1993.

905. **Andersen, C., Balmer, K., and Lagerstrom, P.-O.,** Enantioselective assay of warfarin in blood plasma by liquid chromatography on Chiralcel OC, *J. Chromatogr.,* 615, 159, 1993.

906. **Narayanan, S. R.,** Immobilized proteins as chromatographic supports for chiral resolution, *J. Pharm. Biomed. Anal.,* 10, 251, 1992.

907. **Enquist, M. and Hermansson, J.,** Separation of the enantiomers of β-receptor blocking agents and other cationic drugs using a CHIRAL-AGP column. Binding properties and characterization of immobilized acid glycoproteins, *J. Chromatogr.,* 519, 285, 1990.

908. **Wainer, I. W., Kaliszan, R., and Noctor, T. A. G.,** Biochromatography using immobilized biopolymers: a new approach to the determination of pharmacological properties, *J. Pharm. Pharmacol.,* 45 (Suppl. 1), 367, 1993.

909. **De Vries, J. X. and Schmitz-Kummer, E.,** Direct column liquid-chromatographic enantiomer separation of the coumarin anticoagulants phenprocoumon, warfarin, acenocoumarol and metabolites on an α_1-acid glycoprotein chiral stationary phase, *J. Chromatogr.,* 644, 315, 1993.

910. **Krstulovic, A. M., Fouchet, M. H., Burke, J. T., Gillet, G., and Durand, A.,** Direct enantiomeric separation of betaxolol with applications to the analysis of bulk drug and biological samples, *J. Chromatogr.,* 452, 477, 1988.

911. **Butter, J. J., Kraak, J. C., and Poppe, H.,** Determination of enantiomers of bupivacaine in serum using an online coupled three column liquid-chromatographic system, *J. Pharm. Biomed. Anal.,* 11, 225, 1993.

912. **Evans, J. M., Smith, R. J., and Stemp, G.,** Separation of the enantiomers of some potassium chanel activators using an α_1-acid glycoprotein column, *J. Chromatogr.,* 623, 163, 1992.

913. **Masurel, D. and Wainer, I. W.,** Analytical and preparative high-performance liquid chromatographic separation of the enantiomers of ifosfamide, cyclophosphamide and trofosfamide and their determination in plasma, *J. Chromatogr.,* 490, 133, 1989.

914. **Hermansson, J., Hermansson, I., and Nordin, J.,** Chacterization of Chiral-AGP capillary column coupled to a micro sample-enrichment system with UV and electrospray mass spectrometric detection, *J. Chromatogr.,* 631, 79, 1993.

915. **Takahashi, H., Ogata, H., Shimizu, M., Hashimoto, K., Masuhara, K., Kashiwada, K., and Someya, K.,** Comparative pharmacokinetics of unbound disopyramide enantiomers following oral administration of racemic disopyramide in humans, *J. Pharm. Sci.,* 80, 709, 1990.

916. **Hussein, Z., Chu, S.-Y., and Granneman, G. R.,** Enantioselective determination of DN 2327, a novel non-benzodiazepine anxiolytic, and its active metabolite in human plasma and urine using high-performance liquid chromatography, *J. Chromatogr.,* 613, 113, 1993.

917. **Furlonger, G. F., Fell, A. F., and Kaye, B.,** Design and application of chiral liquid chromatography for drug metabolism studies, *Analyt. Proc.* 30, 94, 1993.

918. **Shibukawa, A., Nagao, M., Terakita, A., He, J., and Nakagawa, T.,** High-performance frontal analysis-high-performance liquid-chromatographic system for the enantioselective determination of unbound fenoprofen concentration in protein binding equilibrium, *J. Liq. Chromatogr.,* 16, 903, 1993.

919. **Oda, Y., Asakawa, N., Yoshida, Y., and Sato, T.,** Direct-injection high-performance liquid-chromatographic analysis of drug enantiomers in plasma with an avidin column coupled online to an ovomucoid column, *J. Pharm. Sci.,* 81, 1227, 1991.

920. **Oda, Y., Mano, N., Asakawa, N., Yoshida, Y., Sato, T., and Nakagawa, T.,** Comparison of avidin and ovomucoid as chiral selectors for the resolution of drug enantiomers by high-performance liquid chromatography, *Analyt. Sci.,* 9, 221, 1993.

921. **Etienne, M.-C., Speziale, N., and Milano, G.,** HPLC of folinic acid diastereoisomers and 5-methyl tetrahydrofolate in plasma, *Clin. Chem.,* 39, 82, 1993.

922. **Nishikata, M., Nakai, A., Fushida, H., Miyake, K., Arita, T., Iseki, K., Miyazaki, K., and Nomura, A.,** Enantioselective pharmacokinetics of homochlorcyclizine. III. Simultaneous determination of (+)- and (−)-homochlorocyclizine in human urine by high-performance liquid chromatography, *J. Chromatogr.,* 612, 239, 1993.

923. **Banks, M. C., Fell, A. F., and McDowall, R. D.,** Assessment of the performance of a new protein-based phase in the chiral liquid chromatography of drugs, *Analyt. Proc.,* 30, 98, 1993.

924. **Fujima, H., Wada, H., Miwa, T., and Haginaka, J.,** Chiral separation of lorazepam on ovomucoid-bonded columns: peak coalescence due to racemization, *J. Liq. Chromatogr.,* 16, 879, 1993.

925. **Schmidt, N., Brune, K., and Geisslinger, G.,** Stereoselective determination of the enantiomers of methadone in plasma using high-performance liquid chromatography, *J. Chromatogr.,* 583, 195, 1992.

926. **Drewe, J. and Kuesters, E.,** High-performance liquid-chromatographic method for the enantiomeric separation of the chiral metabolites of midazolam, *J. Chromatogr.,* 609, 395, 1992.

927. **Andersen, J. V. and Hansen, S. H.,** Simultaneous determination of (*R*)- and (*S*)-naproxen and (*R*)- and (*S*)-6-*O*-desmethylnaproxen by high-performance liquid-chromatography on a Chiral-AGP column, *J. Chromatogr.,* 577, 362, 1992.

928. **Asakura, M., Nagakura, A., Tarui, S., and Matsumura, R.,** Simultaneous determination of the enantiomers of pimobendan and its main metabolite in rat plasma by high-performance liquid chromatography, *J. Chromatogr.,* 614, 135, 1993.

929. **Marchiset-Leca, D. and Leca, F. R.,** Highly-sensitive method for the determination of a new anthracycline: pirarubicin, *Chromatographia,* 35, 435, 1993.

930. **Fujimoto, H., Nishino, I., Ueno, K., and Umeda, T.,** Determination of the enantiomers of a new 1,4-dihydropyridine calcium antagonist in dog plasma by achiral/chiral coupled high-performance liquid chromatography with electrochemical detection, J. *Pharm. Sci.,* 82, 319, 1993.

931. **Haginaka, J., Seyama, C., Yasuda, H., and Takahashi, K.,** Investigation of enantioselectivity and enantiomeric elution order of propranolol and its ester derivatives on an ovomucoid-bonded column, *J. Chromatogr.,* 598, 67, 1992.

932. **Okamoto, Y., Kawashima, M., Aburatani, R., Hatada, K., Nishiyama, T., and Masuda, M.,** Measurement of underivatized propranolol enantiomers in serum using a cellulose-tris(3,5-dimethylphenylcarbamate) high performance liquid chromatographic (HPLC) chiral stationary phase, *Chem. Lett.,* 1237, 1986.

933. **Vandenbosch, C., Massart, D. L., and Lindner, W.,** Evaluation of the enantioselectivity of an ovomucoid and a cellulase chiral stationary phase towards a set of β-blocking agents, *Analyt. Chim. Acta,* 270, 1–12, 1992.

934. **Eap, C. B., Koeb, L., Holsboer-Traschler, E., and Baumann, P.,** Plama levels of trimipramine and metabolites in four patients: determination of the enantiomer concentrations of the hydroxy-metabolites, *Ther. Drug Monitor.,* 14, 380, 1992.

935. **Chu, Y. Q. and Wainer, I. W.,** Determination of the enantiomers of verapamil in serum using coupled achiral-chiral high-performance liquid chromatography, *J. Chromatogr.,* 497, 191, 1989.

936. **Pasteur, L.,** Researches on the molecular asymmetry of natural organic products. Alembic Club Reprints, No 14, reissue edition, F., and S. Livingstone, Edinburgh, 1948.

937. **Josefsson, M., Carlsson, B., and Norlander, B.,** Fast chromatographic separation of (–)-menthyl chloroformate derivatives of some chiral drugs, with special reference to amlodipine, on porous graphitic carbon, *Chromatographia,* 37, 129, 1993.

938. **He, L. and Stewart, T. J.,** High-performance liquid-chromatographic method for the determination of albuterol enantiomers in human serum using solid-phase extraction and chemical derivatization, *Biomed. Chromatogr.,* 6, 291, 1992.

939. **Miller, R. B. and Guertin, Y.,** High-performance liquid-chromatographic assay for the derivatized enantiomers of atenolol in whole blood, *J. Liq. Chromatogr.,* 15, 1289, 1992.

940. **Rondelli, I., Mariotti, F., Acerbi, D., Redenti, E., and Ventura, G. A. P.,** Selective method for plasma quantitation of the stereoisomers of a new aminotetralin by high-performance liquid chromatography with electrochemical detection, *J. Chromatogr.,* 612, 95, 1993.

941. **Zhou, F. X. and Krull, I. S.,** Direct enantiomeric analysis of amphetamine in plasma by simultaneous solid-phase extraction and chiral derivatization, *Chromatographia,* 35, 153, 1993.

942. **Kikura, R., Ishigami, A., and Nakahara, Y.,** Studies on comparison of metabolites in urine between deprenyl and methamphetamine. III. Enantiomeric composition analysis of metabolites in mouse urine by HPLC with GITC chiral reagent, *Jpn. J. Toxicol. Environ. Health,* 38, 136, 1992.

943. **Olsen, L., Bronnum-Hansen, K., Helboe, P., Jorgensen, G. H., and Kryger, S.,** Chiral separation of β-blocking drug substances using derivatization with chiral reagents and normal-phase high-performance liquid chromatography, *J. Chromatogr.,* 636, 231, 1993.

944. **Mayer, S., Spahn-Langguth, H., Gikalov, I., and Mutschler, E.,** Pharmacokinetics of beclobric acid enantiomers and their conjugates after single and multiple oral dosage of racemic beclobrate, *Arzneim. Forsch.,* 43, 40, 1993.

945. **Mayer, S., Mutschler, E., and Spahn-Langguth, H.,** Pharmacokinetic studies with lipid-regulating agent beclobrate: enantiospecific assay for beclobric acid using a new fluorescent chiral coupling component (S-FLOPA), *Chirality,* 3, 35, 1991.

946. **Ludden, T. M., Boyle, D. A., Giesecker, D., Kennedy, G. T., Crawford, M. H., Ludden, L.K., and Clementi, W. A.,** Absolute bioavailability and dose proportionality of betaxolol in normal healthy subjects, *J. Pharm. Sci.,* 77, 779, 1988.

947. **Stagni, G., Davis, P. J., and Ludden, T. M.,** Human pharmacokinetics of betaxolol enantiomers, *J. Pharm. Sci.,* 80, 321, 1990.

948. **Takasaki, W., Asami, M., Muramatsu, S., Hayashi, R., Tanaka, Y., Kawabata, K., and Hoshiyama, K.,** Stereoselective determination of the active metabolites of a new anti-inflammatory agent (CS 670) in human and rat plasma using antibody-mediated extraction and high-performance liquid chromatography, *J. Chromatogr.,* 613, 67, 1993.

949. **Singh, N. N., Jamali, F., Pasutto, F. M., and Coutts, R. T.,** Stereoselective gas chromatographic analysis of etodolac enantiomers in human plasma and urine, *J. Chromatogr.,* 382, 331, 1986.

950. **Brocks, D. R. and Jamali, F.,** Enantioselective pharmacokinetics of etodolac in the rat: tissue distribution, tissue binding, and in vitro metabolism, *J. Pharm. Sci.,* 80, 1058, 1990.

951. **Carlucci, G., Mazzeo, P., and Palumbo, G.,** Indirect stereoselective determination of enantiomers of furprofen in human plasma by high-performance liquid chromatography, *Chromatographia,* 34, 618, 1992.

952. **Wright, M. R., Sattari, S., Brocks, D. R., and Jamali, F.,** Improved high-performance liquid-chromatographic assay method for the enantiomers of ibuprofen, *J. Chromatogr.,* 583, 259, 1992.

953. **Kondo, J., Imaoka, T., Kawasaki, T., Nakanishi, A., and Kawahara, Y.,** Fluorescence derivatization reagent for resolution of carboxylic acid enantiomers by high-performance liquid chromatography, *J. Chromatogr.,* 645, 75, 1993.

954. **Desai, D. M. and Gal, J.,** Reversed-phase high-performance liquid-chromatograpic separation of the stereoisomers of labetalol via derivatization with chiral and non-chiral isothiocyanate reagents, *J. Chromatogr.,* 579, 165, 1992.

955. **Bhatti, M. M. and Foster, R. T.,** Stereospecific high-performance liquid-chromatographic assay of metoprolol, *J. Chromatogr.,* 579, 361, 1992.

956. **Toyo'oka, T., Ishibashi, M., and Terao, T.,** Resolution of carboxylic acid enantiomers by high-performance liquid chromatography with highly sensitive laser-induced fluorescence detection, *J. Chromatogr.,* 625, 357, 1992.

957. **Alessi-Severini, S., Coutts, R. T., Jamali, F., and Pasutto, F. M.,** High-performance liquid-chromatographic analysis of methocarbamol enantiomers in biological fluids, *J. Chromatogr.,* 582, 173, 1992.

958. **Abolfathi, Z., Belanger, P.-M., Gilbert, M., Rouleau, J. R., and Turgeon, J.,** Improved high-performance liquid-chromatographic assay for the stereoselective determination of mexilitine in plasma, *J. Chromatogr.,* 579, 366, 1992.

959. **Lee, E. J. D., Williams, K. M., Graham, G. G., Day, R. O., and Champion, G. D.,** Liquid chromatographic determination and plasma concentration profile of optical isomers of ibuprofen in humans, *J. Pharm. Sci.,* 73, 1542, 1984.

960. **Liang, W. T., Brocks, D. R., and Jamali, F.,** Stereospecific high-performance liquid-chromatographic assay of pirprofen enantiomers in rat plasma and urine, *J. Chromatogr.,* 577, 317, 1992.

961. **Fiset, C., Phillipon, F., Gilbert, M., and Turgeon, J.,** Stereoselective high-performance liquid-chromatographic assay for the determination of sotalol enantiomers in biological fluids, *J. Chromatogr.,* 612, 231, 1993.

962. **Sallustio, B. C., Morris, R. G., and Horowitz, J. D.,** High-performance liquid-chromatographic determination of sotalol in plasma. I. Application to the disposition of sotalol enantiomers in humans, *J. Chromatogr.,* 576, 321, 1992.

963. **Spahn-Langguth, H., Hahn, G., Mutschler, E., Moehrke, W., and Langguth, P.,** Enantiospecific high-performance liquid-chromatographic assay with fluorescence detection for the monoamine oxidase inhibitor tranylcypromine and its applicability in pharmacokinetic studies, *J. Chromatogr.,* 584, 229, 1992.

964. **Aycard, M., Letellier, S., Maupas, B., and Guyon, F.,** Determination of (*R*)- and (*S*)-warfarin in plasma by high-performance liquid chromatography using pre-column derivatization, *J. Liq. Chromatogr.,* 15, 2175, 1992.

965. **Prevot, M., Tod, M., Chalom, J., Nicolas, P., and Petitjean, O.,** Separation of propafenone enantiomers by liquid chromatography with a chiral counter ion, *J. Chromatogr.,* 605, 33, 1992.

966. **Noroski, J., Mayo, D., and Kirschbaum, J. J.,** Liquid-chromatographic resolution of the isomers of tipredane and phenylthioproline using urea-solubilized β-cyclodextrin in the mobile phase, *J. Pharm. Biomed. Anal.,* 10, 447, 1992.

967. **Eto, S., Noda, H., and Noda, A.,** Chiral separation of barbiturates and hydantoins by reversed-phase high-performance liquid chromatography using a 25 or 50 mm short ODS cartridge column via β-cyclodextrin inclusion complexes, *J. Chromatogr.,* 579, 253, 1992.

968. **Eto, S., Noda, H., Minemoto, M., Noda, A., and Mizukami, Y.,** Optical separation of racemic 5-(*p*-hydroxyphenyl)-5-phenylhydantoin by reversed-phase high-performance liquid chromatography by eluents containing β-cyclodextrin, *Chem. Pharmacol. Bull.,* 39, 2742, 1991.

969. **Hsieh, C.-Y. and Huang, J.-D.,** Two-dimensional high-performance liquid-chromatographic method to assay *p*-hydroxyphenylphenylhydantoin enantiomers in biological fluids and stereoselectivity of enzyme induction in phenytoin metabolism, *J. Chromatogr.,* 575, 109, 1992.

970. **Mitchell, P. and Clark, B. J.,** Drug and analogue structural relationships in chiral separations with mobile-phase additives, *Analyt. Proc.,* 30, 101, 1993.

971. **Valtcheva, L., Mohammad, J., Petterson, G., and Hjerten, S.,** Chiral separation of β-blockers by high-performance capillary electrophoresis based on non-immobilised cellulase as enantioselective protein, *J. Chromatogr.,* 638, 263, 1993.

972. **Goodall, D. M., Riley, M. J., Wu, Z., and Wilson, I. D.,** Determination of the ratio of *R*- and *S*-ibuprofen in urine using a solid-phase extraction procedure followed buy high-performance liquid chromatography with dual polarimetric and ultra-violet absorbance detection, *Analyt. Proc.,* 29, 253, 1992.

973. **Soini, H., Riekkola, M.-L., and Novotny, M. V.,** Chiral separations of basic drugs and quantitation of bupivacaine enantiomers in serum by capillary electrophoresis with modified cyclodextrin buffers, *J. Chromatogr.,* 608, 265, 1992.

974. **D'Hulst, A. and Verbeke, N.,** Chiral separation by capillary electrophoresis with oligosaccharides, *J. Chromatogr.,* 608, 275, 1992.

975. **Peterson, T. E.,** Separation of drug stereoisomers by capillary electrophoresis with cyclodextrins, *J. Chromatogr.,* 630, 353, 1993.

976. **Lurie, I. S.,** Micellar electrokinetic capillary chromatography of the enantiomers of amphetamine, methamphetamine and their hydroxyphenethylamine precursors, *J. Chromatogr.,* 605, 269, 1992.

977. **Barker, G. E., Russo, P., and Hartwick, R. A.,** Chiral separation of leucovorin with bovine serum albumin using affinity apillary electrophoresis, *Anal. Chem.,* 64, 3024, 1992.

978. **Altria, K. D., Goodall, D. M., and Rogan, M. M.,** Chiral separation of β-amino alcohols by capillary electrophoresis using cyclodextrins as buffer additives. I. Effect of varying operating parameters, *Chromatographia,* 34, 19, 1992.

979. **Schutzner, W. and Fanali, S.,** Enantiomers resolution in capillary zone electrophoresis by using cyclodextrins, *Electrophoresis,* 13, 687, 1992.

980. **Busch, S., Kraak, J. C., and Poppe, H.,** Chiral separations by complexation with proteins in capillary zone electrophoresis, *J. Chromatogr.,* 635, 119, 1993.

981. **Gareil, P., Gramond, J. P., and Guyon, F.,** Separation and determination of warfarin enantiomers in human plasma samples by capillary zone electrophoresis using a methylated β-cyclodextrin-containing electrolite, *J. Chromatogr.,* 615, 317, 1993.

982. **Wren, S. A. C. and Rowe, R. C.,** Theoretical aspects of chiral separation in capillary electrophoresis. III. Application to β-blockers, *J. Chromatogr.,* 635, 113, 1993.

983. **Otsuka, K. and Terabe, S.,** Optical resolution of chlorpheniramine by cyclodextrin added capillary zone electrophoresis and cyclodextrin modified micellar electrokinetic chromatography, *J. Liq. Chromatogr.,* 16, 945, 1993.

984. **Dzhingova, R. and Kostachinov, K.,** Method for the determination of diphenyl mercury in biological liquid samples, *Radiochem. Radioanal. Lett.,* 51, 399, 1982.

985. **Yalow, R. S. and Berson, S. A.,** Immunoassay of endogenous plasma insulin in man, *J. Clin. Invest.,* 39, 1157, 1960.

986. **Ekins, R. P.,** The estimation of thyroxine in human plasma by an electrophoretic technique, *Clin. Chim. Acta,* 5, 453, 1960.

987. **Murphy, B. E. P., Engelberg, W., and Pattee, C. J.,** Determination of plasma corticoids, *J. Clin. Endocrinol. Metab.,* 23, 293, 1963.

988. **Erlanger, B. F., Borek, F., Beiser, S. M., and Lieberman, S.,** Steroid-protein conjugates. I. Preparation and characterisation of bovine serum albumin with testosterone and with cortisol, *J. Biol. Chem.,* 228, 713, 1957.

989. **Abraham, G. E.,** Solid-phase radioimmunoassay of 17β-estradiol, *J. Clin. Endocrinol.,* 29, 866, 1969.

990. **Niswender, G. D. and Midgley, A. R.,** Hapten radioimmunoassay for steroid hormones, in *Immunological Methods in Steroid Determination,* Peron, F. G. and Caldwell, B. V., Eds., Plenum, New York, 1970, 149.

991. **Wainer, B. H., Fitch, F. W., Rothberg, R. M., and Fried, J.,** Morphine-3-succinyl-bovine serum albumin. Immunogenic hapten-protein conjugate, *Science,* 176, 1143, 1972.

992. **Spector, S. and Parker, C. W.,** Morphine: radioimmunoassay, *Science,* 168, 1347, 1970.

993. **Aherne, G. W., Piall, E. M., Robinson, J. D., Morris, B. A., and Marks, V.,** Two applications of a radioimmunoassay for morphine, in *Radioimmunoassay in Clinical Biochemistry,* Pasternak, C. A., Ed., Heyden, London, 1975, 81.

994. **Robinson, J. D., Morris, B. A., Piall, E. M., Aherne, G. W., and Marks, V.,** The use of rats in the screening of drug-protein conjugates for immunoreactivity, in *Radioimmunoassay in Clinical Biochemistry,* Pasternak, C. A., Ed., Heyden, London, 1975, 113.

995. **Halloran, M. J. and Parker, C. W.,** The preparation of nucleotide-protein conjugates: carbodiimides as coupling agents, *J. Immunol.,* 96, 373, 1966.

996. **Korn, A. H., Feairheller, S. H., and Filachione, E. M.,** Glutaraldehyde: nature of the reagent, *J. Mol. Biol.,* 65, 525, 1972.

997. **Castro, A., Monji, N., Ali, H., Yi, M., Bowman, E. R., and McKennis, H.,** Nicotine antibodies: comparison of ligand specificities of antibodies produced against two nicotine conjugates, *Eur. J. Biochem.,* 104, 331, 1980.

998. **James, V. H. T. and Jeffcoate, S. L.,** Steroids, *Br. Med. Bull.,* 30, 50, 1974.

999. **Van Bree, J. B. M. M., van Nispen, J. W., Verhoef, J. C., and Breimer, D. D.,** Development of specific antisera against 9-desglycinamide-8-arginine vasopressin by site-specific immunization, *J. Pharm. Sci.,* 80, 46–49, 1990.

1000. **Playfair, J. H. L., Hurn, B. A. L., and Schulster, D.,** Production of antibodies and binding reagents, *Br. Med. Bull.,* 30, 24, 1974.

1001. **Lee, J. W., Pedersen, J. E., Moravetz, T. L, Dzerk, A. M., Mundt, A. D., and Shepard, K. V.,** Sensitive and specific radioimmunoassays for opiates using commercially available materials. I: Methods for the determinations of morphine and hydromorphone, *J. Pharm. Sci.,* 80, 284, 1990.

1002. **Jagoda, E., Stouffer, B., Ogan, M., Tsay, H. M., Turabi, N., Mantha, S., Yost, F., and Tu, J.-I.,** Radio-immunoassay for hydroxyphosphinyl-3-hydroxybutanoic acid (SQ 33600), a hypocholesterolaemia agent, *Ther. Drug Monitor.,* 15, 213, 1993.

1003. **Klotz, U.,** Comparison of cyclosporin blood levels measured by radioimmunoassay and TDx assay using monoclonal antibodies, *Ther. Drug Monitor.,* 13, 461, 1991.

1004. **Dasgupta, A., Saldana, S., Kinnaman, G., Smith, M., and Johansen, K.,** Analytical performance evaluation of EMIT II monoclonal amphetamine-methylamphetamine assay: more specificity than EMIT d.a.u. monoclonal amphetamine-methylamphetamine assay, *Clin. Chem.,* 39, 104, 1993.

1005. **Johnson, H. J., Cernosek, S. F., Guitierrez-ˉCernosek, R. M., and Brown, L. L.,** Development of radioimmunoassay procedure for 4-acetamidobiphenyl, a metabolite of the chemical carcinogen 4-aminobiphenyl, in urine, *J. Analyt. Toxicol.,* 4, 86, 1980.

1006. **Mason, P. A., Law, B. A., Pocock, K., and Moffat, A. C.,** Direct radioimmunoassay for detection of barbiturates in blood and urine, *Analyst,* 107, 629, 1982.

1007. **Mohri, Z., Yasuda, T., Kawahara, K., Shono, F., and Yoshitake, A.,** Radioimmunoassay of Halidor (bencyclane fumarate) in human serum, *J. Pharmacobiol. Dyn.,* 5, 25, 1982.

1008. **Jagoda, E., Ogan, M., Stouffer, B., Tsay, H. M., Turabi, N., Mantha, S., Yost, F., and Tu, J.,** Radioimmunoassay for the new antiviral agent 1-β-D-arabinofuranosyl-*E*-5-(2-bromovinyl)uracil [*E*-5-(2-bromovinyl)uracil], *Ther. Drug Monitor.,* 14, 499, 1992.

1009. **Debrabandere, L., Van Boven, M., and Daenens, P.,** Development of a radio-immunoassay for the determination of buprenorphine in biological samples, *Analyst,* 118, 137, 1993.

1010. **Butz, R. F., Smith, P. G., Schroeder, D. H., and Findlay, J. W. A.,** Radioimmunoassay for bupropion in human plasma: comparison of tritiated and iodinated radio-ligands, *Clin. Chem.,* 29, 462, 1983.

1011. **Tu, J.-I., Brennan, J., Stouffer, B., Mantha, S., Turabo, N., and Tsay, H. M.,** Radioimmunoassay of ceronapril, a new angiotensin-converting enzyme inhibitor, and its application to a pharmacokinetic study in healthy male volunteers, *Ther. Drug Monitor.,* 14, 209, 1992.

1012. **Mee, A. V., Wong, P. Y., Sun, C., Oie, L., Elliott, S., Naik, N., Joaquin, B., and Uchimaru, D.,** Monitoring of cyclosporin concentrations by using dry blood-spot samples, *J. Clin. Lab. Anal.,* 5, 74, 1991.

1013. **Johnson, H. J., Cerosek, S. F., Guitierrez-Cernosek, R. M., and Brown, R. L.,** Validation of a radioimmunoassay procedure for NN′-diacetylbenzidine, a metabolite of the chemical carcinogen benzidine in urine, *J. Analyt. Toxicol.,* 5, 157, 1981.

1014. **Butler, V. P., Tse-Eng, D., Lindenbaum, J., Kalman, S. M., Preibisz, J. J., Rund, D. G., and Wissel, P. S.,** Development and application of a radioimmunoassay for dihydrodigoxin, a digoxin metabolite, *J. Pharmacol. Exp. Ther.,* 221, 123, 1982.

1015. **Mann, V., Benko, A. B., and Kocsar, L. T.,** Use of 17-epimethyl-testosterone radioimmunoassay in following excretion of methandienone metabolites in urine, *Steroids,* 37, 593, 1981.

1016. **Warren, J. T., Coker, G. G., Welch, R. M., Fowle, A. S. E., and Findlay, J. W. A.,** Plasma levels of a novel antidysrhythmic agent meobentine sulphate, in human as determined by radioimmunoassay, *J. Pharm. Sci.,* 71, 665, 1982.

1017. **Bizollon, C. A., Rocher, J. P., and Chevalier, P.,** Radioimmunoassay of nicergoline in biological material, *Eur. J. Nuc. Med.,* 7, 318, 1982.

1018. **Wal, J. M. and Bories, G. F.,** In vitro penicillin aminolysis: application to a radioimmunoassay of trace amounts of penicillin, *Analyt. Biochem.,* 114, 263, 1981.

1019. **Owens, S. M., Woodworth, J., and Mayersohn, M.,** Radioimmunoassay for phencyclidine (PCP) in serum, *Clin. Chem.,* 28, 1509, 1982.

1020. **Rowell, F. J., Hui, S. M., and Paxton, J. W.,** Evaluation of a radioimmunoassay for phenothiazines and thioxanthenes using an iodinated tracer, *J. Immunol. Methods,* 31, 159, 1979.

1021. **Fang, K., Koller, C. A., Brown, N., Covington, W., Lin, J.R., and Ho, D. H.,** Determination of plicamycin in plasma by radioimmunoassay, *Ther. Drug Monitor.,* 14, 255, 1992.

1022. **Brown, K., Gardner, J. J., Lockley, W. J. S., Preston, J. R., and Wilkinson, D. J.,** Radioimmunoassay of sodium chromoglycate in human plasma, *Annu. Clin. Biochem.,* 20, 31, 1983.

1023. **Hossein-Nia, M., Surve, A. H., Weglein, R., Gerbeau, C., and Holt, D. W.,** Radioimmunoassays for spirapril and its active metabolite spiraprilate: performance and application, *Ther. Drug Monitor.,* 14, 234, 1992.

1024. **Fong, K.-L. L., Ho, D.-H. W., Bogerd, R., Pan, T., Brown, N. S., Gentry, L., and Bodey, G. P.,** Sensitive radioimmunoassay for vancomycin, *Antimicrob. Agents Chemother.,* 19, 139, 1981.

1025. **Stretcher, B. N.,** State-of-the-art of zidovudine monitoring, *J. Clin. Lab. Anal.,* 5, 60, 1991.

1026. **Peskar, B. and Spector, S.,** Quantitative determination of diazepam in blood by radioimmunoassay, *J. Pharmacol. Exp. Ther.,* 186, 167, 1973.

1027. **Liu, C. T. and Adler, F. L.,** Immunological studies on drug addiction. I. Antibodies reactive with methadone and their use for the detection of the drug, *J. Immunol.,* 111, 472, 1973.

1028. **Tigelaar, R. E., Rapport, R. L., Imman, J. K., and Kupferberg, H. J.,** Radioimmunoassay for diphenylhydantoin, *Clin. Chim. Acta,* 43, 231, 1973.

1029. **Findlay, J. W. A., Warren, J. T., Hill, J. A., and Welch, R. M.,** Stereospecific radioimmunoassaysfor (+)-pseudoephedrine in human plasma and their application to bio-equivalency studies, *J. Pharm. Sci.,* 70, 624, 1981.

1030. **Lelievre, E., Cardona, H., Brillanceau, M. H., Piraube, C., and Sauveur, C.,** Radioimmunoassays for a new angiotensin-converting enzyme inhibitor, zabicipril, and its active metabolite, zabiciprilate, in human plasma, *J. Pharm. Sci.,* 81, 1065, 1992.

1031. **Matsuzawa, Y., Kiyosaki, T., Oki, T., Takeuchi, T., and Umezawa, H.,** Radioimmunoassay for (+)-pseudoephedrine in human plasma and their application to bio-equivalency studies, *Gann,* 73, 229, 1982.

1032. **Skubitz, K. M., Quinn, R. P., and Lietman, P. S.,** Rapid acyclovir radioimmunoassay using charcoal absorption, *Antimicrob. Agents Chemother.,* 21, 352, 1982.

1033. **Okumura, S., Deguchi, T., and Marumo, H.,** Radioimmunoassay of tritium-labelled fortimicins, *Jpn. J. Antibiot.,* 33, 1125, 1980.

1034. **Faraj, B. A. and Ali, F. M.,** Development and application of a radioimmunoassay for tetracycline, *J. Pharmacol. Exp. Ther.,* 217, 10, 1981.

1035. **Cook, C. E., Williams, D. L., Myers, M., Tallent, C. R., Leeson, G. A., Okerholm, R. A., and Wright, G. J.,** Radioimmunoassay for terfenadine in human plasma, *J. Pharm. Sci.,* 69, 1419, 1980.

1036. **Eckert, H. G., Baudner, S., Weimer, K. W., and Wissman, H.,** Determination of tiamenidinein biological specimens by radioimmunoassay, *Arzneim. Forsch.,* 31, 419, 1981.

1037. **Spector, S.,** Disposition of drugs in man by radioimmunoassay, *Pharmacol. Rev.,* 34, 73, 1982.

1038. **Fong, K.-L., Ho, D. H. W., Carter, C. J. K., Brown, N. S., Benjamin, R. S., Freireich, E. J., and Body, G. P.,** Radio-immunoassay for detection and quantitation of bruceantin, *Analyt. Biochem.,* 105, 281, 1980.

1039. **Piall, E. M., Aherne, G. W., and Marks, V.,** Radioimmunoassay for cytosine arabinoside, *Br. J. Cancer,* 40, 548, 1979.

1040. **Wang, Y., Lantin, E., and Sutow, W. W.,** Methotrexate in blood, urine and cerebrospinal fluid of children receiving high doses by infusion, *Clin. Chem.,* 22, 1053, 1976.

1041. **Sethi, V. S., Burton, S. S., and Jackson, D. V.,** Sensitive immunoassay for vincristine and vinblastine, *Cancer Chemother. Pharmacol.,* 4, 183, 1980.

1042. **Yamaoka, A. and Takatori, T.,** Determination of phenobarbitone by radioimmunoassay, *J. Immunol. Methods,* 28, 51, 1979.

1043. **Robinson, J. D., Morris, B. A., Aherne, G. W., and Marks, V.,** Pharmacokinetics of a single dose of phenytoin in man measured by radioimmunoassay, *Br. J. Clin. Pharmacol.,* 2, 345, 1975.

1044. **Gourmel, B., Fiet, J., Collins, R. F., Villette, J. M., and Dreux, C.,** Simple radioimmunoassay of acebutol in plasma, *Clin. Chim. Acta,* 108, 229, 1980.

1045. **Mould, G. P., Clough, J., Morris, B. A., Stout, G., and Marks, V.,** Propranolol radioimmunoassay and its use in study of its pharmacokinetics following low doses, *Biopharm. Drug Dispos.,* 2, 49, 1981.

1046. **Luce, K. R. and Dixon, R.,** Chlordiazepoxide concentration in saliva and plasma measured by radioimmunoassay, *Res. Commun. Chem. Pathol. Pharmacol.,* 27, 397, 1980.

1047. **Bourne, R., Robinson, J. D., and Teale, D. J.,** A simple radioimmunoassay for plasma diazepam and its application to single dose studies in man, *Br. J. Clin. Pharmacol.,* 63, 371P, 1978.

1048. **Dixon, R., Glover, W., and Earley, J.,** Specific radioimmunoassay for flunitrazepam, *J. Pharm. Sci.,* 70, 230, 1981.

1049. **Aderjan, R. and Schmidt, G.,** Screening radioimmunoassay for 1, 4-benzodiazepines in human blood, serum and urine with use of antibodies against oxazepam hemisuccinate, *Z. Rechtsmed.,* 83, 191, 1979.

1050. **Yeager, E. P., Goeblesmann, U., Soares, J. R., Grant, J. D., and Gross, S. J.,** d^9-Tetrahydrocannabinol by g.l.c.-m.s.-validated radioimmunoassays of haemolysed blood or serum, *J. Analyt. Toxicol.* 5, 81, 1981.

1051. **Iwamura, S. and Kambegawa, A.,** Determination of methyl-ergometrine and dihydroergotoxine in biological fluids, *J. Pharmacobio-Dyn.,* 4, 275, 1981.

1052. **Persiani, S., Puanezzola, E., Broutin, F., Fonte, G., and Strolin-Benedetti, M.,** Radioimmunoassay for the synthetic ergoline derivative cabergoline in biological fluids, *J. Immunoassay,* 13, 457, 1992.

1053. **Takahama, S., Ohba, T., Naruchi, T., and Yamamoto, I.,** Development of radioimmunoassay for TG-41, a new antiulcer drug, and its application, *J. Immunoassay,* 13, 163, 1992.

1054. **Suzuki, H., Miki, M., Sekine, Y., Kagemoto, A., Negoro, T., Maeda, T., and Hashimoto, M.,** Determination of a hypoglycemic drug gliclazide, in human serum by radioimmunoassay, *J. Pharmacobio-Dyn.,* 4, 217, 1981.

1055. **Midha, K. K., Loo, J. C. K., Charette, C., Rowe, M. L., Hubbard, J. W., and McGilveray, I. J.,** Monitoring of therapeutic concentrations of psychotropic drugs in plasma by radioimmunoassays, *J. Analyt. Toxicol.,* 2, 185, 1978.

1056. **Midha, K. K., Cooper, J. K., and Hubbard, J. W.,** Radioimmunoassay of fluphenazine in human plasma, *Commun. Psychpharmacol.,* 4, 107, 1980.

1057. **Clarke, B. R., Tower, B. B., and Rubin, R. T.,** Radioimmunoassay of haloperidol in human serum, *Life Sci.,* 20, 319, 1977.

1058. **Midha, K. K., Mackonka, C., Cooper, J. K., Hubbard, J. W., and Yeung, P. F. K.,** Radioimmunoassay for perphenazine in human plasma, *Br. J. Clin. Pharmacol.,* 11, 85, 1981.

1059. **Michiels, M., Hendriks, R., and Heykants, J.,** Radioimmunoassay of new opiate analgesics alfentanil and sulfentanil. Preliminary pharmacokinetic profile in man, *J. Pharm. Pharmacol.,* 35, 86, 1983.

1060. **Bartlett, A. J., Lloyd-Jones, J. G., Rance, M. J., Flockhart, I. R., Dockray, G. J., Bennett, M. R. D., and Moore, R. A.,** Radioimmunoassay of buprenorphine, *Eur. J. Clin. Pharmacol.,* 18, 339, 1980.

1061. **Honigberg, I. L. and Stewart, J. T.,** Radioimmunoassay of hydromorphone and hydrocodone in human plasma, *J. Pharm. Sci.,* 69, 1171, 1980.

1062. **Ling, G. S. F., Umans, J. G., and Inturrisi, C. E.,** Methadone: radioimmunoassay and pharmacokinetics in the rat, *J. Pharmacol. Exp. Ther.,* 217, 147, 1981.

1063. **Ward, D. P. and Trevor, A. J.,** Radioimmunoassay for phencyclidine: application to kinetic analysis in the rat, *Life Sci.,* 27, 457, 1980.

1064. **Virtanen, R., Kanto, J., and Iisalo, E.,** Radioimmunoassay for atropine and hyoscamine, *Acta Pharmacol. Toxicol.,* 47, 208, 1980.

1065. **Orchard, M. A., Blair, I. A., Ritter, J. M., Myatt, L., Jogee, M., and Lewis, P. J.,** Radioimmunoassay at alkaline pH. Method for the quantitative determination of prostacyclin, *Biochem. Soc. Trans.,* 10, 241, 1982.

1066. **Morvay, J., Fotherby, K., and Altorjay, I.,** Rapid radioimmunoassay of ethinyloestradiol in body fluids, *Clin. Chim. Acta,* 108, 147, 1980.

1067. **Nerenberg, C. and Matin, S. B.,** Radioimmunoassay of flunisolide in human plasma, *J. Pharm. Sci.,* 70, 900, 1981.

1068. **Brenner, P. F., Guerrero, R., Cekan, Z., and Diczfalust, E.,** Radioimmunoassay method for six steroids in human plasma, *Steroids,* 22, 775, 1973.

1069. **Alabet, S. M. H., McAllister, W. A. C., Collins, J. V., and Rogers, H. J.,** Comparison of radioimmunoassay and thin-layer chromatographic assay methods for estimation of plasma prednisolone concentrations, *J. Pharmacol. Methods,* 6, 137, 1981.

1070. **Krulik, R, Exner, J., Fuksova, K., Pichova, D., Beitlova, D., and Sikora, J.,** Radio-immunoassay of dibenzazepines and dibenzocycloheptadienes in body fluids and tissues, *Eur. J. Clin. Chem. Clin. Biochem.,* 29, 827, 1991.

1071. **Read, G. F. and Riad-Fahmy, D.,** Determination of tricyclic antidepressant, clomipramine (Anafranil), in plasma by a specific radioimmunoassay procedure, *Clin. Chem.,* 24, 36, 1978.

1072. **Spector, S., Spector, N., and Almeida, M.,** Radioimmunoassay for desmethylimipramine, *Psychopharmacol. Commun.,* 1, 421, 1975.

1073. **Virtanen, R., Salonen, J. S., Scheinin, M., Iisalo, E., and Mattila, V.,** Radioimmunoassay for doxepin and desmethyl-doxepin, *Acta Pharmacol. Toxicol.,* 47, 274, 1980.

1074. **Heptner, W., Badian, M. J., Baudner, S., Christ, O. E., Frazer, H. M., Rupp, W., Weimer, K. E., and Wissmann, H.,** Determination of nomifensine by a sensitive radioimmunoassay, *Br. J. Clin. Pharmacol.,* 4, 123S, 1977.

1075. **Ekins, R. P.,** Radioimmunoasay design, in *Radioimmunoassay in Clinical Biochemistry,* Pasternak, C. A., Ed., Heyden, London, 1975, 3.

1076. **Berson, S. A., Yalow, R. S., Bauman, A., Rothschild, M. A., and Newerly, K.,** Insulin iodine-131 metabolism in human subjects: demonstration of insulin-binding globulin in the circulation of insulin-treated subjects, *J. Clin. Invest.,* 35, 170, 1956.

1077. **Haber, E., Page, L. B., and Richards, F. F.,** Radioimmunoassay employing gel filtration, *Analyt. Biochem.,* 12, 163, 1965.

1078. **Herbert, V., Law, K. S., Gottlieb, C. W., and Bleicher, S. J.,** Coated charcoal immunoassay of insulin, *J. Clin. Endocrinol. Metab.,* 25, 1375, 1965.

1079. **Rosselin, G., Assan, R., Yallow, R. S., and Berson, S. A.,** Separation of antibody-bound and unbound peptide hormones labelled with iodine-131 by talcum powder and precipitated silica, *Nature,* 212, 355, 1966.

1080. **Chard, T., Martin, M., and Landon, J.,** Separation of antibody bound from free peptides using ammonium sulfate and ethanol, in *Radioimmunoassay Methods,* Kirkham, K. E. and Hunter, W. M., Eds., Churchill Livingstone, Edinburgh, 1971, 257.

1081. **Thomas, K. and Ferin, J.,** A new radioimmunoassay for HCG (LH, ICSH) in plasma using dioxane, *J. Clin. Endocrinol. Metab.,* 28, 1667, 1968.

1082. **Hales, C. N. and Randle, P. J.,** Immunoassay of insulin by isotope dilution, *Biochem. J.,* 84, 79P, 1962.

1083. **Catt, K. and Tregear, G. W.,** Solid-phase radioimmunoassay in anti-body coated tubes, *Science,* 158, 1570, 1967.

1084. **Dixon, K.,** Measurement of specific antibodies in human serum, in *Radioimmunoassays in Clinical Biochemistry,* Pasternak, C. A., Ed., Heyden, London, 1975, 15.

1085. **Herbert, V.,** Coated charcoal separation of free labelled hormone from hormone bound antibody, in *Protein and Polypetide Hormones,* Margoulies, M., Ed., Excerpta Medica Foundation, Amsterdam, 1969, 55.

1086. **Catt, K.,** Radioimmunoassay with antibody-coated disks and tubes, *Acta Endocrinol.,* Suppl. 142, 222, 1969.

1087. **Bosworth, N. and Towers, P.,** Scintillation proximity assay, *Nature,* 341, 167, 1989.

1088. **Takeuchi, K.,** Scintillation proximity assay, *Lab. Pract.,* 41, 11, 1992.

1089. **Fenwick, S., Jenner, W. N., Linacre, P., Rooney, R. M., and Wring, S. A.,** Application of the scintillation proximity assay technique to the determination of drugs, *Analyt. Proc.,* 31, 103, 1994.

1090. **Singh, A. K., Granley, K., Misrha, U., Naeem, K., White, T., and Yin, J.,** Screening and confirmation of drugs in urine: interference of hordenine with the immunoassays and thin-layer chromatography methods, *Forensic Sci. Int.,* 54, 9, 1992.

1091. **Yiannakou, L., Loucari-Yiannakou, E., and Souvatzoglou, A.,** Digoxin-like immunoreactivity in cord blood plasma extracts is not only due to endogenous corticosteroids, *Clin. Biochem.,* 24, 475, 1991.

1092. **D'Nicuola, J., Jones, R., Levine, B., and Smith, M. L.,** Evaluation of six commercial amphetamine and methylamphetamine immunoassays for cross-reactivity to phenylpropanolamine and ephedrine in urine, *J. Analyt. Toxicol.,* 16, 211, 1992.

1093. **Jones, R., Klette, K., Kuhlman, J. J., Levine, B., Smith, M. L., Watson, C. V., and Selavka, C. M.,** Trimethobenzamide cross-reacts in immunoassays of amphetamine-methylamphetamine, *Clin. Chim. Acta,* 39, 699, 1993.

1094. **Haak, D., Vecsei, P., Lichtwald, K., Klee., Gless, K. H., and Weber, M.,** Some experiences on radioimmunoassay of glucocorticoids, *Allergolie.* 3, 259, 1980.

1095. **Avendano, C., Alvarez, J. S., Sacristan, J. A., Adin, J., and alsar, M. J.,** Interference of digoxin-like immunoreactive substances with TDx digoxin II assay in different patients, *Ther. Drug Monitor.,* 13, 523, 1991.

1096. **Schlebusch, H.,** Cross-reactivity of endogenous digoxin-like immunoreactive factors in two homogenous immunoassays for determining digoxin, *Labor Med.,* 15, 385, 1992.

1097. **Loo, J. C. K., Gallicano, K. D., McGilveray, I. J., Beaudoin, N., and Jindal, S. L.,** Isolation of cross-reacting compounds to Incstar Cyclo-Trac SP RIA in blood samples obtained from allograft patients on cyclosporin therapy, *J. Clin. Lab. Anal.,* 5, 153, 1991.

1098. **Bermejo, A. M., Ramos, I., Fernandez, P., Lopez-Rivadulla, M., Cruz, A., Chiarotti, M., Fucci, N., and Marsilli, R.,** Morphine determination by gas-chromatography-mass spectrometry in human vitreoaus humour and comparison with radio-immunoassay, *J. Analyt. Toxicol.*, 16, 372, 1992.

1099. **Cone, E. J., Dickerson, S., Paul, B. D., and Mitchell, J. M.,** Forensic drug testing for opiates. V. Urine testing for heroin, morphine and codeine with commercial opiate immunoassays, *J. Analyt. Toxicol.*, 17, 156–164, 1993.

1100. **Kuhnz, W., Louton, T., Back, D. J., and Michaelis, K.,** Radioimmunological analysis of ethimylestradiol in human serum. Validation of the method and comparison with a gas chromatographic/mass spectrometric assay, *Arzneim. Forsch.*, 43, 16, 1993.

1101. **McKay, G., Cooper, S. F., and Midha, K. K.,** Determination of phenothiazines: the present day scene, in *Bioactive Analytes, Including CNS Drugs Peptides and Enantiomers*, Reid, E., Scales, B., and Wilson, I. D., Eds., Plenum Press, New York, 1986, 159.

1102. **Woedtenborghs, R., Geuens. H., Lenoir, H., Janssen, C., and Heykants, J.,** On the selectivity of some recently developed RIA's, in *Analysis for Drugs and Metabolites Including Anti-Infective Agents*, Reid, E. and Wilson, I. D., Eds., Royal Society of Chemistry, Cambridge, 1990, 241.

1103. **Henderson, L. O., Powell, M. K., Hannon, W. H., Miler, B. B., Martin, M. L., Hanzlick, R. L., Vroon, D., and Sexson, W. R.,** Radioimmunossay screening of dried blood spot materials for benzoylecgonine, *J. Analyt. Toxicol.*, 17, 42, 1993.

1104. **Borque, L., Rus, A., Maside, C., and Del Cura, J.,** Automated latex nephelometric immunoassay of theophylline in human serum, *Eur. J. Clin. Chem. Clin. Biochem.*, 3, 307, 1992.

1105. **Armbruster, D. A. and Krolak, J. M.,** Screening for drugs of abuse with the Roche ONTRAK assays, *J. Analyt. Toxicol.*, 16, 172, 1992.

1106. **Tsay, Y.-G. and Palmer, R. J.,** Solid-phase radioimmunoassay of tobramycin, *Clin. Chim. Acta*, 109, 151, 1981.

1107. **Al-Hakiem, M. H. H., White, G. W., Smith, D. S., and Landon, J.,** Direct determination of propranolol in serum or plasma by fluoro-immunoassay, *Ther. Drug Monitor.*, 3, 159, 1981.

1108. **Khanna, P. L. and Ullman, E. F.,** 6-Carboxy-4',5'-dimethoxy-fluorescein: a novel dipole-dipole coupled fluorescence energy-transfer acceptor useful for fluorescence immunoassays, *Analyt. Biochem.*, 108, 156, 1980.

1109. **Al-Hakiem, M. H. H., Nargessi, R. D., Pourfarzaneh, M., White, G. W., Smith, D. S., and Hodgkinson, D. A.,** Fluoro-immunoassay of digoxin in serum, *J. Clin. Chem. Clin. Biochem.*, 20, 151, 1982.

1110. **Al-Hakiem, M. H. H., Simon, M., Mahmoud, S., and Landon, J.,** Fluoro-immunoassay of digitoxin in serum, *Clin. Chem.*, 28, 1364, 1982.

1111. **Kamel, R. S., Landon, J., and Smith, D. S.,** Magnetisable solid-phase fluoroimmunoassay of phenytoin in disposable test-tubes, *Clin. Chem.*, 26, 1281, 1980.

1112. **O'Donnel, C. M., McBride, J., Suffin, S., and Broughton, A.,** Heterogeneous fluorescence immunoassay for gentamicin using a second-antibody separation, *J. Immunoassay*, 1, 375, 1980.

1113. **Al-Hakiem, M. H. H., Smith, D. S., and Landon, J.,** Development of fluoro-immunoassays for determination of individual and combined levels of procainamide and N-acetylprocainamide in serum, *J. Immunoassay*, 3, 91, 1982.

1114. **Kaeferstein, H. and Sticht, G.,** Experiences with a new fluorescence-polarization immunoassay system in the detection of drugs in urine. *GIT Labor-Med.*, 16, 60, 1993.

1115. **Dandliker, W. B., Kelly, R. J., Dandliker, J., Farquhar, J., and Levin, J.,** Fluorescence polarisation immunoassay. theory and experimental method, *Immunochemistry*, 10, 219, 1973.

1116. **Jolley, M. E.,** Fluorescence polarisation for determination of therapeutic drug levels in human plasma, *J. Analyt. Toxicol.*, 5, 236, 1981.

1117. **Winek, C. L., Elzein, E. O., Wahba, W. W., and Feldman, J. A.,** Interference of herbal drinks with urinalysis for drugs of abuse, *J. Analyt. Toxicol.*, 17, 246, 1993.

1118. **Zaninotto, M., Secchiero, S., Paleari, C. D., and Burlina, A.,** Performance of a fluorescence polarization immunoassay system evaluated by therapeutic monitoring of four drugs, *Ther. Drug Monitor.*, 14, 301, 1992.

1119. **Ehrenthal, W., Schoenfeld, U., Gaul, H., Henle, H., and Heubner, A.,** New system and new reagents for the therapeutic drug monitoring, *Labor. Med.*, 15, 453, 1992.

1120. **Cody, J. T. and Schwarzhoff, R.**, Fluorescence polarization immunoassay detection of amphetamine, methamphetamine and illicit amphetamine analogues, *J. Analyt. Toxicol.*, 17, 26, 1993.

1121. **Turner, G. J., Colbert, D. L., and Chowdry, B. Z.**, Broad spectrum immunoassay using fluorescence polarization for the detection of amphetamines in urine, *Ann. Clin. Biochem.*, 28, 588–594, 1991.

1122. **Becker, J., Correll, A., Koepf, W., and Rittner, C.**, Comparative studies on the detection of benzodiazepines in serum by means of immunoassays (FPIA), *J. Analyt. Toxicol.*, 17, 103, 1993.

1123. **Colbert, D. L. and Turner, G. J.**, Experience of cost-effective in-house reagents for the assay of carbamazepine in serum, using the Abbott TDx, *Ther. Drug Monitor.*, 15, 209, 1993.

1124. **Stamp, R. J., Mould, G. P., Muller, C., and Burlina, A.**, Performance of fluorescence polarization immunoassay reagents for carbamazepine, phenytoin phenobarbitone, primidone and valproic acid on a Cobras Fara II Analyzer, *Ther. Drug Monitor.*, 13, 518, 1991.

1125. **Ripple, M. G., Goldberger, B. A., Caplan, Y. H., Blitzer, M. G., and schwarz, S.**, Detection of cocaine and its metabolites in human amniotic fluid, *J. Analyt. Toxicol.*, 16, 328, 1992.

1126. **Chan, G. L., Weinstein, S. S., LeFor, W. W., Spoto, E., Kahana, L., and Shires, D. L.**, Relative performance of specific and non-specific fluorescence polarization immunoassay for cyclosporin in transplant patients, *Ther. Drug Monitor.*, 14, 42, 1992.

1127. **Alvarez, J. S., Sacristan, J. A., and Alsar, M. J.**, Comparison of a monoclonal antibody fluorescent polarization immunoassay with monoclonal antibody radioimmunoassay for cyclosporin determination in whole blood, *Ther. Drug Monitor.*, 14, 78–80, 1992.

1128. **Lindohlm, A., Napoli, K., Rutzky, L., and Kahan, B. D.**, Specific monoclonal radioimmunoassay and fluorescence polarization immunoassay for trough concentration and area-under-the-curve monitoring of cyclosporin in renal transplantation, *Ther. Drug Monitor.*, 14, 292–300, 1992.

1129. **Dusci, L. J., Hackett, L. P., Chiswell, G. M., and Ilett, K. F.**, Comparison of cyclosporin measurement in whole blood by high-performance liquid chromatography, monoclonal fluorescence polarization immunoassay, and monoclonal enzyme-multiplied immunoassay, *Ther. Drug Monitor.*, 14, 327–332, 1992.

1130. **Armijo, J. A., Navarro, F. A., and Angeles de Cos, M.**, Is the monoclonal fluorescence polarization immunoassay for cyclosporin specific? Comparison with specific radioimmunoassay, *Ther. Drug Monitor.*, 14, 333–338, 1992.

1131. **Beutler, D., Molteni, S., Zeugin, T., and Thormann, W.**, Evaluation of instrumental, non-isotopic immunoassays (fluorescence-polarization immunoassay and enzyme-multiplied immunoassay technique) for cyclosporin monitoring in whole blood after kidney and liver transplantation, *Ther. Drug Monitor.*, 14, 424, 1992.

1132. **Panzali, A. and Signorini, C.**, Evaluation of the TDx fluorescence-polarization immunoassay for specific determination of cyclosporin in whole blood, *G. Ital. Chim. Clin.*, 17, 111, 1992.

1133. **Ujhelyi, M. R., Green, P. J., Cummings, D. M., Robert, S., Vlassaes, P. H., and Zarowitz, B. J.**, Determination of free serum digoxin concentrations in digoxin toxic patients after administration of digoxin fab antibodies, *Ther. Drug Monitor.*, 14, 147, 1992.

1134. **Najjar, T. A. O., Matar, K. M., and Alfawaz, I. M.**, Comparison of a new high-performance liquid chromatography method with fluorescence polarization immunoassay for analysis of methotrexate, *Ther. Drug Monitor.*, 14, 142, 1992.

1135. **Chen, Y. and Potter, J. M.**, Fluorescence polarization immunoassay and HPLC assays compared for measuring monoethylglycinexylidide in liver-transplant patients, *Clin. Chem.*, 38, 2426, 1992.

1136. **Cai, W. M., Zhu, G. Z., and Chen, G.**, Free phenytoin monitoring in serum and saliva of epileptic patients in China, *Ther. Drug Monitor.*, 15, 31, 1993.

1137. **Awni, W. M., St. Peter, W. L., Guay, D. R. P., Kenny, M. T., and Matzke, G. R.**, Teicoplanin measurement in patients with renal failure: comparison of fluorescence polarization immunoassay, microbiological assay and high-performance liquid-chromatographic assay, *Ther. Drug Monitor.*, 13, 511, 1991.

1138. **Gennaro, M. C., Abrigo, C., and Biglino, P.**, Quantification of theophylline in human plasma by reversed-phase ion-interaction high-performance liquid chromatography and comparison with the TDx fluorescence polarization immunoassay procedure, *Analyst*, 117, 1071, 1992.

1139. **Uber-Bucek, E., Hamon, M., Pham Huy, C., and Dadoun, H.**, Determination of thevetin B in serum by fluorescebce polarization immunoassay, *J. Pharmacol. Biomed.*, 10, 413, 1992.

1140. **Schneider, R. S.,** Homogenous enzyme immunoassay for opiates in urine, *Clin. Chem.,* 19, 821, 1973.

1141. **Tittle, T. V. and Schaumann, B. A.,** Micro-enzyme-multiplied immunoassay technique plate assay for antiepileptic drugs, *Ther. Drug Monitor.,* 14, 159, 1992.

1142. **Kubotsu, K., Goto, S., Fujita, M., Tuchiya, H., Kida, M., Takano, S., Matsuura, S., and Sakurabayashi, I.,** Automated homogeneous liposome immunoassay systems for anticonvulsant drugs, *Clin. Chem.,* 38, 808, 1992.

1143. **Sauer, M. J., Pickett, R. J. H., and MacKenzie, A. L.,** Determination of clenbuterol residues in bovine liver, urine and eye by enzyme immunoassay, *Analyt. Chim. Acta,* 275, 195, 1993.

1144. **Degand, G., Bernes-Duyckaerts, A., and Maghuin-Rogister, G.,** Determination of clenbuterol in bovine tissues and urine by immunoassay, *J. Agric. Food Chem.,* 40, 70, 1992.

1145. **Grothaus, P. G., Raybould, T. J., Bignami, G. S., Lazo, C. B., and Byrnes, J. B.,** Enzyme immunoassay for the determination of taxol and taxanes in *Taxus* sp. tissues and human plasma, *J. Immunol. Methods,* 158, 5, 1993.

1146. **Reck, B, Dingler, E., and Lohmann, A.,** Development of a sensitive enzyme immunoassay for the determination of vinpocetine in human plasma, *Arzneim. Forsch.,* 42, 1171, 1992.

1147. **Poklis, A., Fitzgerald, R. L., Hall, K. V., and Saady, J. J.,** EMIT-d.a.u. monoclonal amphetamine-methamphetamine assay. II. Detection of methylenedioxyamphetamine (MDA) and methylenedioxymethamphetamine (MDMA), *Forensic Sci. Int.,* 59, 63, 1993.

1148. **Diosi, D. T. and Harvey, D. C.,** Analysis of whole blood for drugs of abuse using EMIT d.a.u. reagents and a Monarch 1000 Chemistry Analyzer, *J. Analyt. Toxicol.* 17, 133, 1993.

1149. **Fyfe, M. J., Chand, P., McCutchen, C., Long, J. S., Walia, A. S., Edwards, C., and Liu, R. H.,** Performance characteristics of phencyclidine assay using Reply Analyzer and Emit d.a.u., Emit 700, and 1:1 (Emit d.a.u.-Emit 700) reagents, *J. Analyt. Toxicol.,* 17, 188–189, 1993.

1150. **Moore, F. M. L., Jarvie, D. R., and Simpson, D.,** Urinaryu amphetamines, benzodiazepines and methadone: cost-effective detection procedures, *Med. Lab. Sci.,* 49, 27, 1992.

1151. **Colbert, D. L.,** Measurement of phenytoin in serum using in-house reagents employing the Syva enzyme-multiplied immunoassay technique principle, *Ther. Drug Monitor.,* 14, 386, 1992.

1152. **Colbert, D. and Landon, J.,** Effects of using alternative antibodies in commercial non-separation enzyme immunoassays for drugs, *Ann. Clin. Biochem.,* 29, 343, 1992.

1153. **Colbert, D. L. and Gooch, J. C.,** In-house opiate enzymoimmunoassay based on the Syva EMIT principle, *Clin. Chem.,* 38, 1483, 1992.

1154. **Van der Weide, J., Luiting, H.-J., and Veefkind, A. H.,** Evaluation of cloned enzyme donor immunoassay for measurement of phenytoin and phenobarbital in serum, *Ther. Drug Monitor.,* 15, 344, 1993.

1155. **Hansen, J. B., Lau, H. P., Janes, C. J., Miller, W. K and Wiebe, D. A.,** Rapid and specific cyclosporin assay for the Du Pont aca discrete clinical analyzer performed directly on whole blood, *J. Clin. Lab. Anal.,* 5, 187, 1991.

1156. **Sabate, I., Gracia, S., Diez, O., Alia, P., and Moliner, R.,** Comparison of cyclosporin in blood concentrations measured by radio-immunoassay and two non-isotopic immuno-assays using mnoclonal antibodies, *Clin. Chem.,* 38, 1187, 1992.

1157. **Poklis, A., Hall, K. V., Eddleton, R. A., Fitzgerald, R. L., Saady, J. J., and Bogema, S.C.,** EMIT-d.a.u. monoclonal amphetamine-methamphetamine assay, I. Stereoselectivity and clinical evaluation, *Forensic Sci. Int.,* 59, 49, 1993.

1158. **Asselin, W. M. and Leslie, J. M.,** Modification of Emit assay reagents for improved sensitivity and cost effectiveness in the analysis of haemolysed whole blood, *J. Analyt. Toxicol.,* 16, 381, 1992.

1159. **Poklis, A.,** Evaluation of EMIT-dau benzodiazepine metabolite assay for urine drug screening, *J. Analyt. Toxicol.,* 5, 174, 1981.

1160. **Pegon, Y., Pourcher, E., and Vallon, J. J.,** Evaluation of the EMIT Tox enzyme immunoassay for toxicological analysis of benzodiazepines in serum, *J. Analyt. Toxicol.,* 6, 1, 1982.

1161. **Wallace, J. E., Harris, S. C., and Shimek, E. L.,** Evaluation of immunoassay for determining diazepam and nordiazepam in serum and urine, *Clin. Chem.,* 26, 1905, 1980.

1162. **O'Connor, J. E. and Rejent, T. A.,** EMIT cannabinoid assay: confirmation by r.i.a., and g.c.-m.s., *J. Analyt. Toxicol.,* 5, 168, 1981.

1163. **Altunkaya, D., Clatworthy, A. J., Smith, R. N., and Start, I. J.,** Urinary cannabinoid analysis: comparison of four immunoassays with gas chromatography-mass spectrometry, *Forensic Sci. Int.,* 50, 15, 1991.

1164. **Gjerde, H.,** Screening for cannabinoids in blood using EMIT: concentrations of delta^9-tetrahydrocannabinol in relation to EMIT results, *Forensic Sci. Int.,* 50, 121, 1991.

1165. **Bannon, P. and Vinet, B.,** EMIT antiepileptic drug assays adapted to an Abbott-VP Analyzer and compared with a gas-chromatographic procedure, *Clin. Biochem.,* 15, 179, 1982.

1166. **Nandedkar, A. K. N., Kutt, H., and Fairclough, G. F.,** Correlation of the EMIT with a gas-liquid chromatographic method for the determination of anti-epileptic drugs in plasma, *Clin. Toxicol.,* 12, 483, 1979.

1167. **Monaco, F. and Piredda, S.,** Carbamazepine 10,11-epoxidedetermined by EMIT carbamazepam reagent, *Epilepsia,* 23, 185, 1982.

1168. **Goldsmith, R. F. and Ouvrier, R. A.,** Salivary anti-convulsant levels in children: comparison of methods, *Ther. Drug Monitor.,* 3, 151, 1981.

1169. **Paxton, J. W.,** Carbamazepine determination in saliva of children: enzyme immunoassay (EMIT) versus high-pressure liquid chromatography, *Epilepsia,* 23, 185, 1982.

1170. **Goldsmith, R. F. and Ouvrier, R. A.,** Salivary anticonvulsant levels in children: comparison of methods, *Ther. Drug Monitor.,* 3, 151, 1981.

1171. **Fenton, J., Schaffer, M., Chen, N. W., and Bermes, E. W.,** Comparison of enzyme immunoassay and gas-chromatography-mass spectrometry in forensic toxicology, *J. Forensic Sci.,* 25, 314, 1980.

1172. **Boyd, D., Shearan, P., Hopkins, J. P., O'Keeffe, M., and Smyth, M. R.,** Matrix solid-phase dispersion linked to immunoassay techniques for the determination of clenbuterol hydrochloride in bovine liver samples, *Anal. Proc.,* 30, 156, 1993.

1173. **McBride, J. H., Kim, S. S., Rodgerson, D. O., Reyes, A. F., and Ota, M. K.,** Measurement of cyclosporin by liquid chromatography and three immunoassays in blood from liver, cardiac, and renal transplant recioients, *Clin. Chem.,* 38, 2300, 1992.

1174. **Reynolds, A. P., Akagi, H., and Hjelm, N. M.,** Reliability of immunoassays of cyclosporin A, *Clin. Chem.,* 38, 1508, 1992.

1175. **Kurze, S., Piegler, E., Schickbauer, T., and Schmid, R. W.,** Performance evaluation of a new EMIT cyclosporine assay in whole blood on a COBAS FARA analyzer, *Labor. Med.,* 15, 466 1992.

1176. **Johnston, A. and Hamer, J.,** Gas chromatography and enzyme immunoassay compared for the analysis of disopyramide in plasma, *Clin. Chem.,* 27, 353, 1981.

1177. **Pape, B. E.,** Enzyme immunoassay of disopyramide in human serum, *Clin. Chem.,* 27, 2038, 1981.

1178. **Kuelpmann, W. R. and Oellerich, M.,** Monitoring of therapeutic serum concentrations of antiepileptic drugs by a newly developed gas-chromatographic procedure and enzyme immunoassay (EMIT): comparative study, *J. Clin. Chem. Clin. Biochem.,* 19, 249, 1981.

1179. **Chinwah, P. and Williams, K.,** Comparison of enzyme immunoassay for gentamicin compared with other methods, *J. Antimicrob. Chemother.,* 6, 561, 1980.

1180. **Oeltgen, P. R., Hamann, S. R., and Blouin, R. A.,** Comparison of gentamicin assays, *Ther. Drug Monitor.,* 2, 423, 1980.

1181. **Fraser, A. D., Bryan, W., and Isner, A. F.,** Urinary screening for α-hydroxy triazolam by FPIA [fluorescence polarization immmunoassay] with confirmation by GC-MS, *J. Analyt. Toxicol.,* 16, 347, 1992.

1182. **Buice, R. G., Evans, W. E., Karas, J., Nicholas, C. A., Sidhu, P., Straughn, A. B., Meyer, A. C., and Crom, W. R.,** Evaluation of immunoassay, radio-assay and radioimmunoasay of serum methotrexate, as compared with liquid chromatography, *Clin. Chem.,* 26, 1902, 1980.

1183. **Von Meyer, L., Kauert, G., and Drasch, G.,** Comparison of enzyme immunochemistry and gas-chromatographic determination of morphine in blood, *Beitr. Gerichtl. Med.,* 39, 113, 1981.

1184. **Couri, D., Suarez del Villar, C., and Toy-Manning, P.,** Simultaneous quantitative assay of phenobarbitone, phenytoin and *p*-hydroxydiphenylphenytoin by h.p.l.c., and a comparison with EMIT, *J. Analyt. Toxicol.,* 4, 227, 1980.

1185. **Kanazawa, T., Sato, Y., Kawakami, K., and Azuma, Y.,** Determination of anticonvulsants in serum. comparison of enzyme immunoassay with high performance liquid chromatography, *Rinsho Byori,* 28, 492, 1980.

1186. **Bessa, R., Magnelli, N. P., Rike, M. A., and Rassi, I. E.**, Phenobarbitone determination: comparison between difference-spectrophotometric and immunoenzyme methods, *Rev. Bras. Pathol. Clin.*, 16, 214, 1980.
1187. **Nandedkar, A. K. N., Kutt, H., and Fairclough, G. F.**, Correlation of the EMIT with a gas-liquid chromatographic method for the determination of anti-epileptic drugs in plasma, *Clin. Toxicol.*, 12, 483, 1979.
1188. **Goldsmith, R. F. and Ouvrier, R. A.**, Salivary anti-convulsant levels in children: comparison of methods, *Ther. Drug Monitor.*, 3, 151, 1981.
1189. **Bannon, P. and Vinet, B.**, EMIT antiepileptic drug assays adapted to an Abbott-VP Analyzer and compared with a gas-chromatographic procedure, *Clin. Biochem.*, 15, 179, 1982.
1190. **Pape, B. E.**, Enzyme immunoassay, liquid chroatography and spectrofluorimetry compared for determination of procainamide and N-acetylprocainamide in serum, *J. Analyt. Toxicol.*, 6, 44, 1982.
1191. **Griffiths, W. C., Dextraze, P., Hayes, M., Mitchell, J., and Diamond, I.**, Assay of serum procainamide and N-acetylprocainamide: comparison of EMIT and reverse-phase high-performance liquid chromatography, *Clin. Toxicol.*, 16, 51, 1980.
1192. **Ha, H. R., Kewitz, G., Wenk, M., and Follath, F.**, Quinidine determination in serum: enzyme immunoassay (EIA) versus h.p.l.c., *Br. J. Clin. Pharmacol.*, 11, 312, 1981.
1193. **Drayer, D. E., Lorenzo, B., and Reidenberg, M. M.**, Liquid chromatography and fluorescence spectroscopy compared with a homogeneous enzyme immunoassay technique for determining quinidine in serum, *Clin. Chem.*, 27, 308, 1981.
1194. **Dextraze, P. G., Foreman, J., Griffiths, W. C., and Diamond, I.**, Comparison of enzyme immunoassay and a high-performance liquid chromatographic method for quantitation of quinidine in serum, *Clin. Toxicol.*, 18, 291, 1981.
1195. **Sivorinovsky, G.**, High-performance liquid chromatography vs. EMIT in therapeutic drug monitoring, with emphasis on theophylline analysis, *Altex Chromatogr.*, 3, 1, 1980.
1196. **Meatherall, R. and Krahn, J.**, Falsely elevated enzyme-multiplied immunoassay serum valproic acid results in 12 patients, *Ther. Drug Monitor.*, 15, 255, 1993.
1197. **Kaneko, S., Homna, H., Kobayashi, H., Sato, T., Koide, N., Haneda, S., Watanabi, J., and Takebe, Y.**, Micro-determination of sodium valproate by gas-liquid chroatography and by homogeneous enzyme-immunoassay technique, *Hirosaki Igaku*, 33, 245, 1981.
1198. **Liu, H., Montoya, J. L., Forman, L. J., Eggers, C. M., Barham, C. F., and Delgado, M.**, Determination of free valproic acid; evaluation of the Centrifree system and comparison between high-performance liquid chromatography and enzyme immunoassay, *Ther. Drug Monitor.*, 14, 513, 1992.
1199. **Olsen, K. M., Gulliksen, M., and Christophersem, A. S.**, Metabolites of chlorpromazine and brompheniramine may cause false-positive urine amphetamine results with monoclonal EMIT d.a.u. immunoassay, *Clin. Chem.*, 38, 611, 1992.
1200. **Burd, J. F., Wong, R. C., Feeney, J. E., Carrico, R. J., and Boguslaski, R. C.**, Homogenous reactant-labelled fluorescent immunoassay for therapeutic drugs exemplified by gentamicin determination in human serum, *Clin. Chem.*, 23, 1402, 1977.
1201. **Burd, J. F., Carrico, R. J., Kramer, H. J., and Denning, C. E.**, Homogenous substrate-labelled fluorescent immunoassay for determining tobramycin concentrations in human serum, in *Enzyme Labelled Immunoassays of Hormones Drugs*, Pal, S. B., Ed., Walter de Gruyter and Co., Berlin-New York, 1978, 387.
1202. **Wong, R. C., Burd, J. F., Carrico, R. J., Buckler, R. T., Thomas, J., and Boguslaski, R. C.**, Substrate-labelled fluorescent immunoassay for phenytoin in human serum, *Clin. Chem.*, 25, 686, 1979.
1203. **Li, T. M., Benovic, J. L., Buckler, R. T., and Burd, J. F.**, Homogenous substrate labelled fluorescence immunoassay for theophylline in serum, *Clin. Chem.*, 27, 22, 1981.
1204. **Krausz, L. M., Hitz, J. B., Buckler, R. T., and Burd, J. F.**, Substrate-labelled fluorescent immunoassay for phenobarbitone, *Ther. Drug Monitor.*, 2, 261, 1980.
1205. **Thompson, S. G. and Burd, J. F.**, Substrate-labelled fluorescent immunoassay for amikacin in human serum, *Antimicrob. Agents Chemother.*, 18, 264, 1980.

1206. **Delbeke, F. T., Teale, P., Debackere, M., and Houghton, E.,** ELISA screening with GC-MS confirmation of the tranquillizer chlorprothixene administered in subtherapeutic doses to horses, *J. Pharm. Biomed. Anal.,* 11, 569, 1993.

1207. **Myagkova, M. A., Zapol'skaya, E. B., Dushkinova, M. V., and Polevaya, O. Y.,** Solid-phase quantitative enzyme immunoassay of _^{9}-tetrahydrocannabinol, *Khim. Farm. Zh.,* 7–8, 101, 1992.

1208. **Eremenko, S. N., Sukhov, I. E., Danilova, N. P., Tolkachev, O. N., Vasilov, R. G., and Morishnikov, A. I.,** Solid-phase immunoassay of papverine, *Khim. Farm. Zh.,* 26, 100, 1992.

1209. **Eisler, M. C., Gault, E. A., Smith, H. V., Peregrine, A. S., and Holmes, P. H.,** Evaluation and improvement of an enzyme-linked immunosorbent assay for the detection of isometamidium in bovine serum, *Ther. Drug Monitor.,* 15, 236, 1993.

1210. **Warty, V., Zuckerman, S., Venkataramanan, R., Lever, J., Fung, J., and Starlz, T.,** FK506 measurement: comparison of different analytical methods, *Ther. Drug Monitor.,* 15, 204, 1993.

1211. **Lewis, L. K., Elder, P. A., and Barrell, G. K.,** Enzyme-linked immunosorbent assay (ELISA) for plasma medroxyprogesterone acetate (MPA), *J. Steroid Biochem. Mol. Biol.,* 42, 179, 1992.

1212. **Kachab, E. H., Wu, W.-Y., and Chapman, C. B.,** Development of an enzyme-linked immunosorbent assay (ELISA) for cephalexin, *J. Immunol. Methods,* 147, 33, 1992.

1213. **Wermeille, M., Moret, E., Siest, J.-P., Ghribi, S., Petit, A.-M., and Wellman, M.,** Determination of dimethindene in human serum by enzyme-linked immunosorbent assay (EIA), *J. Pharm. Biomed. Anal.,* 11, 619, 1993.

1214. **Danilova, N. D. and Vasilov, R. G.,** Production and characterization of anti-theophylline monoclonal antibodies suitable for immunoassay, *Immunol. Lett.,* 28, 79, 1991.

1215. **Danilova, N. P., Beckman, N. I., Ziganshin, A. S., and Vasilov, R. G.,** Monoclonal antibody-based enzyme immunoassay for determination of theophylline in serum and dry blood spots, *Khim. Farm. Zh.,* 7–8, 92, 1992.

1216. **Brandon, D. L., Binder, R. G., Bates, A. H., and Montague, W. C.,** Monoclonal antibody-based ELISA for thiabendazole in liver, *J. Agric. Food Chem.,* 40, 1722, 1992.

1217. **Wal, J.-M., Yvon, M., Pradelles, P., and Grassi, J.,** Enzyme immunoassay of benzyl penicilloyl (BPO) groups using acetylcholinesterase as a label. Application to the study of the BPO-binding sites on albumin, *J. Immunoassay,* 12, 47, 1991.

1218. **Angeletti, R., Paleologo Oriundi, M., Piro, R., Bagnati, R.,** Application of an enzyme-linked immunosorbent assay kit for β-agonist screening of bovine urine in northeastern Italy, *Anal. Chim. Acta,* 275, 215, 1993.

1219. **Leute, R. K., Ullman, E. F., Goldstein, A and Herzenberg, L. A.,** Spin immunoassay technique for determination of morphine, *Nature,* 236, 253, 1971.

1220. **Kaeferstein, H. and Sticht, G.,** Experience gained with the new TRIAGE immunoassay, *Labor. Med.,* 15, 459, 1992.

1221. **Buechler, K. F., Moi, S., Noar, B., McGrath, D., Clancy, M., Shenhav, A., Colleymore, A., Valkirs, G., Lee, T., Bruni, J. F., Walsh, M., Hoffman, R., and Ahmuty, F., et al.,** Simultaneous detection of seven drugs of abuse by the Triage panel for drugs of abuse, *Clin. Chem.,* 38, 1678, 1992.

1222. **Hochhaus, G., Hochhaus, R., Werber, G., Derendorf, H., and Moellmann, H.,** Selective HPLC-RIA for dexamethasone and its pro-drug dexamethasone-21-sulfobenzoate sodium in biological fluids, *Biomed. Chromatogr.,* 6, 283, 1992.

1223. **Ogert, R. A., Kusterbeck, A. W., Wemhoff, G. A., Burke, R., and Ligler, F. S.,** Detection of cocaine using the flow immunosensor, *Analyt. Lett.,* 25, 1999, 1992.

1224. **Palmer, D. A., Edmonds, T. E., and Seare, N. J.,** Flow-injection electrochemical enzyme immunoassay for theophylline using a protein A immunoreactor and *p*-aminophenyl phosphate-*p*-aminophenol as the detection system, *Analyst,* 117, 1679, 1992.

1225. **Zacur, H. A., Linkins, S., Chang, V., Smith, B., Kimball, A. W., and Burkman, R.,** Ethinyloestradiol and norethindrone radioimmunoassay following Sephadex LH-20 column chromatography, *Clin. Chim. Acta,* 204, 209, 1991.

1226. **Stanley, S. M. R., Wilhelmi, B. S., Rodgers, J. P., and Berttschinger, H.,** Immunoaffinity chromatography combined with gas chromatography negative-ion chemical-ionization mass spectrometry for the confirmation of flumethasone abuse in the equine, *J. Chromatogr.,* 614, 77, 1993.

1227. **Irth, H., Oosterkamp, A. J., Van der Welle, W., Tjaden, U. R., and Van der Greef, J.,** Online immunochemical detection in liquid chromatography using fluorescein-labelled antibodies, *J. Chromatogr.,* 633, 65, 1993.

1228. **Ekins, R. P.,** The estimation of thyroxine in human plasma by an electrophoretic technique, *Clin. Chim. Acta,* 5, 453, 1960.

1229. **Jensen, E. V. and Jacobson, H. I.,** Basic guides to the mechanism of estrogen action, R*ec. Prog. Horm. Res.,* 18, 387, 1962.

1230. **Lefkowitz, R. J., Roth, J., and Pastan, I.,** Radioreceptor assay of adrenocorticotrophin hormone: new approach to assay of polypeptide hormones in plasma, *Science,* 170, 633, 1970.

1231. **Simantov, R., Childers, S., and Snyder, S. H.,** Opioid peptides: differentiation by radioimmunoassay and radioreceptor assay, *Brain Res.,* 135, 358, 1977.

1232. **Gilman, A. G.,** Protein binding assay for adenosine 3':5' cyclic monophosphate, *Proc. Natl. Acad. Sci. U.S.A.,* 65, 307, 1970.

1233. **Enna, S. J. and Snyder, S. H.,** A simple, sensitive and specific radioreceptor assay for endogenous GABA in brain tissue, *J. Neurochem.,* 26, 221, 1976.

1234. **Owen, F., Lofthouse, R., and Bourne, R. C.,** A radioreceptor assay for diazepam and its metabolites in serum, *Clin. Chim. Acta,* 93, 305, 1977.

1235. **Creese, I. and Snyder, S. H.,** A simple and sensitive radioreceptor assay for antischizophrenic drugs in blood, *Nature,* 270, 180, 1977.

1236. **Nahorski, S. R., Batta, M. I., and Barnett, D. B.,** Measurement of b-adrenoreceptor antagonists in biological fluids using a radioreceptor assay, *Eur. J. Pharmacol.,* 52, 393, 1978.

1237. **Caselli, G., Ferrari, M. P., Tonon, G., Clavenna, G., and Borsa, M.,** Determination of Nuvenzepine in human plasma by a sensitive [^{3}H]pirenzepine radioreceptor binding assay, *J. Pharm. Sci.,* 80, 173, 1990.

1238. **Kaila, T.,** A sensitive radioligand binding assay for timolol in plasma, *J. Pharm. Sci.,* 80, 296, 1990.

1239. **Murthy, J. N., Chen, Y., Warty, V. S., Venkataramanan, R., Donnelly, J. G., Zeevy, A., and Soldin, S. J.,** Radioreceptor assay for quantifying FK-506 immunosuppressant in whole blood, *Clin. Chem.,* 38, 1307, 1992.

1240. **Dorow, R. G., Seidler, J., and Schneider, H. H.,** A radioreceptor assay to study the affinity of benzodiazepines and their receptor binding activity in human plasma including their active metabolites, *Br. J. Clin. Pharmacol.,* 13, 561, 1982.

1241. **Yamada, S., Matsuoka, Y., Suzuki, N., Sugimoto, N., Kato, Y., and Kimura, R.,** Determination of a novel calcium channel antagonist, mepirodine, in plasma by radioreceptor assay using (+)-[^{3}H]PN 200–110, *Pharm. Res.,* 9, 1227, 1992.

1242. **Yamada, S., Kimura, R., Harada, Y., and Nakayama, K.,** Calcium channel receptor sites for (+)-[^{3}H]PN 200–110 in coronary artery, *J. Pharmacol. Exp. Ther.,* 252, 327, 1990.

1243. **Hunt, P.,** The assay of benzodiazepines by a radioreceptor technique, *Actual. Chimie. Ther.,* 8, 99, 1981.

1244. **Foldesi, I., Falkay, G., and Kovacs, L.,** Simple and sensitive radioreceptor assay of RU 486 [mifepristone] and its biologically active metabolites in human serum, *Acta Pharmacol. Hung.,* 62, 271, 1992.

1245. **Soldin, S. J., Rifai, N., Palaszynski, E. W., Morales, A.and Lipnick, R. N.,** Development of a radioreceptor assay to measure glucocorticoids, *Ther. Drug Monitor.,* 14, 164, 1992.

1246. **Squires, R. F. and Braestrup, C.,** Benzodiazepine receptors in rat brain, *Nature,* 266, 732, 1977.

1247. **Mohler, H. and Okado, T.,** Benzodiazepine receptor: demonstration in the central nervous system, *Science,* 198, 849, 1977.

1248. **Mohler, H., Okada, T., Heitz, P., and Ulrich, J.,** Biochemical identification of the site of action of benzodiazepines in human brain by ^{3}H-diazepam binding, *Life Sci.,* 22, 985, 1978.

1249. **Braestrup, C., Albrechtsen, R., and Squires, R. F.,** High densities of benzodiazepine receptorsin human cortical areas, *Nature,* 269, 702, 1977.

1250. **Mennini, T. and Garattini, S.,** Benzodiazepine receptors:correlation with pharmacological responses in living animals, *Life Sci.,* 31, 2025, 1982.

1251. **Jochemsen, R., Horbach, G. J. M. J., and Breimer, D. D.,** Assay of nitrazepam and triazolam in plasma by a radioreceptor technique and comparison with a gas-chromatographic method, *Res. Commun. Chem. Pathol. Pharmacol.,* 35, 259, 1982.

1252. **Takeuchi, T., Tanaka, S., and Rechnitz, G. A.,** Biotinylated 1012-S conjugate as a probe ligand for benzodiazepine receptors: characterization of receptor binding sites and receptor assay for benzodiazepine drugs, *Analyt. Biochem.,* 203, 158, 1992.

1253. **Tanaka, S., Takeuchi, T., and Rechnitz, G. A.,** Non-isotopic receptor assay for benzodiazepines using a biotin-labelled ligand and biotin-immobilized mictotiter plate, *J. Chromatogr.,* 597, 443, 1992.

1254. **Bruhwyler, J. and Hassoun, A.,** Use of radioreceptor assays for the determination of benzodiazepines in biological samples: a review, *J. Analyt. Toxicol.,* 16, 244, 1992.

1255. **Iversen, L. L.,** Dopamine receptors in the brain, *Science,* 188, 1084, 1975.

1256. **Creese, I., Burt, D. R., and Snyder, S. H.,** Dopamine receptor binding predicts clinical and pharmacological potencies of antischizophrenic drugs, *Science,* 194, 481, 1976.

1257. **Seeman, P., Lee, T., Chau-Wong, M., and Wong, K.,** Antipsychotic drug doses and neuroleptic/dopamine receptors, *Nature,* 261, 717, 1976.

1258. **Fyhrkist, F., Tikkanen, I., Gronhagen-Riska, C., Hortling, L., and Hichens, M.,** Inhibitor binding assay for angiotensin-converting enzyme *Clin. Chem.,* 30, 696, 1984.

1259. **Gronhagen-Riska, C., Tikkanen, I., and Fyhrkist, F.,** Competitive inhibition binding assay (CIBA) of captopril and other ACE inhibitors, *Clin. Chem.,* 162, 53, 1987.

1260. **White, L. O., Holt, H. A., Reeves, D. S., Bywater, M. A., and Bax, R. P.,** Separation and assay of cefotaxime and its metabolites in serum, urine and bile, *Current Chemotherapy and Infectious Disease, Proceedings 11th ICC and 19th ICAAC,* American Society of Microbiology, Washington, D.C., 1980, 153.

1261. **Actor, P., Pitkin, D. H., Lucyszyn, G., Weisbach, J. A., and Brau, J. L.,** Cefatrizine; a new oral cephalosporin: serum levels and urinary recovery in humans after oral or intramuscular administration, comparison study with cephalexin and cefazolin, *Antimicrob. Agents Chemother.,* 9, 800, 1976.

1262. **Rattie, E. S. and Ravin, L. J.,** Pharmacokinetic interpretation of blood levels and urinary excretion data for cefazolin and cephalothin after intravenous and intramuscular administration in humans, *Antimicrob. Agents Chemother.,* 7, 606, 1975.

1263. **Bergan, T.,** Comparative pharmacokinetics of cefazolin, cephalothin, cephacetril, and cephapirine after intravenous administration, *Chemotherapy,* 23, 389, 1977.

1264. **Barza, M., Melethil, S., Berger, S., and Ernst, E. C.,** Comparative pharmacokinetics of cefamandole, cephapirin, and cephalothin in healthy subjects and effect of repeated dosing, *Antimicrob. Agents Chemother.,* 10, 421, 1976.

1265. **Luthy, R., Munch, R., Blaser, J., Bhend, H., and Siegenthaler, W.,** Human pharmacology of cefotaxime (HR756), a new cephalosporin, *Antimicrob. Agents Chemother.,* 16, 127, 1979.

1266. **Sonneville, P. F., Kartodirdjo, R. R., Skeggs, H., Till, A. E., and Martin, C. M.,** Comparative clinical pharmacology of intravenous cefoxitin and cephalothin, *Eur. J. Clin. Pharmacol.,* 9, 397, 1976.

1267. **Harada, Y., Matsubara, S., Kakimoto, M., Noto, T., Nehashi, T., Kimura, T., Suzuki, S., Ogawa, H., and Koyama, K.,** Ceftezole, a new cephalosporin C derivative. II. Distribution and excretion in parenteral administration, *J. Antibiot.,* 29, 1071, 1976.

1268. **Axelrod, J., Meyers, B. R., and Hirschman, S. Z.,** Cephapirin: pharmacology in normal human volunteers, *J. Clin. Pharmacol.,* 12, 84, 1972.

1269. **Rattie, E. S., Bernado, P. D., and Ravin, L. J.,** Pharmacokinetic interpretation of cephradine levels in serum after intravenous and extravascular administration in humans, *Antimicrob. Agents Chemother.,* 10, 283, 1976.

1270. **Peterson, L. R., Bean, B., Fasching, C. E., Korchik, W. P., and Gerding, D. N.,** Pharmacokinetics, protein binding, and predicted extravasular distribution of moxalactam in normal and renal failure patients, *Antimicrob. Agents Chemother.,* 20, 378, 1981.

1271. **Horton, R.,** A comparison of HPLC and bioassay for β-lactam antibiotics, in *Analysis for Drugs and Metabolites, Including Anti-Infective Agents,* Reid, E. and Wilson, I. D., Eds., Royal Society of Chemistry, Cambridge, 1990, 93.

1272. **Smith, R. V. and Stewart, J. T.,** *Textbook of Biopharmaceutic Analysis,* Lea & Febiger, Philadelphia, 1981, 254.

1273. **Barber, M. and Waterworth, P. M.,** Penicillinase-resistant penicillins and cephalosporins, *Br. Med. J.,* ii, 344, 1964.

1274. **Petty, M. A., Reid, J. L., and Miller, S. H. K.,** Plasma converting enzyme activity (C.E.A.): an index of plasma levels of captopril, *Life Sci.,* 26, 2045, 1980.

1275. **Reydel-Bax, P., Redalieu, E., and Rakhit, A.,** Direct determination of angiotensin-converting enzyme inhibitors in plasma by radioenzymatic assay, *Clin. Chem.,* 33, 549, 1987.

1276. **Swanson, B. N., Stauber, K. L., Alpaugh, W. C., and Weinstein, S. H.,** Radioenzymatic assay of angiotensin-converting enzyme inhibitors in plasma and urine, *Analyt. Biochem.,* 148, 401, 1985.

1277. **Gronhagen-Riska, C., Tikkanen, I., and Fyhrkist, F.,** Competitive inhibition binding assay (CIBA) of captopril and other ACE inhibitors, *Clin. Chem.,* 162, 53, 1987.

1278. **Jackson, B., Cubela, R., and Johnston, C. I.,** Measurement of angiotensin converting enzyme inhibitor in serum by radioinhibition binding displacement, *Biochem. Pharmacol.,* 36, 1357, 1987.

1279. **Jacobsen, E. and Rojahn, T.,** Polarographic determination of hexachlorophane, *Analyt. Chim. Acta,* 61, 320 1972.

1280. **Volke, J., Wasilewska, L., and Ryvalova-Kejharova, A.,** Polarographic determination of *p*-fluoro-substituted butyrophenone derivatives, *Pharmazie,* 26, 399, 1971.

1281. **Clifford, J. M. and Smyth, W. F.,** The determination of some benzodiazepines and their metabolites in body fluids. A review, *Analyst,* 99, 241, 1974.

1282. **Smyth, M. R., Smyth, W. F., Palmer, R. F., and Clifford, J. M.,** Acid-based equilibria of some benzhydryl piperazine derivatives, *Analyt. Chim. Acta,* 86, 185, 1976.

1283. **Florence, T. M.,** Polarography of azo compounds and their metal complexes, *J. Electroanalyt. Chem.,* 52, 115, 1974.

1284. **Brooks, M. A., De Silva, J. A. F., and D'Arconte, A. L. M.,** Determination of 2,4-diamino-5-(3,4,5-trimethoxybenzyl) pyrimidine (trimethoprim) in blood and urine by differential pulse polarography, *Analyt. Chem.,* 45, 263, 1973.

1285. **Smyth, W. F., Svehla, G., and Zuman, P.,** Polarography of some sulfur-containing compounds. XV. Polarographic and spectral investigation of acid-base equilibrium in aqueous solutions of substituted 2-thiobarbiturates, *Analyt. Chim. Acta,* 51, 463, 1970.

1286. **Beckett, A. H., Essien, E. E., and Smyth, W. F.,** A polarographic method for the determination of the N-oxide, N-oxide sulphoxide and sulphoxide metabolites of chlorpromazine, *J. Pharm. Pharmacol.,* 26, 399, 1974.

1287. **Hackman, M. R., Brooks, M. A., De Silva, J. A. F., and Ma, T. S.,** Determination of chlordiazepoxide hydrochloride (Librium) and its major metabolites in plasma by differential pulse polarography, *Analyt. Chem.,* 46, 1075, 1974.

1288. **Clifford, J. M.,** Determination of two benzodiazepine anticonvulsants in the plasma of rhesus monkeys when given intravenously, singly or together or in combination with a benzhydryl-piperazine anticonvulsant, in *Assays of Drugs and Other Trace Compounds in Biological Fluids,* Reid, E., Ed., North-Holland, Amsterdam, 1976, 203.

1289. **Brooks, M. A. and Hackman, M. R.,** Trace level determination of 1,4-benzodiazepines in blood by differential pulse polarography, *Analyt. Chem.,* 47, 2059, 1975.

1290. **Kaplan, S. A., Jack, M. L., Weinfeld, R. E., Glover, W., Weissman, L., and Cotler, S.,** Biopharmaceutical and clinical pharmacokinetic profile of bromazepam, *J. Pharmacokinet. Biopharm.,* 4, 1, 1976.

1291. **Smyth, M. R. and Smyth, W. F.,** Polarography as applied to the analysis of certaindrugs and their metabolites in body fluids, in *Assays of Drugs and Other Trace Compounds in Biological Fluids,* Reid, E., Ed., North-Holland, Amsterdam, 1976, 15.

1292. **El Maali, N. A., Ali, A. M., Khodari, M., and Ghandour, M. A.,** Cathode-stripping voltammetric determination of the cephalosporin antibiotic ceftriaxone at the mercury electrode in aqueous and biological media, *Bioelectrochem. Bioenerg.,* 26, 485, 1991.

1293. **Marin, D. and Teijeiro, C.,** Differential-pulse-polarographic determination of the antineoplastic agent cytarabine and its isomer the nucleoside cytidine, *Bioelectrochem. Bioenerg.,* 28, 417, 1992.

1294. **Golabi, D. M. and Nematollahi, D.,** Polarographic determination of doxorubicin hydrochloride and daunorubicin hydrochloride in pharmaceutical preparations and biological media, *J. Pharm. Biomed. Anal.,* 10, 1053, 1992.

1295. **Yan, J. G., Zhu, C. J., Pu, G. G., and Wang, E. K.,** Differential pulse stripping voltammetry of cancerostatic 5-fluorouracil and determination in blood serum, *Bioelectrochem. Bioenerg.* 29, 347, 1993.

1296. **Zapardiel, A., Bermejo, E., Perez Lopez, J. A., Mateo, P., and Hernandez, L.,** Electroanalytical determination of lormetazepam in pharmaceutical preparations and human urine, *Electroanalysis,* 4, 811, 1992.

1297. **Christie, I., Leeds, S., Baker, M., Keedy, F., and Vadgama, P.,** Direct electrochemical determination of paracetamol in plasma, *Anal. Chim. Acta,* 272, 145, 1993.

1298. **Schleyer, E., Lohmann, M., Rolf, C., Gralow, A., Kaufmann, C. C., Unterhalt, M., and Hiddeman, W.,** Column-switching solid-phase trace-enrichment high-performance liquid-chromatographic method for measurement of buprenorphine and norbuprenorphine in human plasma and urine by electrochemical detection, *J. Chromatogr.,* 614, 275, 1993.

1299. **Karlberg, B. and Thelander, S.,** Extraction based on the flow injection principle, Part 1. Description of the extraction system, *Analyt. Chem.,* 98, 1, 1978.

1300. **Landis, J. B.,** Rapid determination of corticosteroids in pharmaceuticals by flow injection analysis, *Analyt. Chim. Acta,* 114, 155, 1980.

1301. **Karlberg, B., Johansson, A., and Thelander, S.,** Extraction based on the flow injection principle. Determination of codeine as the picrate ion-pair in acetylsalicylic acid tablets, *Analyt. Chim. Acta,* 104, 21, 1979.

1302. **Ruzicka, J. and Hansen, E. H.,** Flow injection analysis. Principles, application and trends, *Analyt. Chim. Acta,* 114, 19, 1980.

1303. **Stewart, K. K.,** Flow injection analysis. New tool for old assays. New approach to analytical measurements, *Analyt. Chem.,* 55, 913A, 1983.

1304. **Ruzicka, J.,** Discovering flow injection: journey from sample to a live cell and from solution to suspension, *Analyst,* 119, 1925, 1994.

1305. **Ruzicka, J. and Hansen, E. H.,** Flow injection analysis. I. A new concept of fast continuous flow analysis, *Analyt. Chim. Acta,* 78, 145, 1975.

1306. **Mindegaard, J.,** Flow multi-injection analysis — a system for the analysis of highly concentrated samples without prior dilution, *Analyt. Chim. Acta,* 104, 185, 1979.

1307. **Skeggs, L. T.,** Automatic method for colorimetric analysis, *Am. J. Clin. Path.,* 28, 311, 1957.

1308. **Betteridge, D.,** Flow injection analysis, *Analyt. Chem.,* 19, 832A, 1978.

1309. **Rocks, B. and Riley, C.,** Flow injection analysis: a new approach to quantitative measurements in clinical chemistry, *Clin. Chem.,* 28, 409, 1982.

1310. **Tijssen, R.,** Axial dispersion and flow phenomena in helicaly coiled tubular reactors for flow analysis and chromatography, *Analyt. Chim. Acta,* 114, 71, 1980.

1311. **Evans, M., Palmer, D. A., Miller, J. N., and French, M. T.,** Flow injection fluorescence immunoassay for serum phenytoin using perfusion chromatography, *Analyt. Proc.,* 31, 1994.

1312. **Palmer, D. A., Edmonds, T. E., and Seare, N. J.,** Flow-injection immunosensor for theophylline, *Analyt. Lett.,* 26, 1425, 1993.

1313. **Dieterle, W. and Faigle, J. W.,** Multiple inverse isotope dilution assay for oxprenolol and nine metabolites in biological fluids, *J. Chromatogr.,* 259, 301, 1983.

1314. **Hammer, W. M. and Brodie, B. B.,** Application of isotope derivative technique to assay of secondary amines: estimation of desipramine by acetylation with 3H-acetic anhydride. *J. Pharmacol. Exp. Ther.,* 157, 503, 1967.

1315. **Nisikawa, M., Tatsuno, M., Suzuki, S., and Tsuchihashi, H.,** Analysis of quaternary ammonium compounds in human urine by direct-inlet electron-impact-ionization mass spectrometry, *Forensic Sci. Int.,* 51, 131, 1991.

1316. **Culea, M., Palibroda, N., Chiriac, M., Moldovan, Z., Abraham, A. D., and Frangopol, P. T.,** Isotope-dilution mass spectrometry for procaine determination in biological samples, *Biol. Mass Spectromet.,* 20, 740, 1991.

1317. **Brzezinka, H., Bold, P., and Budzikiewicz, H.,** Screening method for the rapid detection of barbiturates in serum by means of tandem mass spectrometry, *Biol. Mass Spectromet.,* 22, 346, 1993.

1318. **Ismail, I. M. and Wilson, I. D.,** Detection, identification and quantitative analysis of drugs by 1H NMR, in *Bioactive Analytes, Including CNS Drugs Peptides and Enantiomers,* Reid, E., Scales, B., and Wilson, I. D., Eds., Plenum Press, New York, 1986, 337.

1319. **Nicholson, J. K., Sadler, P. J., Tulip, K., and Timbrell, J. A.,** The application of high resolution proton NMR spectroscopy to the detection of drug metabolites in biological samples, in *Bioactive Analytes, Including CNS Drugs Peptides and Enantiomers,* Reid, E., Scayles, B., and Wilson, I. D., Eds., Plenum Press, New York, 1986, 321.

1320. **Threadgill, M. D. and Gescher, A.,** NMR as an aid in studying N-alkylformamides and metabolites, in *Bioanalysis of Drugs and Metabolites, Especially Anti-Inflammatory and Cardiovascular,* Reid, E., Robinson, J. R., and Wilson, I. D., Eds., Plenum Press, New York, 1988, 389.

1321. **Pickford, R., Smith, I. K., and Wilson, I. D.,** Urinanalysis by ^{1}H NMR: application to cephaloridine-treated rats, in *Bioanalysis of Drugs and Metabolites, Especially Anti-Inflammatory and Cardiovascular* Reid, E., Robinson, J. R., and Wilson, I. D., Eds., Plenum Press, New York, 1988, 371.

1322. **Sanins, S. M., Timbrell, J. A., Elcombe, C., and Nicholson, J. K.,** Proton NMR studies on the metabolism and biochemical effects of hydrazine in vivo, in *Bioanalysis of Drugs and Metabolites, Especially Anti-Inflammatory and Cardiovascular* Reid, E., Robinson, J. R., and Wilson, I. D., Eds., Plenum Press, New York, 1988, 375.

1323. **Ghauri, F. Y. K., Wilson, I. D., and Nicholson, J. K.,** ^{19}F- and ^{1}H-studies of the metabolism of 4-trifluoromethylbenzoic acid in the rat, in *Analysis for Drugs and Metabolites, Including Anti-Infective Agents,* Reid, E. and Wilson, I. D., Eds., Royal Society of Chemistry, Cambridge, 1990, 321.

1324. **Wade, K. E., Troke, J., Macdonald, C. M., Wilson, I. D., and Nicholson, J. K.,** ^{19}F NMR studies of the metabolism of trifluoromethylaniline, in *Bioanalysis of Drugs and Metabolites, Especially Anti-Inflammatory and Cardiovascular* Reid, E., Robinson, J. R., and Wilson, I. D., Eds., Plenum Press, New York, 1988, 383.

1325. **Taylor, A., Blackledge, C. A., Nicholson, J. K. Williams, D. A. R., and Wilson, I. D.,** Use of cyclodextrins and their derivatives for chiral analysis by high field nuclear magnetic resonance spectroscopy, *Analyt. Proc.,* 29, 229, 1992.

1326. **Wade, K. E., Wilson, I. D., and Nicholson, J. K.,** The use of ^{15}N-NMR in studying the metabolism of ^{15}N-labelled xenobiotics, exemplified by ^{15}N-anailine, in *Analysis for Drugs and Metabolitesd, Including Anti-Infective Agents,* Reid, E. and Wilson, I. D., Eds., Royal Society of Chemistry, Cambridge, 1990, 325.

1327. **Farrant, R. D., Salman, S. R., Lindon, J. C., Cupid, B. C., and Nicholson, J. K.,** Deuterium NMR spectroscopy of biofluids for the identification of drug metabolites: application to *NN*-dimethylformamide, *J. Pharm. Biomed. Anal.,* 11, 687, 1993.

1328. **De Silva, J. A. F.,** The compromise between sensitivity and specificity in analysing biological fluids for drugs, in *Blood Drugs and Other Analytical Challenges,* Reid, E., Ed., Ellis Horwood, Chichester, 1978, 7.

1329. **Chen, D. Y., Adelhelm, K., Cheng, S. L., and Dovichi, N. J.,** A simple laser-induced fluorescence detector for sulforhodamine 101 in a capillary electrophoresis system: detection limits of 10 yoctomoles or six molecules, *Analyst,* 119, 349, 1994.

1330. **Pearce, J. C., Churchill, A., and Terry, A. C.,** Structured approach to method development in high-performance liquid chromatography, *Chemomem. Intell. Lab. Sys.,* 17, 213, 1992.

1331. **Wrisley, L.,** Use of computer simulations in the development of gradient and isocratic high-performance liquid chromatography methods for analysis of drug compounds and synthetic intermediates, *J. Chromatogr.,* 628, 191, 1993.

1332. **Law, B. and Weir, S.,** Fundamental studies in reversed-phase liquid-solid extraction of basic drugs. III. Sample matrix effects, *J. Pharm. Biomed. Anal.,* 10, 487, 1992.

1333. **Casa, M., Berrueta, L. A., Gallo, B., and Vicente, F.,** Solid-phase extraction conditions for the selective isolation of drugs from biological fluids predicted using liquid chromatography, *Chromatographia,* 34, 79, 1992.

1334. **Starkey, B. J., Green, A. J. E., and Mould, G. P.,** The determination of plasma amiodarone and desmethylamiodorone by HPLC, in *Bioanalysis of Drugs and Metabolites, Especially Anti-Inflammatory and Cardiovascular* Reid, E., Robinson, J. R., and Wilson, I. D., Eds., Plenum Press, New York, 1988, 185.

1335. **De Ridder, J. J.,** Accuracy and precision in GC-MS quantitation, in *Blood, Drugs, and Other Analytical Challenges,* Reid, E., Ed., Ellis Horwood, Chichester, 1978, 153.
1336. **Scales, B.,** Quality control of results and sources of error, in *Blood Drugs and Other Analytical Challenges,* Reid, E., Ed., Ellis Horwood, Chichester, 1978, 43.
1337. **Mullins, E.,** Introduction to control charts in the analytical laboratory, *Analyst,* 119, 369, 1994.
1338. **Burnett, D. and Ayers, G. J.,** Quality control of drug assays, in *Therapeutic Drug Monitoring,* Richens, A. and Marks, V., Eds., Churchill Livingstone, Edinburgh, 1981, 183.
1339. **Boobis, S., Persaud, N., and Richens, A.,** Results of a quality control scheme for the assay of theophylline in serum, *Ther. Drug Monitor.,* 1, 257, 1979.
1340. **Penry, J. K.,** Reliability of serum anti-epileptic drug concentrations and patient management, in *Anti-Epileptic Drugs: Quantitative Analysis and Interpretation,* Pippenger, C. E., Penry, J. K., and Kutt, H., Eds., Raven Press, New York, 1978, 1.
1341. **Pippenger, C. E., Paris-Kutt, H., Penry, J. K., and Daly, D. D.,** Anti-epileptic drug levels quality control program: interlaboratory variability, in *Anti-Epileptic Drugs: Quantitative Analysis and Interpretation,* Pippenger, C. E., Penry, J. K., and Kutt, H., Eds., Raven Press, New York, 1978, 187.
1342. **Pippenger, C. E., Penry, J. K., White, B. G., Daly, D. D., and Buddington, R.,** Interlaboratory variability in determination of plasma anti-epileptic drug concentrations, *Arch. Neurol.,* 33, 351, 1976.
1343. **Ayers, G. J., Goudie, J. H., Reed, K., and Burnett, D.,** Quality control in the simultaneous assay of anticonvulsants using an automated gas-chromatographic system with a nitrogen detector. *Clin. Chim. Acta,* 76, 113, 1977.
1344. **Reeves, D. S. and Bywater, H. J.,** Quality control of serum gentamicin assays — experience of national surveys, *J. Antimicrob. Chemother.,* 1, 103, 1975.
1345. **Turner, W. S., Turano, P., and Badzinski, S.,** An attempt to establish quality control in the determination of plasma chlorpromazine by a multi-laboratory comparison, in *Pharmacokinetics of Psychoactive Drugs: Blood Level and Clinical Response,* Spectrum Publication Inc., New York, 1976, 33.
1346. **Wilson, J. F., Tsanaclis, L. M., Perrett, J. E., Williams, J., Wicks, J. F. C., and Richens, A.,** Performance of techniques for measurement of therapeutic drugs in serum. A comparison based on external quality assessment data, *Ther. Drug Monitor.,* 14, 98–106, 1992.
1347. **Simmonds, R. J. and Wood, S. A.,** Quality control of routine assays at contract facilities, in *Analysis for Drugs and Metabolites, Including Anti-Infective Agents,* Reid, E. and Wilson, I. D., Eds., Royal Society of Chemistry, Cambridge, 1990, 69.
1348. Searle Investigation Task Force Final Report, March 1976. Internal memo of the Department of Health Education and Welfare, Washington D.C., .cw 8.cw 121976.
1349. Pharmacokinetics, *Proceedings of International Harmonization of Non-Clinical Drug Safety Requirements,* Tokyo, Japan, 1990, 5.
1350. **Wright, D. S.,** Analytical method considerations for toxicokinetic studies, in *Drug Toxicokinetics,* Welling, P. G. and de la Iglesia, F. A., Eds., Marcel Dekker, New York, 1993, 1.
1351. **Chamberlain, J.,** *The Analysis of Drugs in Biological Fluids,* CRC Press, Boca Raton, 1985, 168.
1352. **Mathieu, M. P.,** *New Drug Development: A Regulatory Overview,* OMEC International, Washington, D.C., 1987, 29.
1353. **Wu, A. H. B., Johnson, K. G., and Wong, S. S.,** Impact of revised NIDA guidelines for methamphetamine testing in urine, *Clin. Chem.,* 38, 2352, 1992.
1354. **Tsanaclis, L. M. and Wilson, J. F.,** Intra- and inter-laboratory sources of imprecision in drug measurements by different techniques, *Clin. Chem.,* 39, 851, 1993.
1355. **Reid, E., Ed.,** *Assays of Drugs and Other Trace Compounds in Biological Fluids,* North-Holland, Amsterdam, 1976.
1356. **Reid, E., Ed.,** *Blood Drugs and Other Analytical Challenges,* Ellis Horwood, Chichester, 1978.
1357. **Reid, E., Ed.,** *Trace Organic Sample Handling,* Ellis Horwood Limited, Chichester, 1981.
1358. **Reid, E. and Leppard, J. P., Eds.,** *Drug Metabolite Isolation and Determination,* Plenum Press, New York, 1983.
1359. **Reid, E. and Wilson, I. D., Eds.,** *Drug Determination in Therapeutic and Forensic Contexts,* Plenum Press, New York, 1984.

1360. **Reid, E., Scayles, B., and Wilson, I. D., Eds.,** *Bioactive Analytes, Including CNS Drugs Peptides and Enantiomers,* Plenum Press, New York, 1986.

1361. **Reid, E., Robinson, J. R., and Wilson, I. D., Eds.,** *Bioanalysis of Drugs and Metabolites, Especially Anti-Inflammatory and Cardiovascular,* Plenum Press, New York, 1988.

1362. **Reid, E. and Wilson, I. D., Eds.,** *Analysis for Drugs and Metabolitesd, Including Anti-Infective Agents,* Royal Society of Chemistry, Cambridge, 1990.

1363. **Reid, E. and Wilson, I. D., Eds.,** *Bioanalytic Approaches for Drugs, Including Anti-Asthmatics and Metabolites,* Royal Society of Chemistry, Cambridge, 1992.

1364. **Reid, E., Hill, H. M., and Wilson, I. D.,** *Biofluid and Tissue Analysis for Drugs Including Hypolipidaemics,* Royal Society of Chemistry, Cambridge, 1994.

1365. **Hajdu, P. and Chamberlain, J.,** Quality control systems for routine drug analysis, in *Drug Determination in Therapeutic and Forensic Contexts,* Reid, E. and Wilson, I. D., Eds., Plenum Press, New York, 1985, 399.

1366. **Aden Abdi, Y., Villen, T., Gustafsson, L.L., Ericsson, O., and Sjoqvist, F.,** Methodological commentary on the analysis of metriphonate and dichlorvos in biological samples, *J. Chromatogr.,* 612, 336, 1993.

1367. **Unni, L. K.,** Methodological commentary on the analysis of metriphonate and dichlorvos in biological samples, *J. Chromatogr.,* 612, 338, 1993.

1368. **Smith R. V. and Escalona-Castillo, H. J.,** Comparison of esterification methods for the gas chromatographic determination of indoprofen in plasma and urine, *Microchem. J.,* 22, 305, 1977.

1369. **Adam, H. K.,** Stability of drugs in stored biological samples, in *Trace Organic Sample Handling,* Reid, E., Ed., Ellis Horwood Limited, Chichester, 1981, 291.

1370. **Brogan, W. C., Kemp, P.M., Bost, R. O., Glamann, D. B., Lange, R. A., and Hillis, L. D.,** Collection and handling of clinical blood samples to assure the accurate measurement of cocaine concentration, *J. Analyt. Toxicol.,* 16, 152, 1992.

1371. **Karch, S. B.,** Is ecgogenine methyl ester a major in vivo metabolite of cocaine in humans? *J. Analyt. Toxicol.,* 17, 318, 1993.

1372. **Isenschmid, D. S.,** Is ecgogenine methyl ester a major in vivo metabolite of cocaine in humans? A reply *J. Analyt. Toxicol.,* 17, 318, 1993.

1373. **Davey, E. A. and Murray, J. B.,** Hydroysis of diamorphine in aqueous solutions, *Pharmacol. J.,* 203, 737, 1969.

1374. **Nakamura, G. R., Thornton, J. I., and Nogchi, T. T.,** Kinetics of heroin deacetylation in aqueous alkaline solution and in human serum and whole blood, *J. Chromatogr.,* 110, 81, 1975.

1375. **Barrett, D. A., Dyssegaarfd, A. L. P., and Shaw, P. N.,** The effect of temperature and pH on the deacetylation of diamorphine in aqueous solution and in human plasma, *J. Pharm. Pharmacol.,* 44, 606, 1992.

1376. **Scales, B.,** Discussion on subtle gas chromatography, in *Blood Drugs and Other Analytical Challenges,* Reid, E., Ed., Ellis Horwood, Chichester, 1978, 311.

1377. **Edmonds, B. K. and Nierenberg, D. W.,** Serum concentrations of retinol, d-α-tocopherol and β-carotene: effects of storage at $-70°$ for five years, *J. Chromatogr.,* 614, 169, 1993.

1378. **Smith, R. V., Wilcox, R. E., and Humphrey, D. W.,** Stability of apomorphine in frozen plasma samples, *Res. Commun. Chem. Pathol.,* 27, 183, 1980.

1379. **Cowan, D. A.,** Problems in the analysis of N-oxygenated products of the phenothiazines, in *Assays of Drugs and Other Trace Compounds in Biological Fluids,* Reid, E., Ed., North-Holland, Amsterdam, 1976, 193.

1380. **Curry, S. H. and Evans, S.,** A note on the assay of chlorpromazine N-oxide and its sulphoxide in plasma and urine, *J. Pharm. Pharmacol.,* 28, 1467, 1976.

1381. **Ferry, D. G., Ferry, D. M., Moller, P. W., and McQueen, E. G.,** Indomethacin estimation in plasma and serum by electron capture gas chromatography, *J. Chromatogr.,* 89, 110, 1974.

1382. **Curry, S. H. and Whelpton, R.,** Critical factors in drug analysis, in *Trace Organic Sample Handling,* Reid, E., Ed., Ellis Horwood Limited, Chichester, 1981, 347.

1383. **de Silva, J. A. F.,** GLP in analytical method validation, in *Trace Organic Sample Handling,* Reid, E., Ed., Ellis Horwood Limited, Chichester, 1981, 298.

1384. **Degen, P. H.,** The assay of hydralazine: a review, in *Bioanalysis of Drugs and Metabolites, Especially Anti-Inflammatory and Cardiovascular* Reid, E., Robinson, J. R., and Wilson, I. D., Eds., Plenum Press, New York, 1988, 193.

1385. **Zak, S. V., Bartlett, M. F., Wagner, W. E., Gilleran, T. G., and Lukas, G.,** Disposition of hydralazine in man and a specific method for its determination in biological fluids, *J. Pharm. Sci.,* 63, 225, 1974.

1386. **Moreau, J., Palette, C., Cordonnier, P., Naline, E., Advenier, C., and Pays, M.,** Separation and identification of the 4-hydroxyantipyrine sulfoconjugate, *J. Chromatogr.,* 576, 103, 1992.

1387. **Jackson, J. V.,** Extraction methods in toxicology, in *Isolation and Identification of Drugs,* Clarke, E. G. C., Ed., The Pharmaceutical Press, London, 1969, 16.

1388. **Reid, E.,** Sample preparation in the micro-determination of organic compounds in plasma and urine, *Analyst,* 101, 1, 1976.

1389. **Campbell, D. B.,** Separation of drugs from body fluids: (1) low volume extraction and (2) problems associated with plasma protein, in *Assays of Drugs and Other Trace Compounds in Biological Fluids,* Reid, E., Ed., North-Holland, Amsterdam, 1976, 105.

1390. **Whelpton, R. and Curry, S. H.,** Solvent extraction as applied to fluphenazine and its metabolites, in *Assays of Drugs and Other Trace Compounds in Biological Fluids,* Reid, E., Ed., North-Holland, Amsterdam, 1976, 115.

1391. **Curry, S. H. and Whelpton, R.,** Statistics of drug analysis, and the role of internal standards, in *Blood Drugs and Other Analytical Challenges,* Reid, E., Ed., Ellis Horwood, Chichester, 1978, 29.

1392. **Bailey, D. N. and Guba, J. J.,** Gas-chromatographic analysis for chlorpromazine and some of its metabolites in human serum, with use of nitrogen detector, *Clin. Chem.,* 25, 1211, 1979.

1393. **Reid, E.,** Analyzes entailing sample processing: an overview, in *Trace Organic Sample Handling,* Reid, E., Ed., Ellis Horwood Limited, Chichester, 1981, 15.

1394. **Scales, B.,** Problems in getting valid results: an overview, in *Trace Organic Sample Handling,* Reid, E., Ed., Ellis Horwood Limited, Chichester, 1981, 353.

1395. **Brunson, M. K. and Nash, J. F.,** Gas-chromatographic measurement of codeine and norcodeine in human plasma, *Clin. Chem.,* 21, 1956, 1975.

1396. **Wester, R., Noonan, P., Markus, C., Bible, R., Aksamit, W., and Hribar, J.,** Identification of carbamate esters formed during chloroform extraction of tricyclic depressants in urine, *J. Chromatogr.,* 209, 463, 1981.

1397. **Franklin, R. A., Heatherington, K., Morison, B. J., Sherren, P., and Ward, T. J.,** Detection of cyanogen chloride as an impurity in dichloromethane and its significance in drug and metabolite analysis, *Analyst,* 103, 660, 1978.

1398. **Reid, E.,** Metabolite methodolgy: the state of play, in *Drug Metabolite Isolation and Determination,* Reid, E. and Leppard, J. P., Eds., Plenum Press, New York, 1983, 255.

1399. **Leppard, J. P. and Reid, E.,** Assay of blood for metoclopramide, in *Blood Drugs and Other Analytical Challenges,* Reid, E., Ed., Ellis Horwood, Chichester, 1978, 315.

1400. **Jackson, J. V.,** Extraction methods in toxicology, in *Isolation and Identification of Drugs,* Clarke, E. G. C., Ed., The Pharmaceutical Press, London, 1969, 16.

1401. **Ramsey, J. and Campbell, D. B.,** An ultra rapid method for the extraction of drugs from biological fluids, *J. Chromatogr.,* 63, 303, 1971.

1402. **Trevor, A., Rowland, M., and Way, E. L.,** Techniques for studying drug disposition in vivo, in *Fundamentals of Drug Metabolism and Drug Disposition,* La Du, B. N., Mandel, H. G., and Way, E. L., Eds., Williams & Wilkins, Baltimore, 1972, 369.

1403. **De Bree, H.,** A note on taking aliquots from samll volumes of upper phases, in *Drug Metabolite Isolation and Determination,* Reid, E. and Leppard, J. P., Eds., Plenum Press, New York, 1983, 69.

1404. **Mitchell, J. D., Perrigo, B. J., and Mason-Daniel, V. A.,** Falsely negative urine drug assay results due to filtration, *Clin. Chem.,* 38, 2556, 1992.

1405. **Hartvig, P., Freij, G., and Vessman, J.,** Electron capture gas chromatography of terodiline after heptafluorobutyrylation, *Acta Pharmacol. Suecica,* 11, 97, 1974.

1406. **Degen, P. and Riess, W.,** Simplified method for the determination of oxprenolol and other b-receptor blocking agents in biological fluids by gas-liquid chromatography, *J. Chromatogr.,* 121, 72, 1976.

1407. **Reid, E.,** Recommended terminology and practices in gas chromatographic assays, especially GC-ECD, entailing sample work-up, in *Blood Drugs and Other Analytical Challenges,* Reid, E., Ed., Ellis Horwood, Chichester, 1978, 61.

1408. **Chalmers, R. A.,** The quantitative freeze-drying of organic compounds of low relative molecular mass, in *Assays of Drugs and Other Trace Compounds in Biological Fluids,* Reid, E., Ed., North-Holland, Amsterdam, 1976, 121.

1409. **Scoggins, B. A., Maguire, K. P., Norman, T. R., and Burrows, G. D.,** Measurements of tricyclic antidepressants. I. A review of methodology, *Clin. Chem.,* 26, 5, 1980.

1410. **Vandermark, F. L., Adams, R. F., and Schmidt, G. J.,** Liquid chromatographic procedure for tricyclic drugs and their metabolites in plasma, *Clin. Chem.,* 24, 87, 1978.

1411. **Jorgenson, A.,** A gas chromatographic method for the determination of amitriptyline and nortriptyline in human serum, *Acta Pharmacol. Toxicol.,* 36, 79, 1975.

1412. **Sonsalla, P. K., Jennison, T.A., and Finkle, B. S.,** Importance of evaporation conditions and two internal standards for quantitation of amitriptyline and nortriptyline, *Clin. Chem.,* 28, 1401, 1982.

1413. **Reid, E.,** Chromatographic and other assay approaches, in *Assays of Drugs and Other Trace Compounds in Biological Fluids,* Reid, E., Ed., North-Holland, Amsterdam, 1976, 55.

1414. **James, V. H. T. and Wilson, G. A.,** Determination of aldosterone in biological fluids, in *Assays of Drugs and Other Trace Compounds in Biological Fluids,* Reid, E., Ed., North-Holland, Amsterdam, 1976, 149.

1415. **Edelbroeck, P. M., de Hass, E. J. M., and de Wolff, F. A.,** Liquid chromatographic determination of amitriptyline and its metabolites in serum, with adsorption onto glass minimised, *Clin. Chem.,* 28, 2143, 1982.

1416. **De Silva, J. A. F.,** Sample preparation for trace drug analysis, in *Trace Organic Sample Handling,* Reid, E., Ed., Ellis Horwood Limited, Chichester, 1981, 192.

1417. **Schill, G.,** Ion-pair solvent extraction and chromatography, in *Assays of Drugs and Other Trace Compounds in Biological Fluids,* Reid, E., Ed., North-Holland, Amsterdam, 1976, 87.

1418. **Thackeray, P. and Hoar, D.,** A note on assay of blood for traces of anionic and non-ionic surfactants, in *Trace Organic Sample Handling,* Reid, E., Ed., Ellis Horwood Limited, Chichester, 1981, 161.

1419. **Ritter, W.,** Assays where high sensitivity was achieved, especially by HPTLC with automatic spotting and colour reactions, in *Drug Metabolite Isolation and Determination,* Reid, E. and Leppard, J. P., Eds., Plenum Press, New York, 1983, 231.

1420. **Blair, A. D., Forrey, A. W., Christopher, T. G., Maxwell, B., and Cutler, R. E.,** Determination of ethambutol in plasma and urine by chemical ionisation GC-MS using a deuterated internal standard, in *Assays of Drugs and Other Trace Compounds in Biological Fluids,* Reid, E., Ed., North-Holland, Amsterdam, 1976, 231.

1421. **Stewart, M. J.,** Extraction of drugs and metabolites from plasma and urine, in *Drug Metabolite Isolation and Determination,* Reid, E. and Leppard, J. P., Eds., Plenum Press, New York, 1983, 41.

1422. **Goehl, T. J., DeWoody, C. T., and Sundaresan, G. M.,** Sublimation losses of salicylic acid from plasma during analysis, *Clin. Chem.,* 27, 776, 1981.

1423. **Horning, M. G., Gregory, P., Nowlin, J., Stafford, K., Lertratanangkoon, K., Butler, C., Stillwell, W. G., and Hill, R. M.,** Isolation of drugs and drug metabolites from biological fluids by the use of salt-solvent pairs, *Clin. Chem.,* 20, 282, 1974.

1424. **Scales, B.,** Two problem compounds: polarographic and other approaches, in *Assays of Drugs and Other Trace Compounds in Biological Fluids,* Reid, E., Ed., North-Holland, Amsterdam, 1976, 281.

1425. **Bonato, P. S. and Lanchote, V. L.,** Rapid procedure for the purification of biological samples to be analyzed by high-performance liquid chromatography, *J. Liq. Chromatogr.,* 16, 2299, 1993.

1426. **Parkin, J. E.,** Salting-out solvent extraction for pre-concentration of benzalkonium chloride prior to high-performance liquid chromatography, *J. Chromatogr.,* 635, 75, 1993.

1427. **Van der Vlis, E., Irth, H., Tjaden, U. R., and Van der Greef, J.,** Reversed-phase liquid-chromatographic determination of doxorubicin after online trace enrichment on iron(III)-loaded 8-hydroxyquinoline-bonded silica, *Anal. Chim. Acta,* 271, 69, 1993.

1428. **Ruane, R. J., Wilson, I. D., and Tomkinson, G. P.,** The use of secondary ionic interactions for the solid-phase extaction of some 'β-blocking' drugs on C-18 bonded silica, in *Bioanalysis of Drugs and Metabolites, Especially Anti-Inflammatory and Cardiovascular,* Reid, E., Robinson, J. R., and Wilson, I. D., Eds., Plenum Press, New York, 1988, 295.

1429. **Whelpton, R. and Hurst, P. R.**, Solid-phase extraction: personal experiences, in *Bioanalysis of Drugs and Metabolites, Especially Anti-Inflammatory and Cardiovascular*, Reid, E., Robinson, J. R., and Wilson, I. D., Eds., Plenum Press, New York, 1988, 289.

1430. **Hill, H. M., Dehelean, L., and Bailey, B. A.**, Application of solid-phase extraction techniques to the analysis of basic and acidic drugs in biological fluids, in *Bioanalysis of Drugs and Metabolites, Especially Anti-Inflammatory and Cardiovascular*, Reid, E., Robinson, J. R., and Wilson, I. D., Eds., Plenum Press, New York, 1988, 299.

1431. **Casas, M., Berrueta, L. A., Gallo, B., and Vincente, F.**, Solid-phase extraction of benzo-1,4-diazepines from biological fluids, *J. Pharm. Biomed. Anal.*, 11, 277, 1993.

1432. **Casas, M., Berrueta, L. A., Gallo, B., and Vincente, F.**, Solid-phase extraction conditions for the selective isolation of drugs from biological fluids predicted using liquid chromatography, *Chromatographia*, 34, 79, 1992.

1433. **Alliegro, M. A., Dyer, K. D., Cragoe, E. J., Glaser, B. M., and Alliegro, M. C.**, High-performance liquid-chromatographic method for quantitating plasma levels of amiloride and its analogues, *J. Chromatogr.*, 582, 217, 1992.

1434. **Jandreski, M. A. and Vanderslice, W. E.**, Clinical measurement of serum amiodarone and desethylamiodarone by using solid-phase extraction followed by HPLC with a high-carbon reversed-phase column, *Clin. Chem.*, 39, 496, 1993.

1435. **Sarkar, M. A., March, C., and Karnes, H. T.**, Solid-phase extraction and simultaneous high-performance liquid-chromatographic determination of antipyrine and its major metabolites in urine, *Biomed. Chromatogr.*, 6, 300, 1992.

1436. **Stiff, D. D., Frye, R. F., and Branch, R. A.**, Sensitive high-performance liquid-chromatographic determination of chlorzoxazone in plasma, *J. Chromatogr.*, 613, 127, 1993.

1437. **Moore, C., Browne, S., Tebbett, I., and Negrusz, A.**, Determination of cocaine and its metabolites in brain tissue using high-flow solid-phase extraction columns and high-performance liquid chromatography, *Forens. Sci. Int.*, 53, 215, 1992.

1438. **Wei, H., Andollo, W., and Hearn, W. L.**, Solid-phase extraction technique for the isolation and identification of opiates in urine, *J. Analyt. Toxicol.*, 16, 307, 1992.

1439. **Howling, R. and Hutchinson, G.**, Determination of dextromethorphan and its metabolites in human urine using solid-phase extraction and reversed-phase high-performance liquid chromatography, *Analyt. Proc.*, 30, 369, 1993.

1440. **Kamali, F.**, Determination of plasma diazepam and desmethyldiazepam by solid-phase extraction and reversed-phase high-performance liquid chromatography, *J. Pharm. Biomed. Anal.*, 11, 625, 1993.

1441. **Nicholls, G., Clark, B. J., and Brown, J. E.**, Solid-phase extraction and optimized separation of doxorubicin, epirubicin and their metabolites using reversed-phase high-performance liquid chromatography, *J. Pharm. Biomed. Anal.*, 10, 949, 1992.

1442. **Campins-Falco, P., Herraez-Hernandez, R., and Sevillano-Cabeza, A.**, Sensitive determination of ethacrynic acid in urine samples by reversed-phase liquid chromatography with ultra-violet detection using solid-phase extraction techniques for sample clean-up, *Anal. Chim. Acta*, 270, 39, 1992.

1443. **Lensmeyer, G. L., Wiebe, D. A., and Doran, T. C.**, Application of the Empore solid-phase extraction membrane to the isolation of drugs from blood. II. Mexiletine and flecainide, *Ther. Drug Monitor.*, 14, 408, 1992.

1444. **Inagaki, K., Takagi, J., Lor, E., Okamoto, M. P., and Gill, M. A.**, Determination of fluconazole in human serum by solid-phase extraction and reversed-phase high-performance liquid chromatography, *Ther. Drug Monitor.*, 14, 306, 1992.

1445. **Bain, B. M., Harrison, G., Jenkins, K. D., Pateman, A. J., and Shenoy, E. V. B.**, Sensitive radio-immunoassay, incorporating solid-phase extraction, for fluticasone propionate in plasma, *J. Pharm. Biomed. Anal.*, 11, 557, 1993.

1446. **Caturla, M. C. and Cusido, E.**, Solid-phase extraction for the high-performance liquid-chromatographic determination of indomethacin, suxibuzone, phenylbutazone and oxyphenabutazone in plasma, avoiding degradation of compounds, *J. Chromatogr.*, 581, 101, 1992.

1447. **Hennig, B. and Luedecke, F.**, Solid-phase extraction for the determination of isosorbide dinitrate and its metabolites in plasma, *GIT Fachz. Lab.*, 35, 1225, 1991.

1448. **Gaillard, Y., Gay-Montchamp, J. P., and Ollagnier, M.,** Gas-chromatographic determination of meprobamate in serum or plasma after solid-phase extraction, *J. Chromatogr.,* 577, 171, 1992.

1449. **Pierce, T. L., Murray, A. G. W., and Hope, W.,** Determination of methadone and its metabolites by high-performance liquid chromatography following solid-phase extraction in rat plasma, *J. Chromatogr. Sci.,* 30, 443, 1992.

1450. **Pawula, M., Barrett, D. A., and Shaw, P. N.,** Improved extraction method for the HPLC determination of morphine and its metabolites in plasma, *J. Pharm. Biomed. Anal.,* 11, 401, 1993.

1451. **Hartley, R., Green, M., Lucock, M. D., Ryan, S., and Forsythe, W. I.,** Solid-phase extraction of oxcarbazepine and its metabolites from plasma for analysis by high-performance liquid chromatography, *Biomed. Chromatogr.,* 5, 212, 1991.

1452. **Simpson, R. C., Boppana, V. K., Hwang, B. Y.-H., and Rhodes, G. R.,** Determination of oxiracetam in human plasma by reversed-phase high-performance liquid chromatography with fluorimetric detection, *J. Chromatogr.,* 631, 227, 1993.

1453. **Mancinelli, A., Pace, S., Marzo, A., Martelli, E. A., and Passetti, G.,** Determination of pentoxiphylline and its metabolites in human plasma by high-performance liquid chromatography with solid-phase extraction, *J. Chromatogr.,* 575, 101, 1992.

1454. **Marko, V. and Bauerova, K.,** Study of the solid-phase extraction of pentoxiphylline and its major metabolite as a basis of their rapid low concentration gas-chromatographic determination in serum, *Biomed. Chromatogr.,* 5, 256, 1991.

1455. **Campins-Falco, P., Herraez-Hernandez, R., and Sevillano-Cabeza, A.,** Sensitive determination of probenicid in urine samples by reversed-phase liquid chromatography and UV-visible detection using solid-phase extraction techniques for sample clean-up, *Chromatographia,* 35, 317, 1993.

1456. **Martin, P., Leadbetter, B., and Wilson, I. D.,** Immobilized phenylboronic acids for the selective extraction of β-blocking drugs from aqueous solution and plasma, *J. Pharm. Biomed. Anal.,* 11, 307, 1993.

1457. **Micelli, G., Lozupone, A., Quaranta, M., Donadeo, A., Coviello, M., and Lorusso, V.,** Determination of intraperitoneal mitoxantrone in the serum of cancer patients using a high-performance liquid chromatograph coupled to an advanced automated sample processor (AASP), *Biomed. Chromatogr.,* 6, 168, 1992.

1458. **Lecaillon, J. B., Rouan, M. C., Campestrini, J., and Dubois, J. P.,** A fully automated analytical system using solid-phase extraction: application to the determination of carbamazepine and two metabolites in plasma, in *Analysis for Drugs and Metabolitesd, Including Anti-Infective Agents,* Reid, E. and Wilson, I. D., Eds., Royal Society of Chemistry, Cambridge, 1990, 265.

1459. **Chen, X.-H., Franke, J.-P., Ensing, K., Wijsbeek, J., and De Zeeuw, R. A.,** Semi-automated solid-phase extraction procedure for drug screening in biological fluids using the ASPEC system in combination with Clean Screen DAU columns, *J. Chromatogr.,* 613, 289, 1993.

1460. **Soltes, L.,** Enhanced selectivity of the isolation of basic drugs from body fluids by means of extraction minicolumns and sophisticated elution schemes, *Biomed. Chromatogr.,* 6, 43–49, 1992.

1461. **Andresen, A. T., Krogh, M., and Rasmussen, K. E.,** Online dialysis and weak cation-exchange enrichment of dialysate. Automated high-performance liquid chromatography of pholcidine in human plasma and whole blood, *J. Chromatogr.,* 582, 123, 1992.

1462. **Ramsey, J. D.,** Filter paper contaminant, *Chem. Br.,* 17, 57, 1981.

1463. **Jones, C. R.,** Problems arising from assaying low concentrations by HPLC, in *Drug Metabolite Isolation and Determination,* Reid, E. and Leppard, J. P., Eds., Plenum Press, New York, 1983, 33.

1464. **Woodbridge, A. P.,** Prevention of contamination in the residue laboratory, in *Trace Organic Sample Handling,* Reid, E., Ed., Ellis Horwood Limited, Chichester, 1981, 156.

1465. **Roll, B. D., Douglas, J. D., and Peterson, R. V.,** GLC analysis of bis(2-ethylhexyl)phthalate plasticizer in tissue and plasma, *J. Pharm. Sci.,* 63, 1628, 1974.

1466. **Hooper, W. D. and Smith, M. T.,** Novel source of ubiquitous phthalates as analytical contaminant, *J. Pharm. Sci.,* 70, 346, 1981.

1467. **Feyerabend, C. and Russell, M. A. H.,** Assay of nicotine in biological materials: sources of contamination and their elimination, *J. Pharm. Pharmacol.* 32, 178, 1980.

1468. **Ende, M., Pfeifer, P., and Spiteller, G.,** Contamination by the use of Extralut columns, *J. Chromatogr.,* 183, 1, 1980.

1469. **Cairns, E. R.,** Sampling procedures for blood ^{9}-tetrahydrocannabinol analysis and interference by anti-oxidant, *J. Analyt. Toxicol.,* 4, 160, 1980.

1470. **Rapaka, R. S., Roth, J., Goehl, T. J., and Prasad, V. K.,** Anticoagulants interfere with analysis for frusemide in plasma, *Clin. Chem.,* 27, 1470, 1981.

1471. **Cotham, R. H. and Shand, D. G.,** Spuriously low plasma propranolol concentration resulting from blood collection methods, *Clin. Pharmacol. Ther.,* 18, 535, 1975.

1472. **Borga, O., Piafsky, K. M., and Nilsen, B. G.,** Plasma protein binding of basic drugs. I. Selective displacement from alpha-1-acid glycoprotein by tris(2-butoxyethyl) phosphate, *Clin. Pharmacol. Ther.,* 22, 539, 1977.

1473. **Piafsky, K. M. and Borga, O.,** Inhibitor of drug-protein binding in "vacutainers", *Lancet 2, 963, 1976.*

1474. **Kessler, K. M., Leech, R. C., and Spavin, J. F.,** Blood collection techniques, heparin and quinidine binding, *Clin. Pharmacol. Ther.,* 25, 204, 1979.

1475. **Stargel, W. W., Roe, C. R., Routledge, P. A., and Shand, D. G.,** Importance of blood-collection tubes in plasma lidocaine determinations, *Clin. Chem.,* 25, 617, 1979.

1476. **Moore, G., Nation, R., Wilkinson, G. R., and Schenker, S.,** Pharmacokinetics of meperidine in man, *Clin. Pharmacol. Ther.,* 19, 486, 1976.

1477. **Midha, K. K., Loo, J. C. K., and Rowe, M. L.,** The influence of blood sampling techniques on the distribution of chlorpromazine and tricyclic antidepressants between plasma and whole blood, *Res. Comm. Psychol. Psychiat. Behav.,* 4, 193, 1979.

1478. **Midha, K. K., Cooper, J. K., Lapierre, Y. D., and Hubbard, J. W.,** Fluphenazine, trifluoperazine and perphenazine in Vacutainers, *Can. Med. Assoc. J.,* 124, 263, 1981.

1479. **Midha, K. K., Loo, J. C. K., and Rowe, M. L.,** The influence of blood sampling techniques on the distribution of chlorpromazine and tricyclic antidepressants between plasma and whole blood, *Res. Commun. Psychol. Psychiat. Behav.,* 4, 193, 1979.

1480. **Veith, R. and Perera, C.,** Tricyclic antidepressants, *N. Engl. J. Med.,* 300, 504, 1979.

1481. **Kessler, K. M., Ho-Tung, P., Steele, B., Silver, J., Pickoff, A., Sheshandri, N., and Myerburg, R. J.,** Simultaneous quantitation of quinidine, procainamide, and *N*-acetylprocainamide in serum by gas-liquid chromatography and with a nitrogen-phosphorus selective detector, *Clin. Chem.,* 28, 1187, 1982.

1482. **Shah, V. P., Knapp, G., Skelly, J. P., and Cabana, B. E.,** Interference with measurement of certain drugs in plasma by a plasticizer in Vacutainer tubes, *Clin. Chem.,* 28, 2327, 1982.

1483. **Westerlund, D. and Nilsson, L. B.,** Some matrix effects on the extraction of hydrophobic amines and a quaternary ammonium compound from serum and plasma, in *Trace Organic Sample Handling,* Reid, E., Ed., Ellis Horwood Limited, Chichester, 1981, 270.

1484. **Westerlund, D., Nilsson, L. B., and Jaksh, Y.,** Straight-phase ion-pair chromatography of zimelidine and similar divalent ions, *J. Liq. Chromatogr.,* 2, 373, 1979.

1485. **Siek, T. J. and Rieders, E. F.,** Formation of 2-(4-chlorophenoxy)-2-methylpropanamide from a clofibrate metabolite of urine, *J. Analyt. Toxicol.,* 5, 194, 1981.

1486. **Hornke, I., Fehlhaber, H.-W., Girgand, M., and Jantz, H.,** Metabolism of nomifensine: isolation and identification of the conjugates of nomifensine-^{14}C from hyman urine, *Br. J. Clin. Pharmacol.,* 9, 255, 1980.

1487. **Upton, R.A., Buskin, J. N., Williams, R. L., Holford, N. H. G., and Riegelman, S.,** Negligible excretion of unchanged ketoprofen, naproxen, and probenicid, in urine, *J. Pharm. Sci.,* 69, 1254, 1980.

1488. **Blankaert, N., Compernolle, F., Leroy, P., van Houtte, R., Fevery, J., and Heirwegh, K. P. M.,** The fate of bilirubin-IXα glucuronide in cholestasis and during storage *in vitro.* Intramolecular rearrangement to positional isomers of glucuronic acid, *Biochem. J.,* 171, 203, 1978.

1489. **Eggers, N. J. and Doust, K.,** Isolation and identification of probenecid acyl glucuronide, *J. Pharm. Pharmacol.,* 33, 123, 1981.

1490. **Hignite, C. E., Ischantz, C., Lemons, S., Wiese, H., Arzanoff, D. L., and Hoffman, D. H.,** Glucuronic acid conjugates of clofibrate: four isomeric structures, *Life Sci.,* 28, 2077, 1981.

1491. **Sinclair, K. A. and Caldwell, J.,** The pH-dependent intramolecular rearrangement of glucuronic acid conjugates of xenobiotics, *Biochem. Soc. Trans.,* 9, 215, 1981.

1492. **Beckett, A. H. and Cowan, D. A.,** Pitfalls in drug metabolism methodology, in D*rug Metabolism in Man,* Gorrod, J. W. and Beckett, A. H., Eds., Taylor and Francis, London, 1978, 237.

1493. **Patel, R. B. and Welling, P. G.,** On-column reduction of promethazine metabolite during gas chromatography, *Clin. Chem.,* 27, 1780, 1981.

1494. **Wright, M. R. and Jamali, F.,** Limited extent of sterochemical conversion of nonsteroidal anti-inflammatory drugs induced by derivatization methods employing ethyl chloroformate, *J. Chromatogr.,* 616, 59, 1993.

1495. **Hornbeck, C. L., Carrig, J. E., and Czarny, R. J.,** Detection of a GC-MS artifact peak as methamphetamine, *J. Analyt. Toxicol.,* 17, 257, 1993.

1496. **Dell, D.,** Views on method validation, in *Analysis for Drugs and Metabolites, Including Anti-Infective Agents,* Reid, E. and Wilson, I. D., Eds., Royal Society of Chemistry, Cambridge, 1990, 9.

1497. **Burrows, J. L.,** Evaluation of calibration routines in pharmaceutical and bioanalytical separations, with reference to Monte-Carlo simulations, *J. Pharm. Biomed. Anal.,* 11, 523, 1993.

1498. **Brown, R. K., Zenie, F. H., and Johnston, G.,** Robotics in chemical laboratories, *Chemistry Br.,* 22, 640, 1986.

1499. **Barlow, G. B.,** Automation of column chromatography, in *Trace Organic Sample Handling,* Reid, E., Ed., Ellis Horwood Limited, Chichester, 1981, 221.

1500. **Pacha, W. and Eckert, H.,** Automated analysis in pharmacokinetic studies, in T*race Organic Sample Handling,* Reid, E., Ed., Ellis Horwood Limited, Chichester, 1981, 240.

1501. **Stockwell, P. B.,** Automated sample pre-treatment, in *Trace Organic Sample Handling,* Reid, E., Ed., Ellis Horwood Limited, Chichester, 1981, 227.

1502. **Leppard, J. P. and Reid, E.,** A multi-sample TLC applicator, in *Trace Organic Sample Handling,* Reid, E., Ed., Ellis Horwood Limited, Chichester, 1981, 262.

1503. **Leppard, J. P., Harrison, A. D. R., and Nicholas, J. D.,** An automatic applicator for thin film chromatography, *J. Chromatogr. Sci.,* 14, 438, 1976.

1504. **Lecaillon, J. B., Souppart, C., Dubois, J. P., and Delacroix, A.,** The routine use, and column-stability implications, of several column-switching HPLC methods for determining drugs in plasma and urine, in *Bioanalysis of Drugs and Metabolites, Especially Anti-Inflammatory and Cardiovascular* Reid, E., Robinson, J. R., and Wilson, I. D., Eds., Plenum Press, New York, 1988, 225.

1505. **Rieck, W. and Platt, D.,** High-performance liquid-chromatographic method for the determination of alprazolam in plasma using the column-switching technique, *J. Chromatogr.,* 116, 259, 1992.

1506. **Pan, H.-T., Kumari, P., Lim, J., and Lin, C.-C.,** Determination of a cephalosporin antibiotic, ceftibuten, in human plasma with column-switching high-performance liquid chromatography with ultra-violet detection, *J. Pharm. Sci.,* 81, 663, 1992.

1507. **Weigmann, H. and Hiemke, C.,** Determination of clozapine and its major metabolites in human serum using automated solid-phase extraction and subsequent isocratic high-performance liquid chromatography with ultra-violet detection, *J. Chromatogr.,* 583, 209, 1992.

1508. **Mader, R. M., Rizovski, B., Steger, G. G., Rainer, H., Proprentner, R., and Kotz, R.,** Determination of methotrexate in human urine at nanomolar levels by high-performance liquid chromatography with column switching, *J. Chromatogr.,* 613, 311, 1993.

1509. **Yamashita, K., Motohashi, M., and Yashika, T.,** Automated high-performance liquid-chromato-graphic method for the simultaneous determination of cefotiam and {delta}³-cefotiam in human plasma using column switching, *J. Chromatogr.,* 577, 174, 1992.

1510. **Fischer, J., Kelly, M. T., Smyth, M. R., and Jandera, P.,** Determination of ivermectin in bovine plasma by column-switching LC using online solid-phase extraction and trace enrichment, *J. Pharm. Biomed. Anal.,* 11, 217, 1993.

1511. **Haertter, S. and Hiemke, C.,** Column-switching and high-performance liquid chromatography in the analysis of amitriptyline, nortriptyline and hydroxylated metabolites in human plasma or serum, *J. Chromatogr.,* 116, 273, 1992.

1512. **Dolezalova, M.,** On-line solid-phase extraction and high-performance liquid-chromatographic determination of nortriptyline and amitriptyline in serum, *J. Chromatogr.,* 579, 291, 1992.

1513. **Farthing, D., Karnes, T., Gehr, T. W. B., March, C., Fakhry, I., and Sica, D. A.,** External-standard high-performance liquid-chromatographic method for quantitative determination of furosemide in plasma by using solid-phase extraction and online elution, *J. Pharm. Sci.,* 81, 569, 1992.

1514. **Allsopp, M. A., Sewell, G. J., and Rowland, C. G.,** Column-switching liquid chromatography assay for the analysis of carboplatin in plasma ultrafiltrate, *J. Pharm. Biomed. Anal.,* 10, 375, 1992.

1515. **Edholm. L.-E.,** Coupled columns in bioanalytical work, *J. Pharm. Biomed. Anal.,* 4, 181, 1986.

1516. **Anderson, N. G.,** Basic principles of fast analyzers, *Am. J. Clin. Pathol.,* Suppl. 778, 53, 1970.

1517. **Shaw, W. and McHan, J.,** Adaptation of EMIT procedures for maximum cost effectiveness to two different centrifugal analyzer systems, *Ther. Drug Monitor.,* 3, 185, 1981.

1518. **Broughton, A. and Ross, D. L.,** Drug screening by enzymatic immunoassay with the centrifugal analyzer, *Clin. Chem.,* 21, 186, 1975.

1519. **Urquart, N., Godolphin, W., and Campbell, D. J.,** Evaluation of automated enzyme immunoassays for five anticonvulsants and theophylline adapted to a centrifugal analyzer, *Clin. Chem.,* 25, 785, 1979.

1520. **Studts, D., Haven, G. T., and Kiser, E. J.,** Adaptation of microvolume EMIT assays for theophylline, phenobarbital, phenytoin, carbamazepine, primidone, ethosuximide, and gentamicin to a CentrifiChem chemistry analyzer, *Ther. Drug Monitor.,* 5, 335, 1983.

1521. **Haven, G. T.,** Phenytoin and phenobarbital measurement by centrifugal analyzer, *Clin. Chem.,* 22, 2057, 1976.

1522. **Lasky, F. D., Ahuja, K. K., and Karmen, A.,** Enzyme immunoassays with the miniature centrifugal fast analyzer, *Clin. Chem.,* 23, 1444, 1977.

1523. **Finley, P. R., Williams, R. J., Lichti, D. F., and Byers, J. M.,** Assay of phenobarbital: adaptation of "EMIT" to the centrifugal analyzer, *Clin. Chem.,* 23, 738, 1977.

1524. **Brinkman, U. A. T., de Jong, G. J., and Gooijer, C.,** The use of fluorescence and chemiluminescence techniques for sensitive and selective detection in HPLC, in *Bioanalysis of Drugs and Metabolites, Especially Anti-Inflammatory and Cardiovascular* Reid, E., Robinson, J. R., and Wilson, I. D., Eds., Plenum Press, New York, 1988, 321.

1525. **Lingeman, H., Van de Nesse, R. J., Brinkman, U. A. T., Gooijer, C., and Velthorst, N. H.,** Laser-induced fluorescence as a detection mode in column liquid chromatography, in *Analysis for Drugs and Metabolites, Including Anti-Infective Agents,* Reid, E. and Wilson, I. D., Eds., Royal Society of Chemistry, Cambridge, 1990, 355.

INDEX

Colorimetric measurements, 84–88
9-chloroacridine, 88
acetaminophen, 84–85
aromatic aldehydes, 88
colorimetric assays for drugs in biological
fluids, 86
colorimetric tests for drug substances, 85
Folin-Ciocalteau reagent, 88
isoniazid, 88–90
phenylbutazone, 84
procainamide, 87
Colorimetric tests for drug substances, 85
Colorimetry, 84
Column chromatography preparation units,
260–261
Column packings, 140–143
Combination drugs, 62–64
Competitive protein-binding assay, 165, 172–173
Complexation, 154
Compliance, 31–33
Computers, in analysis of drugs, 259
Conjugate hydrolysis, 38
Conjugate selection, 166–167
amphetamines, 166–167
bovine serum albumin, 166–167
lysergic acid, 166–167
prostaglandins, 166–167
Schotten-Baumann reaction, 166–167
Conjugates, 68–69
analysis of, 69
Contaminants, sources of, 253–255
Coprescribed drugs, 61
Corticosteroids, 108, 114, 185
Corticosterone-binding globulin, 165
Critical micelle concentration (CMC), 143–144
Cross-reaction of metabolites, 174
Cyclodextrins, 156–157
Cyclosporin, 141, 148

D

Dansyl chloride, 148, 210
Dansyl reagents, 102–104
Degradation, 62–65
Derivative spectroscopy, 91–93
first derivative, 92
second derivative, 92
Derivatives for UV detection, 146–147
Derivatization, 110, 147
postcolumn, 147
precolumn, 147
Derivatization with chiral reagents, 159–160
Detectors
electron capture, 126–129
flame ionization, 126

HPLC
electrochemical, 150–151
fluorescence, 147–150
mass spectrometry, 151–152
ultraviolet absorption, 145–147
mass spectrometry, 130–134
nitrogen specific, 129–130
Dexamethasone, 149
Dextran-coated charcoal, 164
Diastereoisomers, 155, 159–160
Diazepam, 185–186
Diazomethane, 121
Difference spectra, 88–91
Differential pulse polarography, 196–197
Diffusion or zone of inhibition assay, 192
Digitoxin, 8
Digoxin, 7, 8, 174
Disopyramide, 8
Disrupting biological samples, 37
Documentation, 223
Dropping mercury electrode, 150
Drug abuse, 5
alcohol, 5
marijuana, 5
tobacco, 5
Drug analysis in biological fluids, methods for
analytical profile, 208–212
application, 223
centrifugal analyzers, 267
chemical ionization-mass spectrometry, 134
chiral GC, 137
colorimetric assays, 86
documentation, 223
electron-capture detector, 127–128
evaluation of, 219–222
extraction cartridges, 252
fluorescence, 148
fluorescence polarization, 179
fluorescence spectrum, 210
gas chromatography, 124, 127, 129, 131–134,
137, 211
gas chromatography-mass spectrometry, 133
high-performance liquid chromatography,
144, 146, 148–149, 153, 155, 157–161
mass spectrometry, 131, 153, 211
mobile phases, 144
nuclear magnetic resonance, 212
physicochemical characteristics, 209
polarization immunoassay, 179
polarography, 197
pre-extraction cartridges, 252
publication of, 223–224
radioimmunoassay, 170–171
radioreceptor assay, 185
screening assays, 146

P